Capacity mobilisation and capacity building are central to achieving the goals of the climate regime of the Paris Agreement. However, capacity-building initiatives have largely failed as they were hitherto not sufficiently focused on local agendas and self-reliance. This timely book provides useful guidance for change in this respect. Putting local actors first, it provides fresh perspectives to the debate about the economic co-benefits of climate action, especially business development, jobs, and technological learning. Focusing on insights from renewable electrification experiences in East Africa, the authors provide compelling arguments and guidance for ambitious policies and capacity-building initiatives to capture the gains from the greening of energy systems. It is a must-read for scholars and policymakers who are interested in local determination and participation in the transition to low-carbon energy regimes.

Youba Sokona, Vice-Chair of the Intergovernmental Panel on Climate Change (IPCC)

Innovation and Renewable Electrification in Kenya (IREK) project's book on *Building Innovation Capabilities for Sustainable Industrialisation* is a great insight of the capability development in renewable energy processes in East Africa and the challenges faced in renewable energy access and adoption in today's fast changing and evolving East African market. The book provides excellent advice on how to build effective and resilient renewable energy systems that create manufacturing jobs and whose deployments generate highly skilled service employment that can contribute to the economic growth of the East African states. A book well worth reading by academic researchers as well as policy makers and practitioners with an interest in renewable energy pathways and related discussions on capabilities for sustainable industrialisation.

Dr. Edward Mungai, CEO Kenya Climate Innovation Centre

The effective development and diffusion of green technologies is critical to development in sub-Saharan Africa. The major strength of this book is that it combines a macro-level perspective of national and international dynamics with a micro-level focus on renewable energy projects to unpack this process. The authors show that innovation at various levels is a central prerequisite for development and diffusion that attains both environmental and economic benefits associated with the diffusion of solar and wind power technologies. Drawing on a novel framework combining technological capabilities, global value chains, and innovation systems, this book will make an important contribution to theoretical debates about pathways of diffusion in green sectors. More importantly, the application of this framework to detailed renewable electrification case studies allows the authors to provide deep insights and concrete advice for policies aimed at creating economic development from sustainability transition in low- and middle-income countries.

Xiaolan FU, Director of the Technology and Management Centre for Development at Oxford University

This book addresses two aspects of long-term development in African countries. One of these is about the expansion of renewable electricity production. The other, much broader, is about change in the sectoral structure of production in the economy – a source of increased employment, productivity growth, and higher incomes, but also, beyond that, a basis for achieving the multi-dimensional goals of sustainable industrialisation. Drawing on a wealth of case studies, the authors suggest that the expansion of renewable electrification, sometimes impressive, has made only limited contributions to structural change. Behind this, they argue, lie limited activities to develop the necessary technological and managerial capabilities. They throw down a challenge to explore new kinds of policy action in different kinds of context.

Martin Bell, Emeritus Professor, Science Policy Research Unit, University of Sussex

The great potential of Africa can only be realised through big scale job creation based upon industrialisation and technological development. This book helps us understand, how this can be combined with economic, social, and ecological sustainability. Through a series of case studies, it gives insights in how green electrification, based on technologies developed outside Africa, can be implemented in such a way that it helps building local capabilities fundamental for sustainable industrialisation. It is a book worth reading for scholars and policy makers.

Bengt-Åke Lundvall, Emeritus Professor, Aalborg University and Lund University

BUILDING INNOVATION CAPABILITIES FOR SUSTAINABLE INDUSTRIALISATION

This book argues that renewable electrification in developing countries provides important opportunities for local economic development, but new pathways are required for turning these opportunities into successful reality.

Building Innovation Capabilities for Sustainable Industrialisation offers a novel input into the debate on development of capabilities for sustainable industrialisation and delivers key insights for both researchers and policy makers when it comes to the question of how to increase the economic co-benefits of renewables expansion. The chapters in the book use a tailored analytical framework in their studies of renewable electrification efforts in Kenya and other countries in sub-Saharan Africa. They draw on a mix of project, sector, and country level case studies to address questions such as: What capabilities are developed through on-going renewable electrification projects in developing economies? How can the expansion of renewable electrification be supported in a way that also encourages sustainable economic development? What role do international linkages (South-South and North-South) play and what role should they play in the greening of energy systems in developing economies? The authors provide a new understanding of how green transformation and sustainable industrialisation can be combined, highlighting the opportunities and constraints for local capability building and the scope for local policy action.

This book will be of great interest to students and scholars of development studies, energy studies, sustainability and sustainable development, as well as practitioners and policy makers working in development organisations and national governments.

Rasmus Lema, DPhil in Development Studies (Sussex), Associate Professor, Department of Business and Management, Aalborg University and Visiting

Professor, College of Business and Economics, University of Johannesburg. Principal Investigator of the IREK project.

Margrethe Holm Andersen, PhD in Social Science (Aalborg), Senior Advisor in Innovation and Development, Department of Politics and Society, Aalborg University. Member of the IREK research team and the IREK management group.

Rebecca Hanlin, PhD in Science and Technology Studies (Edinburgh), Non-Resident Fellow at the African Centre for Technology Studies (ACTS), Innovation and Development Specialist for AfricaLics, Visiting Fellow at a number of universities in Europe and Africa. Member of the IREK research team and the IREK management group.

Charles Nzila, PhD in Applied Science (Gent), Senior Lecturer at Moi University. Co-Project Investigator of the IREK project and member of the IREK management group.

Pathways to Sustainability Series

This book series addresses core challenges around linking science and technology and environmental sustainability with poverty reduction and social justice. It is based on the work of the Social, Technological and Environmental Pathways to Sustainability (STEPS) Centre, a major investment of the UK Economic and Social Research Council (ESRC). The STEPS Centre brings together researchers at the Institute of Development Studies (IDS) and SPRU (Science Policy Research Unit) at the University of Sussex with a set of partner institutions in Africa, Asia and Latin America.

Titles in this series include:

Transformative Pathways to Sustainability
Learning Across Disciplines, Cultures and Contexts
The Pathways Network

The Politics of Knowledge in Inclusive Development and Innovation
Edited by David Ludwig, Birgit Boogaard, Phil Macnaghten and Cees Leeuwis

Building Innovation Capabilities for Sustainable Industrialisation
Renewable Electrification in Developing Economies
Edited by Rasmus Lema, Margrethe Holm Andersen, Rebecca Hanlin and Charles Nzila

For more information about this series, please visit: www.routledge.com/Pathways-to-Sustainability/book-series/ECPSS

BUILDING INNOVATION CAPABILITIES FOR SUSTAINABLE INDUSTRIALISATION

Renewable Electrification in Developing Economies

Edited by Rasmus Lema, Margrethe Holm Andersen, Rebecca Hanlin and Charles Nzila

Routledge
Taylor & Francis Group
LONDON AND NEW YORK

earthscan
from Routledge

First published 2022
by Routledge
2 Park Square, Milton Park, Abingdon, Oxon OX14 4RN

and by Routledge
605 Third Avenue, New York, NY 10158

Routledge is an imprint of the Taylor & Francis Group, an informa business

British Library Cataloguing-in-Publication Data
A catalogue record for this book is available from the British Library

Library of Congress Cataloging-in-Publication Data
Names: Lema, Rasmus, editor. | Andersen, Margrethe Holm, editor. | Hanlin, Rebecca, editor. | Nzila, Charles, editor.
Title: Building innovation capabilities for sustainable industrialisation: renewable electrification in developing economies / edited by Rasmus Lema, Margrethe Holm Andersen, Rebecca Hanlin and Charles Nzila.
Description: Milton Park, Abingdon, Oxon; New York, NY: Routledge, 2022. | Includes bibliographical references and index.
Identifiers: LCCN 2021027873 (print) | LCCN 2021027874 (ebook) | ISBN 9780367516208 (paperback) | ISBN 9780367516246 (hardback) | ISBN 9781003054665 (ebook)
Subjects: LCSH: Sustainable development–Developing countries. | Renewable energy sources–Developing countries. | Electrification–Developing countries.
Classification: LCC HC59.72.E5 B85 2022 (print) | LCC HC59.72.E5 (ebook) | DDC 333.7/091724–dc23
LC record available at https://lccn.loc.gov/2021027873
LC ebook record available at https://lccn.loc.gov/2021027874

ISBN: 978-0-367-51624-6 (hbk)
ISBN: 978-0-367-51620-8 (pbk)
ISBN: 978-1-003-05466-5 (ebk)

DOI: 10.4324/9781003054665

Typeset in Bembo
by Deanta Global Publishing Services, Chennai, India

CONTENTS

ILLUSTRATIONS

Figures

Tables

CONTRIBUTING AUTHORS

Rasmus Lema (editor)
DPhil in Development Studies (Sussex), Associate Professor, Department of Business and Management, Aalborg University and Visiting Professor, College of Business and Economics, University of Johannesburg. Principal Investigator of the IREK project.

As an expert in Innovation and Development, Rasmus has 20 years of experience with research work in developing countries (including China, India, and Kenya). His research is focused on capability-building, international technology collaboration, and innovation policy in the context of the green transformation.

Margrethe Holm Andersen (editor)
PhD in Social Science (Aalborg), Senior Advisor in Innovation and Development, Department of Politics and Society, Aalborg University. Member of the IREK research team and the IREK management group.

Margrethe has more than 25 years of experience in research, consultancy, planning, and implementation of development cooperation, working on a broad range of issues including capacity development, agriculture, climate change and environmental management, community development, gender issues, civil society, and fragile states. Current research interests include capacity building in renewable energy, inclusive innovation, health systems strengthening, and research capacity building.

Rebecca Hanlin (editor)
PhD in Science and Technology Studies (Edinburgh), Non-Resident Fellow at the African Centre for Technology Studies (ACTS), Innovation and Development Specialist for AfricaLics, Visiting Fellow at a number of universities in Europe

and Africa. Member of the IREK research team and the IREK management group.

Rebecca is a science, technology, and innovation policy expert with over 20 years of experience working in developing countries (Cuba, Tanzania, Kenya, Nigeria) with an emphasis on promoting innovation and business development opportunities for small and medium sized businesses in Africa. Main area of research is innovation and development with a focus on renewable energy, health systems, and social innovation.

Charles Nzila (editor)
PhD in Applied Science (Gent), Senior Lecturer at Moi University. Co-Project Investigator of the IREK project and member of the IREK management group.

Charles is an expert in technology and sustainability assessments, renewable energy, and energy poverty with an emphasis on the promotion of sustainable energy, cleaner production, and life cycle engineering in Africa. He has over 13 years of research experience and a strong background in applied research and sustainable development. Main research interests include renewable energy and enhancing energy accessibility in Kenya, clean technology, sustainable development, and life cycle analysis.

Ann Kingiri
PhD in Development Policy and Practice (The Open University, UK), Director, of the Science, Technology, Innovation, Knowledge and Society (STIKS) programme at the African Centre for Technology Studies (ACTS), visiting fellow at Johannesburg University, Innovation Policy Specialist for AfricaLics. Member of the IREK research team.

Ann has an interdisciplinary training background (biological sciences, environmental science, innovation, and development policy) and vast experience in networking and advocacy in multicultural settings involving diverse development and policy actors in the public and private sectors. Areas of expertise include policy analysis, climate change, and sustainable development.

Faith H. Wandera
MSc in Environmental Planning and Management (Kenyatta), Senior Deputy Director of Renewable Energy at Ministry of Energy (Kenya), PhD candidate at Moi University. Member of the IREK research team.

Faith is a renewable energy policy expert with emphasis on bioenergy development with more than 25 years of experience in the energy sector during which she has worked with a wide range of renewable energy technologies, including solar PV, solar thermal, small hydro, wind, and biogas. Faith's PhD dissertation focuses on dissemination of small wind turbine technology and possibilities for enhanced use of SWT in providing remote rural communities in Kenya with access to electricity.

Cecilia Gregersen

MSc in Business and Development Studies (Copenhagen Business School), PhD fellow at Aalborg University. Member of the IREK research team.

Cecilia has experience as a policy officer in the Strengthening European External Action programme at the European Centre for Development Policy Management in the Netherlands. Her main research area is the nature and potential for learning and the emergence of innovation capabilities through global technology collaborations in the wind sector. Cecilia's PhD thesis explores why, how, and through which mechanisms capabilities may emerge, and how the embeddedness of a technology transfer project relates to the innovation system in the Kenyan wind sector.

Mbeo Calvince Ogeya

Research Fellow, the Energy and Climate Change Programme, Stockholm Environmental Institute (SEI), Africa, PhD student at Jaramogi Oginga Odinga University of Science and Technology.

Mbeo has worked as an energy consultant, mentor, and national policy advisor and has 11 years' research experience in renewable energy technologies and systems modelling, with extensive skills in LEAP energy systems modelling, engineering design, solar photovoltaic, bioenergy, and energy efficiency. Main areas of interest are climate change, natural resources, and air quality.

Philip Osano

PhD in Geography (McGill), Centre Director of Stockholm Environmental Institute (SEI), Africa.

Philip is an environmental policy expert with diverse research on biodiversity and ecosystem governance; land use change; pastoral and rangeland management; climate change adaptation; agricultural policy; water and land management; and integrated environmental planning. Main research interests are in environmental policy, development, and international affairs.

Josephat Mongare Okemwa

MSc in Economics (Kenyatta), Research assistant at African Centre for Technology Studies (ACTS), working with the IREK project.

Josephat has research experience on financial inclusion, agriculture, and health with a particular focus on applying methods and insights from the behavioural sciences to questions of economic growth. Main research interests are science, technology, and innovation towards sustainable development and the economics of development.

Ulrich Elmer Hansen

PhD in Innovation Studies (DTU), Senior Researcher at DTU Management Engineering, a UNEP/DTU partnership. Member of the IREK research team.

Ulrich is an expert in economic geography and development economics and has a strong interest in learning, technology capability, innovation, catching-up, and renewable energy. Research areas include globalisation of R&D, industrial upgrading in global value chains and technological capability-building at firm and industry level, the diffusion of solar PV and wind turbine technologies in East Africa, and the development of biomass power plants in China.

Michael Korir

PhD in Entrepreneurship and Business Management (Kenyatta), teaching professor at the Department of Management Science, School of Business and Economics, Moi University. Member of the IREK research team.

Michael's areas of expertise include innovation, entrepreneurship, and strategic management, including competitive strategy issues.

Birgitte Gregersen

MSc in Economics (Aalborg), Associate Professor, Department of Business and Management, Aalborg University, member of the Innovation, Knowledge and Economic Dynamics (IKE) research group.

Birgitte has researched and published within the field of technical change and employment, IT in the public sector, public technology procurement, studies of national systems of innovation, university–industry linkages, innovation policy, and sustainable development. Current research is centred around systems of innovation with a special focus on institutions and learning capabilities from a sustainable development perspective.

Joni Karjalainen

MSc in Political Science (Helsinki), PhD Student, Project Researcher, University of Turku, Finland Futures Research Centre, focused on the long-term transformation of the energy system.

Joni's research work includes projects such as Great Electrification in Peer-to-Peer Society, Neo-Carbon Energy – Transformative Energy Futures 2050, Access to Sustainable Energy for All, and Kenya and Tanzania Beyond 2015. Main research interests include the dynamics that affect the adoption of renewable energy technologies and the emergence of solar energy business in East and West Africa.

Rob Byrne

PhD in Science and Technology Policy (Sussex), Research Fellow in the Sussex Energy Group and Lecturer in Science Policy Research Unit (SPRU), at the University of Sussex. Member of the STEPS Centre, Climate Strategies, and the IREK Scientific Advisory Board.

Rob has an extensive background in practical engineering work, academic social science research, and policy analysis. His research expertise is mainly in sustainable energy access, drawing on academic perspectives in development studies, geography, political economy, innovation studies, socio-technical transitions, and strategic niche management. Co-author (with David Ockwell) of the book Sustainable Energy for All – Innovation, technology and pro-poor green transformations (EarthScan, 2017).

Padmasai Lakshmi Bhamidipati

PhD in Sustainable Energy at the Technical University Denmark (DTU), and Advisor in Energy and Climate at UNEP-DTU Partnership.

Lakshmi works at the intersection of research, policy and implementation mainly in the areas of sustainable energy transitions, climate mitigation, and local community engagement in developing countries. Her other work includes: enabling diffusion of climate technologies, supporting private sector engagement in renewable energy, and researching socio-economic and developmental impacts of utility-scale energy projects, mainly in Sub-Saharan Africa and South Asia.

Julian Kirchherr

DPhil in Geography and the Environment (Oxford), Associate Partner, McKinsey & Company, and Principal Investigator (Circular Economy), Copernicus Institute of Sustainable Development, Utrecht University.

Julian's main research area is sustainable energy transitions in emerging economies and the role the private sector can play in these. Recent research includes work on the environmental and socio-economic impacts of the Chinese-led Belt and Road Initiative (BRI), Chinese hydropower projects, and the role of circular start-up hubs in the transitioning towards a circular economy.

Nina Kotschenreuther

BA in Nordic Philology (Copenhagen), IREK Editorial Assistant, the African Centre for Technology Studies (ACTS).

Nina has more than 20 years of experience developing impact and research communication as editorial assistant, with a particular experience with research publications within the field of innovation studies.

PREFACE

Many low- and middle-income countries are currently seeking to increase electrification using renewables. The considerable investments in renewable energy projects expected in the coming years provide new opportunities for developing local capabilities in designing, constructing, operating, and maintaining renewable electrification plants, thereby contributing to structural change. But there is a real risk that these opportunities do not become reality. Renewable energy supply mechanisms are often designed, constructed – and to some extent also operated and maintained – using foreign equipment, foreign financing, and foreign workers. The degree to which low- and middle-income countries will capture the economic gains from renewable electrification is therefore a key issue and one that is so far understudied, not least in the context of low- and middle-income countries in Africa.

This book brings together insights based on case studies focusing mainly on solar and wind energy in Kenya, Tanzania, Ethiopia, and South Africa. Drawing on both qualitative and quantitative evidence, it contributes to discussions on how renewable electrification processes can be shaped to optimise economic 'co-benefits' in the form of e.g., increased job creation, development of local supply chains, and technological learning. It argues that activities to develop the technological and managerial capabilities needed for economic co-benefits to materialise remain limited. Furthermore, it argues that new policy initiatives are needed to ensure that processes of renewable electrification contribute significantly to the multi-dimensional objectives of sustainable industrialisation.

ACKNOWLEDGEMENTS

This book is a product of a research project on Innovation and Renewable Electrification in Kenya (IREK). The project was funded by the Danish Ministry of Foreign Affairs and brought together researchers from the African Centre for Technology Studies (ACTS), Moi University, and Aalborg University (AAU) for over six years of collaboration.

We would like to thank all the researchers involved in the project. This includes Ann Kingiri and Josephat Mongare Okemwa (ACTS); Michael Korir, Faith Wandera, and Dominick Samoita (Moi University); Cecilia Gregersen, Arne Remmen and Poul Alberg Østergaard (Aalborg University); and Ulrich Elmer Hansen (DTU/UNEP).

We would also like to thank members of the scientific advisory committee and other scholars and practitioners for their advice, their presentations, and comments on the working papers that laid the foundation for various chapters of the book. Thanks are due to Bengt-Åke Lundvall (Aalborg University), Edward Mungai (Kenya Climate Innovation Centre), Feng Zhao (FTI Consulting), Helene Ahlborg (Chalmers University of Technology, Sweden), Joni Karjalainen (University of Turku), Padmasai Lakshmi Bhamidipati (UNEP/DTU), Mathilde Brix Pedersen (UNEP/DTU), Mbeo Calvince Ogeya (Stockholm Environmental Institute), Morten Blomqvist (Danish Ministry of Foreign Affairs), Oliver Johnson (Stockholm Environmental Institute), Robert Byrne (SPRU/ Sussex Energy Group), and Roberta Rabellotti (University of Pavia). Special thanks are reserved to Nina Kotschenreuther for all the support she has provided over the years and for initial editing of the chapters in this book.

All of those who agreed to be interviewed during field research and who took time to participate in the various stakeholder workshops also deserve a big note of thanks.

We express special thanks to Kenya's Parliamentary Committee on Energy 2018 for accepting a submission from the IREK research team focusing on the importance of building local capabilities as part of local content requirements in Kenya's Energy Bill, which became Law in 2019. Discussions with stakeholders on policy measures for renewable electrification and sustainable industrialisation have been very useful for our learning process, and Dr. Ann Kingiri and support staff from ACTS deserve specific mention in this regard.

Finally, the support for research (on which major parts of this book are based) is from the Danish Ministry of Foreign Affairs (Grant: DFC 14-09AAU) and is gratefully acknowledged. On behalf of the entire IREK team, we are also appreciative of the excellent administrative support accorded to the project team staff by management and support staff at ACTS and Moi University in Kenya and at Aalborg University and Danida Fellowship Centre (DFC) in Denmark.

1

RENEWABLE ELECTRIFICATION AND SUSTAINABLE INDUSTRIALISATION

Rebecca Hanlin, Margrethe Holm Andersen, Rasmus Lema, and Charles Nzila

Abstract

This book argues that debates about renewable electrification must move beyond their predominant focus on access to clean energy. Increased access to electricity makes important contributions to sustainable development but it does not produce the full range of co-benefits which can arise from green energy investments. The book argues that policy makers need to start focusing more heavily on questions of the development of local activities and capabilities in designing, constructing, and operating renewable electricity infrastructure. A key issue is the degree to which sustainable access to clean energy will be sustainable when these renewable energy supply mechanisms are often designed, constructed, operated, and maintained predominantly with foreign equipment, foreign financing, and foreign workers. This is what this book sets out to examine and discuss in the context of green industrialisation discourses. This chapter outlines the background to the sustainable industrialisation debate. It also specifies the objectives and provides an overview of the book and its key themes.

Introduction

There is a wealth of literature which has shown how access to electricity can create new business opportunities and increase economic activities in formerly non-electrified communities (e.g., Peters and Sievert, 2016). This is not the focus of this book. Rather, it is focused overall on the contribution which renewable electrification can make in meeting sustainable industrialisation goals. Renewable electrification includes both the creation of access to electricity to formerly non-electrified communities as well as transformation of existing energy systems with renewables.

DOI: 10.4324/9781003054665-1

We argue that the debate about the benefits of renewable electrification needs a push. It must move beyond (a) the ability to create sustainable development benefits as a result of increased electricity access and (b) the ability to contribute to climate change mitigation. These are important goals in their own right, but there is a need to include (c) the local economic benefits that can potentially arise from the renewable electrification process itself. The debate needs to start focusing more heavily on questions about the localisation of economic activities and development of local capabilities for designing, constructing, and supplying renewable electrification infrastructure such as solar parks, windfarms, and hybrid grids.

The concern should be not only with distribution of economic activities but also with the learning gains that may arise in connection with these activities. The economic activities involved in renewable electrification are temporal in nature, but the learning gains can have a more lasting effect on the change of economic development paths. Accordingly, we argue that building up capabilities for economic change – or innovation capabilities – constitutes an important missing link in ensuring the transition to a more sustainable development in developing economies. The book brings together new insights on the development of local capabilities and suggests policy measures that can support and accelerate the development of capabilities required for sustainable industrialisation in Kenya and countries facing challenges similar to those of Kenya. Such capabilities are vital for the transformation required at a time when the need for access to electricity for development in low and lower middle-income countries continues to increase while warnings against the potentially disastrous effects of continued high emissions of CO_2 and the need to bring down CO_2 emissions dominate international and domestic debates.

The three main objectives of the book are therefore:

- To establish a new conceptual framework for analysing and understanding linkages between renewable electrification and sustainable industrialisation in developing economies.
- To contribute to the empirical understanding of how capabilities for sustainable industrialisation are being developed (or not) through renewable energy projects in low- and lower middle-income countries in Africa, based on in-depth studies mainly, but not only, from Kenya.
- To contribute to the development of transformative innovation policies, i.e., policies that support green transition in a manner that also takes into consideration aspects of distribution and directions of development.

This introductory chapter creates a backdrop and then sets the scene for the book. First, we introduce key discourses and debates about sustainable industrialisation. Second, we focus on renewable electrification processes and challenges. We then bring these together and outline the key focus of the book by highlighting how renewable electrification may produce 'co-benefits' that may contribute to sustainable industrialisation processes. Finally, we outline the chapters and key themes of the book.

Sustainable industrialisation

Industrialisation is a process of structural change and traditionally, the term has referred to a shift in the sectoral composition of economies starting from agriculture, moving into manufacturing, and eventually into knowledge-intensive services (Kuznets, 1973; Gabardo, Pereima, and Einloft, 2017). Industrialisation as a result is synonymous with economic development driven by manufacturing and high-value services (Szirmai, Naudé, and Alcorta, 2013). These activities create pathways for developing countries to grow their economies because they provide employment and linkages to other parts of the economy, and they can boost consumption and generate more foreign currency through value-added exports (Opuku and Yan, 2019).

Indeed, it is often argued that African countries have no option but to industrialise through manufacturing (Oyelaran-Oyeyinka and Adesina, 2020) or at least that industry needs to play an essential role in economic development, for example with an increase in agro-industry (Lundvall and Lema 2014).[1] Yet, analysts have increasingly questioned whether current low- and lower middle-income countries can and should develop along pathways similar to those that were historically prevalent in the triad of North America, Europe, and East Asia. Rodrik (2018, p. 17) suggests Africa can only achieve development if it is based on 'a growth model that is different from earlier miracles based on industrialisation' and there is an increasing focus on the possibility for African countries to develop economically through the services sector, without having to go through a manufacturing phase of industrial development (Newfarmer, Page, and Tarp, 2018). As such, there is increasing agreement that economic development may involve some degree of traditional manufacturing activities and agricultural and commodity processing, but may also involve entirely new types of industry and services. The key question that we are concerned with in this book is whether green energy has a role to play in this respect.

Greening of industrial development

The prospects for a new economy are captured by terms such as 'green growth', 'green economy', and 'green industrialisation', which have all been advanced in policy circles. The various terms have been coined and promoted by different agencies, including the World Bank, the UN, and global think tanks as well as national governments.

A key notion in these concepts is that of 'low carbon development', which requires a decoupling of economic growth from environmental impact (United Nations Environment Programme, 2014; Jackson, 2017). In general, green growth concepts focus on the idea of an 'economic development that is based on sustainable use of non-renewable resources and that fully internalizes environmental costs, including most critically those related to climate change' (Rodrik, 2014, p. 469). This involves the use of 'green technologies', i.e., technologies that have the least possible impact on the environment and are possible through an industrialisation process that is 'green' in process and outcome (UNIDO, 2009; Organisation for Economic Co-operation and Development, 2011; World Bank, 2012).

The term 'green industrialization' as defined by the United Nations Industrial Development Organization (UNIDO, 2009, 2011) is particularly relevant for this book. It defines 'green industrialisation' as having two main dimensions: (a) the greening of industries within themselves; and (b) the creation of green enterprises – i.e., enterprises which offer environmental goods and services. It is this latter dimension which most directly highlights the idea that greening of industrial development is associated with costs, but also with opportunities. But significant questions have remained about the distribution of costs and potential opportunities.

Combining greening and social inclusion

In recent years, the mainstream development discourses have been characterised by explicit attempts to bring environmental and social objectives together with economic development aims. For example, in the run up to the development of the Sustainable Development Goals (SDGs), UNIDO (2014) started using the term 'sustainable industrial development' and has since moved to the term 'inclusive and sustainable industrialisation' (UNIDO, 2017). UNIDO (2017) define this form of industrialisation as including three elements (UNIDO, 2017): (a) creating shared prosperity – offering equal opportunities and an equitable distribution of benefits to all, (b) advancing economic competitiveness, and (c) safeguarding the environment – addressing the need to decouple generated prosperity of industrial activities from excessive natural use and negative environmental impacts.

The discourse has thus shifted from 'green' to multifaceted 'sustainable' industrialisation. Focusing on sustainable industrialisation, according to UNIDO, can provide African countries a way of increasing employment, lowering energy costs, and reducing the pressure on infrastructure in cities as well as ensuring prosperity is shared across all those in society (UNIDO, 2020).

However, despite the widespread use of the term 'sustainable industrialisation' (e.g., Sampath, 2016) there is no agreed or readily available definition of what 'sustainable' means. Or put differently: the term 'sustainable' has been used to denote many different meanings, ranging from environmental sustainability (UN, 1987) to financial sustainability or more broadly the ability to sustain a certain type of production by a firm or existence of a company over a longer period of time. Whichever mode or combination of approaches are pursued, there is strong support for sustainable industrialisation at global level through the introduction of SDG 9 focused on promoting inclusive and sustainable industrialisation and at a regional level through the African Union's Action Plan for the Accelerated Industrial Development of Africa (African Union, 2021).

In the next chapter we will specify how we take on board the relevant parts of these discourses and define sustainable industrialisation for operational use in our empirical work. Here it suffices to say that while we seek to incorporate the multi-criteria nature of the term, we use it in a more narrow way, which is tailored to our analysis of the process of renewable electrification.

In this book, the general use of the term 'sustainable industrialisation' can be defined broadly as having a double meaning. First, it is *environmentally friendly industrialisation*; or to be more precise, it gives rise to an increase in industrial activities (typically manufacturing and related services, but also e.g., agroindustry or high-value services associated with the digital economy) which do not conflict with the principles of sustainable development as they are commonly defined.[2] Second, it is a type of industrial development which can be *maintained in the long run*; industrial activities are enduring and rooted firmly in the local economy. This also implies contributing to social inclusiveness through creation of jobs and incomes in local communities. In the broad sense, 'sustainable industrialisation' is industrial development, which is sustainable in both the environmental, social, and general sense of the term. Although the book does not have an explicit focus on, e.g., justice and gender equality, it addresses key issues related to how renewable electrification processes may help reduce inequalities, e.g., through provision of local jobs, increased local capabilities, and other co-benefits from renewable energy projects.

Therefore, in this book, we are concerned with a defined subset of sustainable industrial development activities, namely those that contribute directly to the process of restructuring which contribute to green industrialisation (as defined by UNIDO) and to bringing the economy within the planetary boundaries (Rockstrom et al., 2009). Thus, the sustainable industrialisation activities discussed in this book align with low-carbon development objectives, as they are related to strategies that mitigate emissions to avoid dangerous climate change while at the same time achieving economic and social development (Lema, Iizuka and Walz, 2015).

Renewable electrification challenge(s)

We are concerned in this book with how to shape the process of renewable electrification in ways that maximise the contribution to sustainable industrialisation. Before we proceed to outline the research questions, contents, and main themes of the book, it is worthwhile to elaborate on the main drivers of renewable electrification, the context for empirical studies and some of the main capability challenges.

Access to electricity and climate change mitigation

There are currently 1.2 billion people – one in six people – in the world without access to electricity (Rivas Saiz, 2018). The problem is particularly severe in low- and middle-income countries in sub-Saharan Africa. The latest figures (Blimpo and Cosgrove-Davis, 2019) state that 43% of the population of sub-Saharan Africa have access to electricity (half the global access figure of 84%), with access in rural areas a staggeringly low 25%. Ensuring access to electricity thus remains one of the greatest development challenges of our time (Jacobson, 2007; Lay, Ondraczek, and Stoever, 2013).

National governments and multilateral agencies have put in place strategies for creating access to electricity, and today renewable electrification plays an increasingly prominent role. Many of these policy initiatives are framed within the context of the Sustainable Energy for All initiative (SEforALL, 2020) and activities related to SDG 7, which aims to 'Ensure access to affordable, reliable, sustainable and modern energy for all'. While aimed at increasing access, these strategies also have the associated aim of accelerating the transition towards increased use of renewable energy. Such a transition is important because it would help increase the number of people with access to electricity but also because it may help stabilise production and delivery of energy (IRENA, IEA and REN21, 2018) and last but not least because of the current warnings about the need to reduce worldwide CO_2 emissions (IPCC, 2018).

Context: East Africa and Kenya specifically

The East African Community (EAC) states are rich in renewable energy resources such as solar, wind, hydro, biomass, and geothermal energy. A large proportion of this potential remains untapped, yet it can be efficiently exploited to boost the generation of energy and cope with the strong load growth besides enhancing electrification of the partner states. All EAC partner states have adopted quite promising electricity access targets whereby the goals by Kenya and Uganda to achieve 100% access to grid-connected power by 2030 and 2040 respectively are regarded as the most ambitious in the East Africa region. These targets require a paradigm shift, since notwithstanding their impressive outlook, they do raise rather pertinent capability questions as well as whether achieving them is both realistic and possible considering the past trends.

Of all these East African countries, Kenya, with 90% of its energy mix sourced from renewable energies (Reuters, 2019), is clearly a country to watch: it is at the forefront of ensuring increased supply of electricity to a growing population using renewable energy. Meanwhile, the country still faces many of the same challenges faced by other low- and lower middle-income countries.

As with other countries in sub-Saharan Africa, Kenya faces the challenge of increasing demand for modern energy services in the face of its high population growth. In 2000 only 15% of the Kenyan population had access to electricity, by 2013 this had raised considerably. Figures from 2018 indicate that access to energy in Kenya may have increased to 75% which places it well above average for sub-Saharan Africa which was 47.7% (World Bank, 2020).

The increased percentages in energy access relate to implementation of various national electrification projects undertaken by Kenya Power Authorities such as the Last Mile Connectivity Project and the Global Partnership on Output-Based Aid (GPOBA) slum electrification project. Many of the newly connected households have a low level of energy consumption, however (two units on average per month), and by March 2017 many newly connected households were reported as not yet having used the pre-loaded units that come with the meter when installed (Kenya Power and Lighting Company, 2017). This indicates that having access to electricity (being formally connected) and having the need for electricity and/or the resources to pay for electricity used from the national grid may not be the same.

Kenya Vision 2030 (2008) identifies energy as key in achieving its goals for economic development and poverty alleviation. The more recent Big Four Agenda of the Kenyan Government emphasises the importance of intensifying energy diversification, promoting transition from traditional fuels to modern sources of fuels, adopting energy efficiency technologies, promoting off-grid options, and attaining a more efficient energy mix with a reduced share of thermal power generation as critical ways forward (KIPPRA, 2018). Vision 2030 estimates that for Kenya to achieve its goal, its electricity generation must grow from the current level of 1,500 MW to 19,200 MW by 2030. In 2019, Kenya's installed electricity capacity was 2,818.9 MW (KNBS, 2020).

To increase electrification and adapt to climate changes, Kenya's Energy Act of 2019 (and the previous 2006 Act) strongly emphasises development of other renewable energy sources to diversify the national energy mix. Among the renewable energy sources given high priority are solar and wind, but these sources of energy still only represent a small fraction of installed capacity produced through renewable energy. The proportion of electricity generation by source in 2019 shows that geothermal remains the major source of electricity in Kenya accounting to 45% of total generation (hydro 28%, wind 13%, thermal oil 11%, and others 3%). Wind generation rose from 375.6 GWh in 2018 to 1,562.7 GWh in 2019 following full operationalisation of the Lake Turkana Wind Power Plant, becoming the third largest source of electricity generation in 2019. Solar generation increased from 13.7 GWh in 2018 to 92.3 GWh in 2019, attributed to the commissioning of the Garissa Solar Power Plant (KNBS, 2020).

The renewable electrification process, localisation, and capabilities

Traditionally, much of the physical technology, skills, knowledge, and companies involved in renewable energy projects in Africa have been bought from outside. While this has changed from always being from developed to developing country (e.g., German solar panels to Kenya), the rise of manufacturing capabilities in the so-called BRICS (Brazil, Russia, India, China, and South Africa) implies there is now also increasing South-South technology transfer. In any case, 'technology transfer' is a complex issue (Ockwell and Mallett, 2013). It relates to a much bigger issue relating to the degree to which countries should rely on external technology or develop their own local innovation capabilities (Baker and Savacool, 2017). However, without local capabilities, there is a big risk that renewable electrification remains overly dependent on external actors and hence does not become sustainable – and a big risk that the possible benefits in terms of increased local employment, local content, and new business opportunities are not realised. This, of course, is a risk of missed economic opportunities in the short run, but also a risk of not gaining experience from engagement in these activities for the purpose of better local anchoring of future activities.

In fact, it is possible to see the capabilities that are needed and built through interaction of renewable energy plants at various stages of the production and deployment of renewable energy solutions. There are capabilities required in manufacturing core technology and parts for the plants themselves (whether these are wind turbine blades or solar panels, or the cement, nuts, and bolts needed to fix the turbines and panels). Then there are capabilities related to the construction, operation, and maintenance of a renewable electrification plant. These are not just construction level capabilities, but also important planning and financing capabilities. The operation and maintenance of the plants once commissioned also require particular capabilities as does the distribution of the power to the consumer for their consumption.

More concretely, capabilities can be built in the design, build, and operation (DBO) of solar, wind, and other renewable energy plants. The capabilities that are built are not just those for individuals and firms directly involved in the DBO of renewable energy plants but include allied supplier firms and their individuals as well. This is because firms directly involved in renewable energy plant DBO activities rely on others, for example to provide supplies and transport of those supplies, to conduct outsourced work, such as initial ground clearance, and to provide support facilities.

The renewable electrification process and potential co-benefits

This book discusses the potential of using renewable electricity expansion and transition efforts to contribute to future and wider sustainable industrial development aims. We are interested in sustainable industrialisation through electrification that is not only green but is also more durable and more inclusive (see above). A key element of inclusivity is around who is involved and how they are involved in the electrification process, and durability relates to the nature of capabilities i.e., whether they are sufficiently rooted and relevant for subsequent activities in the sustainable development process.

In order to dissect and specify these aims, we draw on vocabulary from the multilateral policy arena and focus on 'co-benefits'. In the broadest sense, these are the positive benefits related to the reduction of greenhouse gases (IPCC 201?). More concretely, mitigation with green energy, such as renewable electrification, can have a range of additional benefits. For example, a key environmental co-benefit is the reduction of air pollution. Economic co-benefits are manifold and include, for instance, the reduction of costly energy imports. However, we limit our focus to what we see as the most important co-benefits when it comes to sustainable industrialisation: localised economic activity and capability-building. As mentioned above, in key policy discussions these benefits tend to be overshadowed by the emphasis on local benefits that arise from access to clean energy.

As such, the argument and focus of this book can be illustrated through Figure 1.1. The figure shows that renewable electrification, in addition to the reduction of carbon emissions, creates the primary benefit of increased electricity production. In turn, this allows for the development of new businesses and improved productivity, potentially in high productivity growth industries such as manufacturing and high-value services. It creates new access in formerly non-electrified communities, which can enable various forms of production and services. These outcomes are illustrated by arrows at the top of the diagram in Figure 1.1. We seek to emphasise the potential industrial gains available through the *process* of renewable electrification. This is essentially a process of creating and installing new green energy infrastructure. We argue that this process provides important opportunities to derive activities and capabilities that are highly relevant to sustainable industrialisation. These are the arrows at the bottom of the diagram in Figure 1.1.

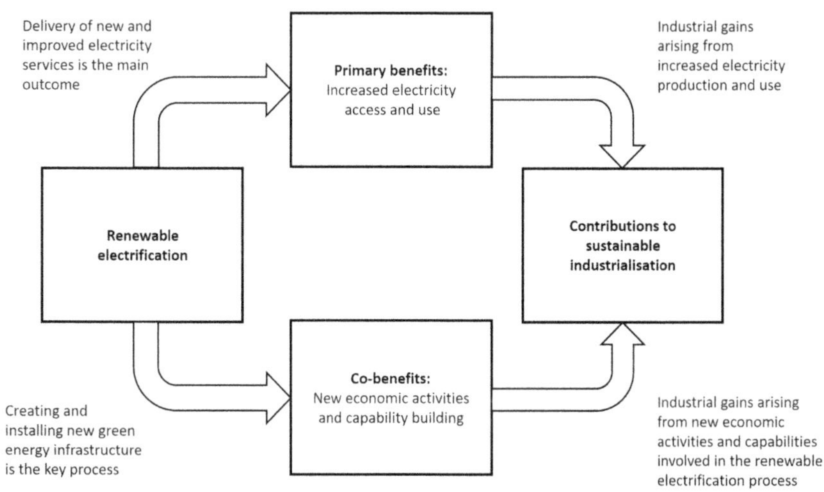

FIGURE 1.1 The importance of co-benefits. *Source: authors.*

This book focuses on how renewable electrification processes can be shaped in ways that maximise their contribution to sustainable industrialisation. It does this by analysing the technological capability-building in the wind and solar renewable energy sub-sectors in Kenya with the inclusion of some cases from other countries in East Africa.

Outline of the book

The book starts by presenting a conceptual framework for understanding and analysing renewable electrification processes with specific emphasis on learning processes at different levels, development of (technological) capabilities, and outcomes in the form of capabilities that enable increased employment, local content, and business opportunities (Andersen and Lema, 2022; this volume).

The book subsequently presents a number of case studies and insights into how the systems of production and learning of solar and wind energy in Kenya (and a few other African countries, i.e., Tanzania and Ethiopia) are developing and how capabilities for renewable electrification are being built (or not). It includes examples of how international linkages influence such processes and presents insights on how policies for renewable electrification and green transformation in Kenya may be shaped to help foster the desired development.

Drawing on selected case studies, including a number of projects in Kenya, using transition theory, the book discusses the successes and failures of diffusion of renewable energy for rural transformation (Mbeo et al., 2022; this volume).

Disaggregating sectoral systems of production and innovation in renewable electrification pathways, Hansen et al. (2022; this volume) argue that 'size' is an important determinant of the appropriateness of renewable energy technological trajectories in sub-Saharan Africa. Distinguishing between small-scale (mini-grids) and large-scale (grid-connected) deployment paths in renewable energy, they find that innovation and diffusion dynamics differ more between small and large than between wind and solar.

In the following chapter the use and application of the Technological Innovation Systems (TIS) framework to analyse the Small Wind Turbine (SWT) innovation system in Kenya is presented (Wandera, 2022; this volume). While many studies have been conducted on the diffusion of SWT in developed countries, relatively fewer studies on this topic have been conducted in developing countries, especially in East Africa.

Based on a survey of existing renewable electrification projects in Kenya (Nzila and Korir, 2022; this volume) and on research into a selected number of renewable energy projects deemed critical for the understanding of how the project level (design and processes) influences the up-take of renewable energy (Hanlin and Okemwa, 2022; this volume), the book presents insights into the current status of capabilities for renewable electrification in Kenya.

The book further investigates how 'learning from importing' of renewable energy products and services may help build local capabilities for renewable electrification, placing particular emphasis on governance structures and knowledge-exchanges between buyers (in Kenya) and lead firm suppliers in more advanced economies. Lessons on the creation of 'learning spaces' in the Lake Turkana Wind Project (Kenya) and the Adama Wind Project (Ethiopia) (Gregersen and Gregersen, 2022; this volume) are presented along with lessons on capability development and accumulation of innovative capabilities in off-grid solar energy companies in Kenya and Tanzania (Karjalainen and Byrne, 2022; this volume).

The role of South-South technology collaboration in renewable energy is investigated with the purpose to explore to what extent and under what conditions renewable investments have economic co-benefits in terms of spill-overs and linkage development effects. One peculiarity of African renewable energy sectors is the rapid increase and likely future growth of Chinese involvement in large-scale renewable energy projects. We investigate to what extent economic co-benefits arise when Chinese investors develop renewable energy projects (Bhamidipati et al., 2022; this volume).

The book includes a chapter focused on Kenyan energy policies and legislation, including the increasing attention to local content issues that are closely linked to discussions regarding development of local capabilities. In light of current discussions about the need for building capabilities to ensure a green and sustainable development path in developing economies and the need for more emphasis on transformative innovation policies, we look at stakeholders' perspectives on local content requirements (Kingiri and Okemwa, 2022; this volume).

Finally, the book concludes by providing an overall assessment of successes and challenges in developing capabilities required for renewable electrification in Kenya and other countries with similar characteristics regarding electrification and/or engineering, design, and project management capabilities. Based on this assessment, the book offers suggestions for key stakeholders, such as universities, private sector actors, and policy makers (both domestic and foreign) in how to capitalise on the opportunities available for capability-building and industrial development through renewable electrification. Implications for pathways to transition to a green and low-carbon development are a key focus area of this chapter. Furthermore, the chapter suggests a number of outstanding questions and issues that would warrant future research (Lema et al., 2022; this volume).

Key themes in the book

As mentioned, the key overriding topic throughout the book is how to shape the development of renewable energy pathways to maximise co-benefits in terms of sustainable industrialisation through development of a broad range of (industrial) capabilities. In considering the role of learning and capabilities in supporting innovation in the renewable electrification sector in Africa, this book highlights three key themes or future research areas that need more attention. We come back to these in the final chapter.

Project design, organisation, and linkages

Renewable electrification activities – their design and construction – are most often conducted as 'projects'. These projects may have highly differentiated 'anatomies' which in turn may create different types and degrees of co-benefits. In this book we ask questions about the degree to which local actors are involved in renewable electrification projects. How much 'local content' is provided and where in the project is it located? Are local actors providing mission critical inputs, or inputs which are strategically less important? We also ask questions such as which linkages are involved, what is the nature of linkages between local and foreign actors, and to what extent do they include elements of knowledge transfer and capacity-building? To what extent (if at all) are issues of local content and capability-building reflected in upfront project planning, and are policies in place to facilitate their realisation?

Deployment model and choice of technology

The different natures that projects take may reflect wider 'modes' of renewable electrification. With regards to the type or shape of the technology, it is about whether renewable energy is produced in a centralised manner for 'the grid' (i.e., is used to power a set of households or businesses, often through a national power provider) or whether renewable energy is being used to power homes and businesses 'off-grid', i.e., in a decentralised and independent manner. It also relates to the importance of the type and size of the technology being used. Size of technology relates to small or large-scale projects or plants.

Predominantly, in the countries we look at in this book, small scale tends to be more off-grid in technology choice while on-grid tends to be more large scale.[3] In this book, we investigate how such characteristics matter for the associated (potential) opportunities for local industrial activity and capability-building in and around the project. We attempt to answer the following questions around the location of learning and capabilities building: what are the patterns of learning, capabilities, and outcomes across large and small-scale deployment models? How do such patterns differ between different renewable energy technologies such as wind and solar PV?

Policies and political actors at the national and global level

We consider in this book the degree to which global policies and schemes may influence renewable electrification at the local level, e.g., as development banks, donor agencies, and other political actors get involved in financing. Energy policies at the national and local level are obviously important, but the nature of the electrification process is also influenced by industrial and trade policy, educational policy, etc. A key broad proposition for this book is that deliberate and active policies are required for renewable electrification to contribute to sustainable industrialisation. Such benefits do not (necessarily) emerge as a fortunate by-product of renewable electrification; they are likely to be absent or very constrained unless they are specifically planned for in local or global policies. Thus, the chapters in this book look at the following questions: what are the main policy areas which influence learning and capability development for sustainable industrialisation (e.g., energy, industry, trade policy, etc.)? To what extent are deliberate policies implemented to facilitate capability development in this field? What opportunities and obstacles are important for ensuring optimal learning and capability development in renewable energy projects?

In summary

Through the key themes outlined above, this book engages substantially with debates about directions of development by examining distinct pathways of electrification with differing transformative potentials when it comes to wider benefits (beyond electricity) from electrification processes. We are particularly focused on looking at the processes of change – how pathways evolve over time (Leach et al., 2010). The case studies and chapters in this book investigate how the system of actors and technologies changes and adapts over time in response to events, new actors' entry, changes in policy, and technology. We focus, more specifically, on how policies on electrification and related policies on local content, certification of electrical contractors, etc. lead to specific dynamics in the system that lead to particular technological pathways and/or lock-in to certain types of actors and processes. We also focus on the relationship between different technologies and how the actors interact with them as well as how system dynamics are changed by these interactions. Many of the chapters consider how inclusion in these systems leads to greater or lesser opportunities for inclusion of local actors in 'the system' – whether this is at an infrastructure project level; at the level of 'renewable electrification' as a sectoral system of innovation; or at the level of 'industrialisation debates' nationally, at county level, or at community level. They discuss to different extents what this inclusion means in terms of skills development, employability, and long-term opportunities for business. We are interested therefore in both 'green and just transformations' (Scoones, Leach, and Newell, 2015).

The book thus addresses local sustainable industrialisation 'outcomes' arising from renewable electrification processes and examines underlying explanatory factors such as the 'social choice' underlying particular directionalities in techno-economic progress and associated circumstances such as unequal distribution of technological capability and economic power between different actors and change agents.

To ensure increased integration of renewable energy in the electrification processes in Kenya and other developing economies, the development of local capabilities of various kinds are key. This book focuses on innovation capabilities – by which we mean both technological and other types of capabilities related, e.g., to evaluating the social, financial, and technical feasibility of different projects as well as planning, implementation, operation, and maintenance. In terms of technological capabilities, both capabilities to use and operate given forms of technologies in a specific context and capabilities to create and implement innovations in production to change the forms and configurations of current technologies are important.

By explicitly focusing on the early experience gained regarding capability-building in renewable energy projects in an African context and the provision of new primary data collected through both quantitative surveys and qualitative studies of selected renewable energy projects, the book contributes to setting a new direction of future innovation and industrialisation policies and practices, while taking into consideration issues of diversity and distribution of benefits in the renewable electrification processes.

Theoretically, we combine research fields (notably development studies, business and strategic management studies, innovation studies, and global value chain thinking) that are rarely combined. As many of the green technology value chains are highly globalised, we find that this unique combination contributes to a better understanding of ongoing renewable electrification processes and better opportunities for identification of policies that can effectively support development of capabilities for sustainable industrialisation.

Acknowledgements

Support from the Danish Ministry of Foreign Affairs, Grant: DFC 14-09AAU for research on which this chapter is based is gratefully acknowledged.

Notes

1 As such, Naudé (2019) argues that African countries have three different options for industrialisation. First, '*acquiring traditional manufacturing capabilities*', i.e., classical movement from agriculture to manufacturing over time. Second, '*fostering sectors with the characteristics of manufacturing*' or movement towards industrialisation through service sector (ICT, finance, tourism, etc.). Third, '*resurgent entrepreneurship-led industrialisation*' which is about development of niche high-productivity growth businesses, e.g., manufacturing parts of 3D printers or the aviation industry.
2 Sustainable development meets the needs of the present without compromising the needs of future (WCED, 1987).
3 This may be changing as more companies, for example in Kenya, are now installing their own medium-sized solar PV systems to reduce their dependency on the national grid.

References

African Union (2021) *AIDA – Accelerated Industrial Development for Africa* [Online]. Available at: https://au.int/en/ti/aida/about (accessed 15/08/21).

Aiginger, K. and Rodrik, D. (2020) 'Rebirth of industrial policy and an agenda for the twenty-first century', *Journal of Industry, Competition and Trade*, 20, pp. 189–207. https://doi.org/10.1007/s10842-019-00322-3

Andersen, M.H. and Lema, R. (2022) 'Towards a conceptual framework: Renewable electrification and sustainable industrialisation', in *Building Innovation Capabilities for Sustainable Industrialisation: Renewable Electrification in Developing Economies*. New York: Routledge. https://doi.org/10.4324/9781003054665-2

Archibugi, D. and Coco, A. (2005) 'Measuring technological capabilities at the country level: A survey and a menu for choice', *Research Policy*, 34, pp. 175–194. https://doi.org/10.1016/j.respol.2004.12.002

Baker, L. and Sovacool, B.K. (2017) 'The political economy of technological capabilities and global production networks in South Africa's wind and solar photovoltaic (PV) industries', *Political Geography*, 60, pp. 1–12. https://doi.org/10.1016/j.polgeo.2017.03.003

Battat, J., Frank, I. and Shen, X. (1996) *Suppliers to Multinationals: Linkage Programs to Strengthen Local Companies in Developing Countries*. Washington, DC: The World Bank.

Bhamidipati, P.L. et al. (2022) 'Chinese green energy projects in Sub-Saharan Africa: Are there co-benefits?', in *Building Innovation Capabilities for Sustainable Industrialisation: Renewable Electrification in Developing Economies*. New York: Routledge. https://doi.org/10.4324/9781003054665-10

Blimpo, M.P. and Cosgrove-Davies, M. (2019) *Electricity Access in Sub-Saharan Africa: Uptake, Reliability, and Complementary Factors for Economic Impact*. Africa Development Forum Series. Washington, DC: World Bank.

Gabardo, F.A., Pereima, J.B. and Einloft, P. (2017) 'The incorporation of structural change into growth theory: A historical appraisal', *EconomiA*, 18(3), pp. 392–410. https://doi.org/10.1016/j.econ.2017.05.003

Government of Kenya (2008) *Kenya Vision 2030*. [Online]. Retrieved from: http://vision2030.go.ke/ (Accessed: 20/08/20).

Gregersen, C. and Gregersen, B. (2022) 'Interactive learning spaces: Insights from two wind power megaprojects', in *Building Innovation Capabilities for Sustainable Industrialisation: Renewable Electrification in Developing Economies*. New York: Routledge. https://doi.org/10.4324/9781003054665-8

Hanlin, R. and Okemwa, J. (2022) 'Interactive learning and capability-building in critical projects', in *Building Innovation Capabilities for Sustainable Industrialisation: Renewable Electrification in Developing Economies*. New York: Routledge. https://doi.org/10.4324/9781003054665-7

Hansen, U.E. et al. (2022) 'Centralised and decentralised deployment models: Is small beautiful?', in *Building Innovation Capabilities for Sustainable Industrialisation: Renewable Electrification in Developing Economies*. New York: Routledge. https://doi.org/10.4324/9781003054665-4

Intergovernmental Panel on Climate Change (IPCC) (2018) *Global Warming of 1.5 C*. Paris: Intergovernmental Panel on Climate Change.

IRENA, IEA and REN21 (2018), 'Renewable Energy Policies in a Time of Transition'. IRENA, OECD/IEA and REN21.

Jackson, T. (2017) *Prosperity without Growth*. (2nd Ed.). Abingdon: Routledge.

Jacobson, A. (2007) 'Connective power: Solar electrification and social change in Kenya', *World Development*, 35(1), pp. 144–162. https://doi.org/10.1016/j.worlddev.2006.10.001

Karjalainen, J. and Byrne, R. (2022): 'Moving forward? Building foundational capabilities in Kenyan and Tanzanian off-grid solar PV firms', in *Building Innovation Capabilities for Sustainable Industrialisation: Renewable Electrification in Developing Economies*. New York: Routledge. https://doi.org/10.4324/9781003054665-9

Kenya National Bureau of Standards (KNBS) (2020) *Economic Survey*. Nairobi: Kenya National Bureau of Standards.

Kenya Power and Lighting Company (2017) *Kenya Power Confirms 5.9 Million Customers Connected to the Grid* [Online]. Available at: https://kplc.co.ke/content/item/1951/kenya-power-confirms-5.9-million-customers-connected-to-the-grid (Accessed: 23/06/2020).

Kingiri, A. and Okemwa, J. (2022) 'Local content and capabilities: Policy processes and stakeholders in Kenya', in *Building Innovation Capabilities for Sustainable Industrialisation: Renewable Electrification in Developing Economies*. New York: Routledge. https://doi.org/10.4324/9781003054665-11

KIPPRA (2018) 'Realizing the "Big Four" agenda through energy as an enabler', *Policy Monitor*, 9(3 January-March 2018). Nairobi: Kenya Institute for Public Policy Research and Analysis.

Kuznets, S. (1973) 'Modern economic growth: Findings and reflections', *America Economic Review*, 63, pp. 247–258.

Lay, J., Ondraczek, J. and Stoever, J. (2013) 'Renewables in the energy transition: Evidence on solar home systems and lighting fuel choice in Kenya', *Energy Economics*, 40, pp. 350–359. https://doi.org/10.1016/j.eneco.2013.07.024

Leach, M., Scoones, I. and Stirling, A. (2010) *Dynamic Sustainabilities: Technology, Environment, Social Justice*. London: Earthscan.

Lema, R., Iizuka, M. and Walz, R. (2015) 'Introduction to Low-Carbon Innovation and Development: Insights and Future Challenges for Research', *Innovation and Development*, 5(2), pp. 173–187.

Lema, R. et al. (2022) 'Renewable electrification pathways and sustainable industrialization: Lessons learned and their implications', in *Building Innovation Capabilities for Sustainable Industrialisation: Renewable Electrification in Developing Economies*. New York: Routledge. https://doi.org/10.4324/9781003054665-12

Lundvall, B. and Lema, R. (2014) 'Growth and structural change in Africa: Development strategies for the learning economy', *African Journal of Science, Technology, Innovation and Development*, 6, pp. 455–466. https://doi.org/10.1080/20421338.2014.979660

Naudé, W. (2019) *African Countries Can't Industrialise? Yes They Can*. [Online]. Available at: https://www.merit.unu.edu/african-countries-cant-industrialise-yes-they-can/ (Accessed: 24/06/2020).

Newfarmer, R., Page, J. and Tarp, F. (2018) *Industries without Smokestacks: Industrialization in Africa Reconsidered*. Oxford: Oxford University Press.

Nzila, C. and Korir, M. (2022) 'Are the capabilities for renewable electrification in place? A Kenyan firm-level survey', in *Building Innovation Capabilities for Sustainable Industrialisation: Renewable Electrification in Developing Economies*. New York: Routledge. https://doi.org/10.4324/9781003054665-6

Ockwell, D. and Mallett, A. (2013) 'Low carbon innovation and technology transfer' in Urban, F. and Nordensvärd, J. (eds.) *Low Carbon Development: Key Issues. Key Issues in Environment and Sustainability*. Abingdon: Routledge, pp. 109–128.

Ogeya, M.C., Osano, P., Kingiri, A. and Okemwa, J. (2022) 'Challenges and opportunities for the expansion of renewable electrification in Kenya', in *Building Innovation Capabilities for Sustainable Industrialisation: Renewable Electrification in Developing Economies*. New York: Routledge. https://doi.org/10.4324/9781003054665-3

Organisation for Economic Co-operation and Development (2011) *Towards Green Growth: A Summary for Policy Makers May 2011*. Paris: Organisation for Economic Co-operation and Development.

Opoku, E.E.O. and Yan, I.K.M. (2019) 'Industrialization as driver of sustainable economic growth in Africa', *The Journal of International Trade & Economic Development*, 28(1), pp. 30–56. https://doi.org/10.1080/09638199.2018.1483416

Oyelaran-Oyeyinka, B. and Adesina, A.A. (2020) *Structural Transformation in African Development: Industrialization and Urbanization Pathway*. London: Anthem Press.

Peters, J. and Sievert, M. (2016) 'Impacts of rural electrification revisited: The African context', *Journal of Development Effectiveness*, 8(3), pp. 327–345. https://doi.org/10.1080/19439342.2016.1178320

Reuters (2019) *Renewables* Top *90% of Kenyan Power with New 50 MW Solar Plant*. 13 December 2019. [Online]. Available at: https://af.reuters.com/article/topNews/idAFKBN1YH1JC-OZATP (Accessed: 24/06/2020).

Rivas Saiz, M. (2018) *More than One Billion People Do Not Have Access to Electricity. What Will It Take to Get Them Connected?* World Economic Forum. [Online]. Available at: https://www.weforum.org/agenda/2018/08/milagros-rivas-saiz-electricity-access-sdg7/ (Accessed: 24/06/2020).

Rockström, J., Steffen, W., Noone, K., Persson, Å., Chapin, III, F.S., Lambin, E., Lenton, T.M., Scheffer, M., Folke, C., Schellnhuber, H., Nykvist, B., De Wit, C.A., Hughes, T., van der Leeuw, S., Rodhe, H., Sörlin, S., Snyder, P.K., Costanza, R., Svedin, U., Falkenmark, M., Karlberg, L., Corell, R.W., Fabry, V.J., Hansen, J., Walker, B., Liverman, D., Richardson, K., Crutzen, P. and Foley, J. (2009) 'Planetary boundaries: Exploring the safe operating space for humanity', *Ecology and Society*, 14(2), p. 32. [Online]. Available at: http://www.ecologyandsociety.org/vol14/iss2/art32/ (Accessed: 02/09/20).

Rodrik, D. (2014) 'Green industrial policy', *Oxford Review of Economic Policy*, 30(3), pp. 469–491. https://doi.org/10.1093/oxrep/gru025

Rodrik, D. (2018) 'An African growth miracle?', *Journal of African Economies*, 27(1), 10–27.

Sampath, P.G. (2016) 'Sustainable industrialization in Africa: Toward a new development agenda' in Oyelaran-Oyeyinka, B. and Sampath, P.G. (eds.) *Sustainable Industrialization in Africa: Toward a New Development Agenda*. Basingstoke: Palgrave Macmillan, pp. 1–19.

Scoones, I., Leach, M. and Newell, P. (2015) *The Politics of Green Transformation*. London: Routledge.

SEforALL 2020: Sustainable Energy for All, Vienna. www.se4all.org

Szirmai, A., Naudé, W. and Alcorta, L. (2013) *Pathways to Industrialization in the Twenty-First Century: New Challenges and Emerging Paradigms*. Oxford, UK: Oxford University Press Online.

UN (1987) *Report of the World Commission on Environment and Development: Our Common Future*. [Online]. Available at: https://sustainabledevelopment.un.org/content/documents/5987our-common-future.pdf (Accessed: 20/05/21).

United Nations Environment Programme (2014) *Decoupling 2: Technologies, Opportunities and Policy Options*. A Report of the Working Group on Decoupling to the International Resource Panel. Nairobi: United Nations Environment Programme.

United Nations Industrial Development Organization (UNIDO) (2009) *A Greener Footprint for Industry: Opportunities and Challenges of Sustainable Industrial Development*. Vienna: UNIDO.

UNIDO (2011) *Green Industry: Policies for Supporting Green Industry*. Vienna: UNIDO.

UNIDO (2014) *Inclusive and Sustainable Industrial Development (ISID): Creating Shared Prosperity*. Vienna: UNIDO.

UNIDO (2017) *Industrial Development Report 2018. Demand for Manufacturing: Driving Inclusive and Sustainable Industrial Development*. Vienna: UNIDO.

UNIDO (2020) *Creating Shared Prosperity*. [Online]. Available at: https://www.unido.org/our-focus/creating-shared-prosperity (Accessed: 24/06/2020).

Wandera, F. (2022) 'Understanding the diffusion of small wind turbines in Kenya. A technological innovation systems approach', in *Building Innovation Capabilities for Sustainable Industrialisation: Renewable Electrification in Developing Economies*. New York: Routledge. https://doi.org/10.4324/9781003054665-5

World Bank (2012) *Inclusive Green Growth Policies for Real World Challenges*. [Online]. Available at: https://www.worldbank.org/en/news/feature/2012/05/09/inclusive-green-growth-policies-real-world-challenges (Accessed: 23/06/2020).

World Bank (2020) *Access to Electricity (% of Population): Kenya*. [Online]. Available at: https://data.worldbank.org/indicator/EG.ELC.ACCS.ZS?locations=KE (Accessed: 23/06/2020).

World Commission on Environment and Development (WCED) (1987) *Our Common Future*. London: Oxford University Press.

2

TOWARDS A CONCEPTUAL FRAMEWORK

Renewable electrification and sustainable industrialisation

Margrethe Holm Andersen and Rasmus Lema

Abstract

Renewable electrification is advancing at unprecedented rates across the developing world. The key purpose of this chapter is to define a conceptual framework for understanding and analysing how the up-take of renewable energy, i.e., the pathways and processes of electrification based on renewables, may support sustainable industrialisation. Our theoretical point of departure includes a combination of different types of innovation system theory, global value chain thinking, and project-based approaches with clear links to the literature about technology transfer understood as an interactive rather than a linear process. We suggest that renewable energy projects can be seen as embedded in different levels of innovation systems and global value chains combining what we refer to as 'the nested view'. Our conceptual framework furthermore suggests that intra- and inter-active learning processes related to renewable technologies may lead to accumulation of key capabilities that are in turn essential for the creation of new jobs and business opportunities (outcomes) and which may be useful for further sustainable industrialisation processes and development (long-term impact). The framework has been developed in connection with research on solar and wind energy projects in sub-Saharan Africa and how processes of electrification can be shaped to maximise co-benefits in terms of industrialisation that is green, inclusive, and durable. We argue that increased awareness about the implications of different pathways and the need for an engaged, deliberate learning approach taking into account the consequences of different choices for development of (innovative) capabilities is central for making use of the window of opportunity that the current increase in investments in renewable energy constitutes.

DOI: 10.4324/9781003054665-2

Introduction

Until recently, industrialisation and (renewable) energy were mostly discussed as belonging to separate development policy domains. The connection between the two was one-way and was confined to the energy inputs needed to uphold manufacturing and services activities associated with industrialisation. This is no longer the case, however. In large developing countries like China and India, renewables are becoming part of the industrialisation strategy itself (Lema, Iizuka, and Walz, 2015; Mathews and Tan, 2013). Production of renewables creates manufacturing jobs, and their deployments generate highly skilled service employment. Of the almost 9.5 million jobs in sectors producing renewable energy worldwide, 60% are in Asia, and global renewable employment creation continues to shift towards Asia (IRENA, 2016). The Asian giants thus show a way of reframing industrialisation strategies based on renewables (Mathews, 2018).[1] The trend is currently continuing and spreading to other parts of the world (Mazzucato, 2018).

This chapter provides a framework for investigating the two overall research questions that shape this book: do renewable electrification efforts provide opportunities for sustainable industrialisation? If so, what are the opportunities and how can different renewable energy pathways be shaped in a way that supports the realisation of such opportunities?

In order to understand the potential of renewable electrification for sustainable industrialisation and the conditions required to turn such opportunities into reality, we will combine three analytical perspectives: Global Value Chains (GVCs), Innovation Systems thinking, and Project Focused Approaches. The intention here is not to make a broad and encompassing review of the literature in these fields, but rather to map out how the three different theoretical frameworks/approaches can be combined to analyse links between (renewable) electrification processes and sustainable industrialisation.

The connection between industrialisation and renewable energy, however, takes different forms in different parts of the developing world. Although the need for deployment of renewables is greatest in low and middle-income countries, the production and innovation capabilities required to integrate energy and industrial development strategies tend to be weak. Hence, it can be argued that building up innovation capabilities for renewable electrification constitutes an important missing link in ensuring the transition to a more sustainable development in such countries.

In order to explore this missing link, we investigate literature that may help us understand the linkages between learning, capabilities, and outcomes understood as capabilities (technological as well as organisational and project related). In other words, we want to understand how renewable electrification 'works'. On this basis, the chapter presents an analytical framework for analysing whether renewable electrification efforts provide opportunities for sustainable industrialisation, breaking down the capability-related co-benefits in the renewable electrification processes into three key elements: learning, development of capabilities, and the resulting (expected) outcomes. We also discuss connections between these elements – and how they may link up with sustainable industrialisation efforts. Finally, we conclude the chapter by providing some initial answers to the overall research questions and outline how the framework relates to subsequent chapters in the book.

Understanding renewable electrification processes

Renewable electrification is essentially equivalent to the production, deployment, and use of renewable energy. It involves several steps across three sets of chains: a production, a deployment chain, and a user chain. The production chain focuses on the production of core elements in renewable energy, such as wind turbines and solar photovoltaics (PV), and includes product engineering and design, component manufacturing, and equipment assembly. The deployment chain focuses on how such key technologies are put to use in specific contexts and countries and includes planning, finance, construction, connection, operation, maintenance, distribution, and consumption. Finally, the user chain relates to distribution and consumption of energy – and the technologies used to secure this, e.g., national electricity grids, systems for distribution energy in mini-grids, and technologies such as mobile money used to support the distribution of energy through small-scale solar PV systems for individual households.

In each step of these chains, there are multiple actors involved (Lema, Rabellotti, and Sampath, 2018b) ranging from foreign technology suppliers involved in the production of core technologies, to regional or local companies importing core technologies, or, assisting in assembling systems near local markets to companies engaged in transporting elements of renewable energy systems to the project site, preparing the project site, and connecting the energy producing systems to national grids or mini-grids. Different energy providers (public, community based, or private) are involved in delivery of the electricity to end users (institutions, firms, households) through connections to the national grid, mini-grids, or indeed through small solar PV systems targeting individual households.

The interactions between the different actors at the different steps of the value chains differ and may – as we will see in the empirical examples provided in this book – provide more or less efficient and effective possibilities for learning and building different types of capabilities. Such options are also influenced by the conditions under which firms produce and deploy different types of renewable energy technologies, including international policy regimes and national policies, strategies, rules, and regulations. The differences in the value chains also influence the extent to which they may contribute to more inclusive innovation and development pathways.

Another issue critical for understanding how renewable energy projects work is the size of the technology, which is typically also associated with the form or shape of the technology in use (Hansen et al., 2018): basically, there are three models or pathways:

1. *Large-scale grid-connected renewable energy projects*: based, for example, on solar or wind energy and requires extension of grids into areas that currently do not have access.
2. *Mini-grids using renewable energy that are not connected to the national grid*: self-contained grids established in rural villages, using micro hydro, solar, and micro wind – or a mix of these.
3. *Off-grid approach*: electricity generation tied to the individual household or factory – typically solar rooftop solutions or pico-solar systems.

The three different pathways provide different opportunities not least for engaging local firms and actors. It is also evident that electrification in low and lower-middle-income countries, e.g., in sub-Saharan Africa, will involve all three pathways. But important questions remain about the balance between the various pathways, what options they provide for learning and capability-building – and how they can be set up to maximise inclusiveness and economic development (Lema et al., 2018b; Leach et al., 2010). One underlying issue in this chapter is to create the foundation for comparing the main pathways and to study capability-development dynamics within them as well as the implication of these dynamics for fostering broader sustainable industrialisation.

Theoretical starting points

The *global value chain approach* focuses on the way in which value is added to at different stages of the production process and how these are located in different countries. It has been used to analyse successful catching up, e.g., of Asian economies (Gereffi and Korzeniewicz, 1994), but also – more recently – in critical analyses of the linkages between global value chains and sustainability governance (Ponte, 2019). The *national innovation system* concept was introduced by Lundvall (1985a, b) and further elaborated in Lundvall (1992) and Lundvall et al. (2002). It has been used to understand the systemic nature of innovation in the context of economic development (Lundvall et al., 2009). The two concepts of global value chains and national innovation systems have developed in parallel and both are increasingly used to analyse economic development (Bolwig et al., 2010; Ponte and Ewert, 2009). They are also increasingly being linked to the global challenges of climate change and analyses of how different actors may or may not help solve such challenges (Ponte, 2019; Mazzucato, 2018).

The innovation system perspective points to the need to establish and develop domestic linkages while the value chain perspective is concerned with alignment of and power relationships between global lead firms and domestic actors.

We see it as a promising line of research to combine them. Although initial theoretical attempts have been made in this regard (Pietrobelli and Rabellotti, 2011; Lema, Rabellotti, and Sampath, 2018a), few prior studies have brought these approaches together in an operational way for empirical analysis.[2] There is, therefore, a need to develop a framework that can facilitate this – and a big need for investigating further and in a more disaggregated manner, the way interactions between the global value chains and the local and national innovation systems interact.

The third element of the analytical framework presented in this chapter and used to different extents by authors of the specific chapters in the book is the project-focused approaches. A focus on projects and the way they are organised is key to an improved understanding of how renewable electrification processes play out on the ground, what learning takes place in different contexts, and what capabilities are (or are not) generated (see e.g., Hanlin and Okemwa, this volume).

Innovation systems and renewable electrification

The innovation systems approach often analyses national systems of production and innovation to explain variations in innovation performance across countries. Scholars in the field emphasise the role of different types of learning and suggest that both intra- and interactive learning are key to increasing innovation performance (Lundvall, 1985a; Jensen et al., 2007). Analyses of national systems of innovation can help to identify 'system failures' that lead to less advantageous outcomes. In this chapter we do not seek to compare innovation systems in sub-Saharan Africa and more advanced nations that can be considered 'lead markets' in wind (e.g., Denmark) and solar PV (e.g., Germany and China). We rather focus on what opportunities, if any, renewable energy projects and different pathways for renewable electrification may entail for sustainable industrialisation *within* sub-Saharan Africa. Our analysis is concerned with broader issues and implications of innovation, including the choice of technologies and distribution of benefits from different types of innovations (Stirling, A, 2009; STEPS, 2010), but with a particular focus on these issues based on an analysis of renewable energy projects in the region (cf. also Ockwell and Byrne, 2016).

The focus on solar and wind energy sectors makes the 'sectoral system' approach relevant since it has specialised in sectoral comparisons (e.g., Malerba and Nelson, 2011). The sectoral approach needs to be combined with insights from work that has focused on 'system building' in the case of developing innovation systems (Lundvall, 2007). Research has also shown that the sectoral system approach can benefit from the introduction of a more disaggregated perspective. This has led to the introduction of an approach where a distinction is made between both different types of (sub) technologies and different sizes of projects (large, grid-connected wind, and solar power plants vs. wind or solar powered mini-grids) (Hansen et al., 2018; Hansen et al., this volume). Finally, research using the Technological Innovation System thinking as a framework (Kebede, Mitsufuji, and Islam, 2015; Wandera, 2018) indicates that this framework can also be used to shed light on issues and missing links in the up-take of renewable energy in sub-Saharan Africa.

Stronger innovation systems can facilitate enhanced up-take and use of new technologies, including renewables (Lema, Rabellotti, and Sampath, 2018b), but they do not necessarily develop by themselves.

At the national level there are numerous relevant policy domains which cross across different renewable energy sectors, not least those concerned with feed-in tariffs, electricity generation licences and permits, etc. There are cross-ministry development plans which (in principle) synchronise regulation across ministries. There are state policy and regulatory bodies, utilities, transmission systems operators, and education systems. In the private sector there are local equipment manufacturers and assemblers, wholesale importers and distributors, logistics firms, sectoral trade organisations, and many more. There are vertical value chain links as well as horizontal links within the systems that provide various types of inputs to the electrification processes.

In subsequent sections of this chapter we draw in an eclectic way on various types of innovation systems thinking. We find that both national, sectoral, and technology specific innovation systems thinking may help create a better understanding of processes required to increase up-take of renewables, such as solar and wind. More importantly, different systems-traditions have different strength and weaknesses. For example, the Technological Innovation Systems (TIS) approach has developed a static but easily applicable framework for identifying inducing and blocking mechanisms for technological development and diffusion. This has often been used to study renewable energy in Europe. However, it has also been used in studies of renewable energy diffusion in the African context (Kebede and Mitsufuji, 2017; Wandera, 2018).

In summary, we find that various approaches to innovation systems thinking (National Innovation Systems [NIS], Sectoral Innovation Systems [SIS], and TIS) are complementary and can be used to inform our conceptual framework for analysing the development of capabilities through renewable electrification activities and projects – capabilities which may in turn have potential for development of sustainable industrialisation more generally.

Global value chains

The Global Value Chain (GVC) approach typically analyses relationships between lead firms from advanced economies and suppliers in developing economies. From the developing country perspective, it is typically about 'learning from exporting' (upgrading in GVCs). Our focus is the reverse: 'learning from importing' renewable energy products and services. In this context we are concerned with governance structures and knowledge exchanges between buyers and lead firm suppliers in more advanced economies. There are clear connections to literature on technology transfer (Bell, 2007, 2012; Ockwell and Mallet, 2013) and to more recent work focused on technology transfer and local capability formation (Lema, Rabellotti, and Sampath, 2018b; Hansen et al., this volume).

Global value chain thinking (Gereffi, 2014; Humphrey and Schmitz, 2002; Kaplinsky and Morris, 2000) may contribute to a better understanding of learning opportunities in renewable energy projects and activities (low carbon development) in various ways. First, it moves beyond a narrow focus on technology producers (exporters) and users (importers) and thereby helps map out the many actors in the value chains; their role in technological transfer processes and in interactive learning related to development of local capabilities required both for renewable energy and sustainable industrialisation in a broader perspective. Second, the global value chain thinking may help enhance our understanding of issues related to power relations in technology transfer and capability development – and not least the role of powerful lead firms in specific (global) value chains.

The global level governance structures and knowledge exchanges include such domains as trade regulation such as World Trade Organisation (WTO), global standards by the International Organisation for Standards (ISO), and institutional infrastructures concerned with the Technology Mechanism linked to the UN Climate Change process (and, before it, the Clean Development Mechanism (CDM)) within the United Nations Framework Convention on Climate Change (UNFCCC). Global level governance structures and knowledge exchanges also include large equipment producers (think Vestas and Yingli Solar) and project developers, and investment funds, consultancy firms, NGOs, and providers of overseas development assistance. At the global level there are, not least, inter-national linkages formed 'vertically' between producers, project developers, financers, and consultancies in China and Europe and renewable electrification firms/organisation in, for instance, Kenya.

The various networks and linkages established form opportunities for learning, but they may also limit learning opportunities for local stakeholders, for instance when a lead company does not want to share key knowledge related to the technology they are bringing along. In other words, the chain governance influences the possibilities for extracting more economic value through learning in the South (Lema, Rabellotti, and Sampath, 2018b, p. 5). The technology transfer chains linked to e.g., renewable energy, are in a way *reverse value chains* that may (or may not, depending on how organised) help develop local capabilities.

Opportunities for learning exist both in the production, deployment, and use steps of the value chain (cf. Figure 2.1 above) and may be driven both by actors from the North and the South. Empirically, however, there is very little evidence of development of local capability formation and industrial development in renewable energy industries in sub-Saharan Africa irrespective of where the technology comes from (Lema, Rabellotti, and Sampath, 2018b; Bellini, 2017) and so far most of the outcomes in terms of jobs and income opportunities

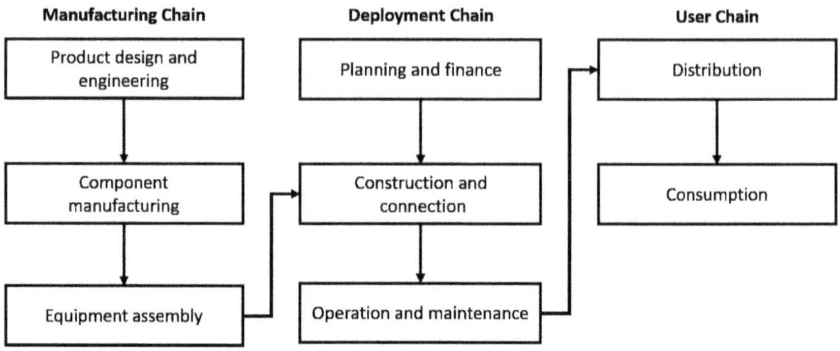

FIGURE 2.1 Steps in the production, deployment and use of renewable energy. Source: drawing based on Lema, Quadros, and Schmitz (2015b) and Lema et al. (2016).

linked to renewable energy projects seem to be linked to the deployment steps, i.e., installation (Lema, Rabellotti, and Sampath, 2018b; Hansen et al., 2018). The capability-building found in this part of the value chains is less 'hardware' oriented than in the manufacturing part of the value chain as it is often focused on servicing and involves operating skills and know-how. But evidently, such skills and capabilities are key to ensuring durable and sustainable use of the new technologies installed and avoid a whole new generation of large (and small) white elephants in the form of renewable energy facilities that end up not functioning.

In short, we consider the value chain approach fruitful for a more systematic investigation of learning opportunities and capability development in renewable electrification (RE) value chains as well as for revealing opportunities for sustainable industrialisation potentially arising from renewable electrification projects. We also find that a more explicit focus on opportunities related to the RE value chains may help identify critical learning and capability gaps that need to be addressed – and which are in turn critical for the required strengthening of local and national-sectoral innovation systems.

Project-focused approaches

The innovation system and value chains approaches are complementary since they emphasise different aspects of economic interaction and because they focus on different units of analysis (Jurowetzki, Lema, and Lundvall, 2018). They are helpful in guiding comprehensive analyses of the circumstances and relationships that will structure the provision of sustainable energy. In particular, they also combine when one seeks to address *user-producer interaction* (as will be discussed below).

But in a certain respect they are both insufficient – individually and when combined – i.e., in terms of scope or unit of analysis. This is because *renewable electrification is essentially a project-based activity.* It implies that a project-lens can help to examine specific cases of renewable energy deployment and trace interactions within the project and beyond it (nationally and globally).

There is a wealth of literature to draw upon, including literature on project management and innovation (Brady and Hobday, 2011). Most of this literature is, however, focused on advanced economies and/or very large-scale projects. Still, there is an innovation and development literature on design and engineering which seems relevant since it has explicit connections to both national systems and global value chains literature. This includes unpublished work by Abdelkadar Djeflat and colleagues on the need to develop design engineering capacity and innovation in North Africa[3] and the pioneering work by Bell (2007) on design and engineering in infrastructure and industrial sectors. The key point here is that there are different stages in all projects, ranging from project development over procurement or manufacturing of the technologies to installation, operation, and maintenance. And that each of these steps includes opportunities for learning.

Bell (2007) distinguishes between (a) owner-driven, (b) contractor-driven, or (c) jointly driven project structures. Owner-driven project modes – where the owner-operator (e.g., the Kenya Electricity Generating Company PLC, which is the leading electric power generating company in East Africa) takes charge of coordination and execution – were dominant until about 20 years ago. This is no longer the case because of increasing outsourcing of design and engineering activities to competent contractors, many of which are multinational firms. Owner-driven models might still, however, be commonplace in *large and grid-connected* facilities where the owner is a utility or a private investor. The Lake Turkana Wind Power Project in Kenya seems to follow such an owner-driven model. Contractor-driven models may be prevalent in case of *mini-grids* where there is community-based ownership or smaller private ownership. Whether this is the case is a question for empirical analysis. In both cases there is a substantial element of involvement of Multi National Corporations (MNC), even if these are sometimes small design and engineering firms, e.g., small and medium solar system providers from Germany. Jointly managed projects involve extensive involvement of both owners and specialist contractors. The Garissa Solar Project may be seen as an example of this in the sense that the project is owned by the Kenyan Rural Electrification and Renewable Energy Cooperation (REREC), but has been developed and installed by specialist contractors from China with some inputs being provided by local suppliers.

There is also a literature on 'strategic niche management' (Geels, 2002) which may be relevant. From a science and technology systems perspective, this literature looks at how small technological niches emerge and grow to influence overall technological trajectories. Authors from this tradition have begun investigating 'sustainability experiments' (projects) as niches in developing countries (Berkhout et al., 2010). Some of this literature has been specifically concerned with rural electrification in developing countries (Drinkwaard, Kirkels, and

Romijn, 2010; Romijn, Raven, and de Visser, 2010). Such literature seems particularly useful as a starting point for thinking about the relationship between projects and larger sectoral systems.

At the project level there are contracts specifying agreed rules, and roadmaps and project design documents that stipulate who does what and how. The project comprises the various actors involved in installing and operating wind or solar PV technology, including principally firms and other actors involved in installation and operation of renewable energy projects. These may include owners, project development firms, equipment producers, operators, maintenance firms, etc. Production and consumption of electricity is physically co-located in the case of mini-grids, but it is separated in the case of grid-connected facilities. In other words, end users are not necessarily an 'actor' in these projects.

While innovation systems linkages may be 'durably' and slowly evolving, interactions in projects are often more 'temporal' and/or 'sequential': a project is typically time-bound and linked to the development of a new product or service e.g., a power plant run by either hydro-power, solar, or wind. When a project is completed, a large number of actors involved will typically move on to a new project. Projects typically cover the full project cycle from initial idea and design through construction and may or may not extend into operation and management phase. This poses particular challenges in countries where general technical and organisational capabilities may not be sufficiently developed. For instance, there is often particularly intense interaction between project participants during the phase of project design, engineering, and installation, while interaction during operation (and maintenance) depends to a large extent on whether service contracts are entered into and whether suppliers of energy sources have a responsibility to assist in making sure the projects remain operational. Different actors may focus on becoming good in one particular part of the project cycle or might span across several different parts of the project cycle. This is also linked to who drives the project (contractor or operator).

A nested view

Figure 2.2 below seeks to provide a simple illustration of how the three levels of analysis can be combined in a 'nested view'. Economic geographers working on global production networks (Henderson et al., 2002) have proposed such a view, but the model is here adapted to the specific topic of concern, i.e., electrification processes in low and lower-middle-income countries.

Nested view models have typically derived from analyses of producer-driven (see below) value chains manufacturing and integrating modular consumer goods such as automobiles and electronics (see e.g., Coe, Dickens, and Hess, 2008; Sturgeon, Van Biesebroeck, and Gereffi, 2008). They have therefore tended to focus primarily on the production landscape, while largely by-passing the interface with consumption (the individual or collective consumer). The value-added chains, which are analysed, tend to start from suppliers in developing countries,

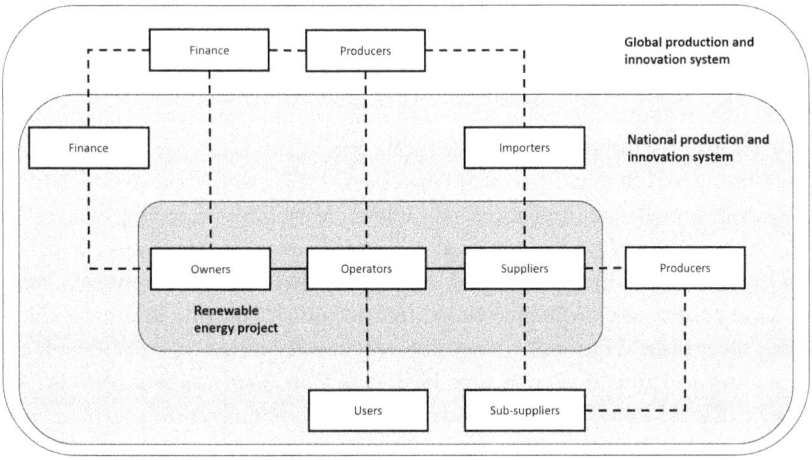

FIGURE 2.2 Nested view. Source: authors.

providing inputs for large lead firms which cater for consumers globally, not least in advanced economies.

The renewable electrification setting is markedly different. The purpose here is therefore to draw only loosely on the notions proposed by Coe, Dicken, and Hess (2008) and others to conceptualise nested relationships in sectors, which are essentially concerned with the provision of (energy) services in the context of renewable electrification in developing countries. This means that the renewable energy project, including the interface with professional or collective consumers (e.g., utilities, big firms or institutions, local communities), constitutes the 'micro level' in the nested view adopted here. It situates renewable energy projects within the national-sectoral and global context.

This view thus comprises three main levels of analysis:

- Global solar/wind industry and governance mechanisms (global level)
- National-sectoral system of innovation production (national level)
- Renewable energy project (project level)

The task is to examine each level both in separate studies focusing on key actors and institutions in and around (a) global value chains, (b) national-sectoral innovation systems, and (c) critical projects, and to trace and examine the interactions between them.

By doing so we aim to explore how learning and capability-building is related to each of the three levels. One key proposition is that the governance of the global value chains (how they are organised, who dominates which parts of the chain, and what power relations are embedded in these) impact the types of learning and capability-building that takes place or may take place. Some MNCs may for instance give high priority to training of local staff because it helps them reduce costs and strengthens their business model, while other international actors may be more hesitant to share knowledge with local actors. Similarly, international policies and financing mechanisms (think e.g., bilateral and regional funding agencies such as the Nordic Investment Fund and the Danish Fund for Investment in Developing Countries) may for instance encourage local content and local learning and capability-building to different degrees and thereby influence the extent to which local actors (firms, knowledge institutions) are able to 'learn from importing'.

The characteristics of the national-sectoral systems of innovation and the actors in these, including policy makers, the financing system, importers, and other local actors, also influence the extent to which learning and capability-building takes place. National energy policies may, for instance, encourage local contents and learning – or may not be focused on such opportunities at all. Similarly, if local knowledge institutions (e.g., vocational training centres and universities) provide local technicians and engineers with a solid level of basic knowledge, e.g., in wiring or other fields, this can enhance possibilities that local staff can benefit from interaction with external actors and enhance their capabilities. A diverse range of pathways (small/large and using different sources of energy) may thus help foster more inclusive development.

Towards an analytical framework

A key issue in this chapter is the current trends in renewable energy production (notably solar and wind), the pathways linked to these trends, and in particular how different sources of technology and their associated characteristics influence possibilities for local competence-building and technology adaptation. These issues are crucial for strategies aimed at using green technologies for sustainable industrialisation and transition in a broad range of low and lower-middle-income countries in sub-Saharan Africa and beyond. They are also crucial because different innovation pathways towards electrification represent different directions with implications for distribution (winners/losers) and diversity (ensuring that multiple options are considered). We argue that increased awareness about the implications of different pathways and the need for an engaged, deliberate learning approach taking into account the consequences of different choices for development of (innovative) capabilities is central.

The purpose of this section is to outline a basic analytical framework for analysing these issues. The framework – presented in Figure 2.3 below – breaks down the renewable electrification process into three key elements: learning, development of capabilities, and the resulting (expected) outcomes – with long-term impact in the form of sustainable industrialisation and other socio-economic benefits resulting from this process. In the following, we will elaborate on each of these three elements considered key for the 'framing' of our analysis of how the renewable electrification process plays out in the countries we are dealing with in the empirical analyses conducted throughout this book.

A basic analytical framework

Our basic analytical framework (Figure 2.3 below) is essentially a theory of change prepared on the basis of the combined reading of the literature discussed above. It is based on the assumption that interactive (in projects, national innovation systems, and in global value chains) as well as intra-active (or organisational) learning (in firms but also in other organisations) are vital to the development of (technological) capabilities. The development of enhanced technological capabilities is in turn expected to lead to outcomes such as increased employment, local business opportunities, and content of contracts, new firms, and more 'inclusive' and relevant electrification processes.

The long-term impact of such outcomes is expected to include increased access to electricity, increased energy security, and related socio-economic benefits such as improved health, increased education levels, and enhanced incomes. Sustainable industrialisation, i.e., industrialisation that is more durable, greener, and more inclusive, will also come about as a long-term impact if renewable electrification processes are properly managed and made use of.

Obviously, there is no automatic link between each of these elements, and a number of factors not made explicit in the model will influence the extent to which the outcomes and long-term impact are actually achieved. Such factors may have to do with the willingness of foreign firms to share knowledge with local actors and with local actors' existing levels of education and capabilities to take up new jobs generated as RE projects are implemented in a country.

Even so, the underlying idea of the basic analytical framework is that understanding the way inter- and intra-active learning unfolds in specific renewable

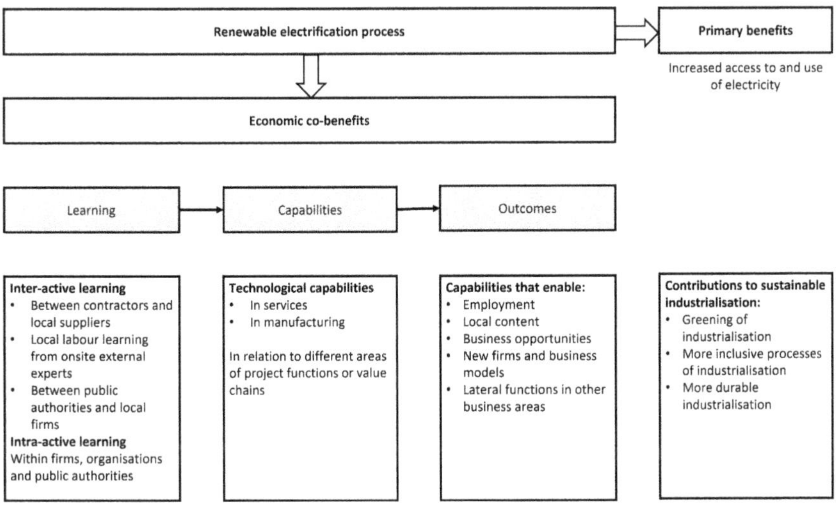

FIGURE 2.3 Renewable electrification processes: learning, capabilities, and outcomes. Source: authors.

energy projects, national innovation systems, and in specific sectoral value chains forming part of the global value chains in solar and wind is key to understanding which capabilities are developed.

Likewise, it is an underlying idea of the framework that developing (technological) capabilities is key to ensuring desired outcomes in the form of employment, local content, business opportunities, and establishment of new firms as well as to more 'inclusive' and relevant electrification processes. Our primary focus in the chapter is on the two first boxes of the model dealing with learning and building (technological) capabilities.

Understanding what capabilities are available in a specific country and the learning processes and processes linked to building different forms of capabilities may also help us understand why some capabilities are not created and what could be done to facilitate the generation of these. In this book the question of capability development is reviewed on the basis of both quantitative surveys and more qualitative work (e.g., case studies of critical projects).

Finally, feedback mechanisms are important: as local staff, firms, projects, and knowledge institutions as well as government bodies build up knowledge in RE, there is a possibility that they become better equipped to take advantage of new knowledge and technologies brought into a particular country e.g., by foreign firms. Virtuous circles may be generated that help increase the stock of capabilities and hence the absorption capacity.

Learning

Learning is a fundamental prerequisite for sustainable economic growth and development (Lema, Rabellotti, and Sampath, 2018b; Cimoli, Dosi, and Stiglitz, 2009). Learning is understood here as the accumulation of relevant capabilities; we are informed by the increasing body of literature which emphasises the importance of local production and innovation capabilities for effective low carbon development. Figure 2.3 illustrates that there are many learning mechanisms – i.e., ways in which learning can take place. Some of these are intra-active (understood as internal to the firm) and some of these are interactive (i.e., include interaction with actors outside the firm).

Inside firms, learning may take place through on-the-job training, through in-house training courses or seminars, or through internal knowledge exchange platforms (intranet), etc. The learning may be more or less formalised – and often will also include use of knowledge from outside e.g., in the form of presentations of new research. In other words, intra- and interactive learning are often combined, and research evidence suggests that the most innovative firms are those that are able to combine practical and interactive learning (Doing-Using-Interacting or DUI) with more scientific insights (Science, Technology, and Innovation or STI) (Jensen et al., 2007). Without neglecting the importance of intra-active learning, the analytical framework developed in this chapter places particular emphasis on inter-active learning understood as learning in interaction with others at three different levels: renewable energy projects (often involving several or even many different firms), the national-sectoral innovation systems (with particular emphasis on solar and wind), and the global value chains in solar and wind as they unfold, develop, and relate to the context experienced in Kenya and other low and lower-middle income countries.

The learning taking place may be more or less strongly tied to specific sectoral innovation systems and to the development of specific technologies, such as e.g., solar PV (see Karjalainen and Byrne, this volume). As such, the degree to which the learning is specific for a sector has implications for subsequent parts of the theory of change, e.g., the extent to which lateral transfer and reuse of capabilities occurs.

What types of learning are particularly important to ensure sustainable industrialisation benefits from renewable electrification? This is a question for empirical research, but the following types (or arenas) of interactive learning have been identified in the extant literature (Lema, Rabellotti, and Sampath, 2018b):

Interactive learning between contractors and local suppliers

Service providers and local firms play a key role in the RE value chains and the importance of their interactions with their clients, especially the main contractor of a wind or solar PV project; this main contractor is essentially a professional user (as opposed to an end-user; see Lundvall, 1985a, b). Professional users have more defined needs in terms of what products and services they require and as such, a good level of interaction – which focuses on learning the needs and wants of the other – between suppliers and the main contractor will ensure a more efficient project and should reduce delays. This is important on both sides, not only for building up competences of local suppliers and their reputation in the market, but also reducing the 'lock-in' of dominant sourcing policies of lead firms in a project setting (Hanlin and Hanlin, 2012). Both are needed if strong backward and forward linkages are to be created within GVCs. These need to be encouraged by governments so as to create the development of dynamic capabilities in firms and through these a stronger more diversified economic base (Morris, Kaplinsky, and Kaplan, 2012; Lundvall and Lema, 2016).

In a situation where learning takes place in a series of single projects – each with a new constellation of users and producers – a major issue is what institutional setting allows for an accumulation of capabilities. Firms that take on the role of coordinating large projects such as the Lake Turkana Wind Power Project (LTWP) in Kenya e.g., through an Engineering, Procurement, Construction, and Management (EPCM) contract (Hanlin and Okemwa, this volume) are one possible way of accumulating knowledge and capabilities, in particular if they are able to successfully bid for a series of projects and retain (key) staff. But small organisations with civil engineering competences in addition to government entities may also play a role here if they have an explicit focus on fostering accumulation and dissemination of learning.

Local labour learning to 'use' the new installations – operating and maintaining

As indicated above, less codified knowledge and experiential learning that comes through 'doing, using and interacting' (Jensen et al., 2007) is an important (but often not well recognised) way of learning. Local staff need to understand how different parts of the technology within a solar PV or wind project work and to be provided with the opportunity to take over the 'operations and maintenance' part of the project lifecycle – getting some of the practical experience required to maintain the systems. Requirements for increased local content in new small or large renewable energy projects being established in a country (e.g., LTWP in Kenya or the wide range of solar energy projects in e.g., Tanzania and Kenya) may increase such options. It requires, however, not just a new mindset from lead firms in the projects, but also government support of the relevant training and education needed to ensure there are technicians and/or engineers available locally to conduct such work. In some cases, lead firms (whether large or small) may organise training both on-site and in other countries to ensure local staff develop the knowledge required for operating and maintaining the systems installed. They are more likely to do so, if they can reduce costs by hiring more staff locally while keeping in place systems that ensure the necessary quality control. In other cases, lead firms may prefer to bring in and use their own staff (Hansen, 2019).

Public authorities and private companies learning to manage major projects

Public authorities are key in terms of regulating and supporting renewable electrification through promotion of training schemes etc. However, sometimes governments are also *de facto* the 'lead firms' in projects: commissioning, managing, and/or running RE projects once construction is completed. This requires a change in mindset for government departments as they need to start behaving like lead firms and act in a more commercially oriented manner than might otherwise be the case.[4] Public authority entities may very well gain from liaison with foreign engineering firms and knowledge institutions such as universities abroad in learning how to manage projects. As noted by Hanlin and Okemwa (this volume) private firms in Kenya (and the region at large) also need to improve their capabilities in project management as such capabilities are vital to the successful implementation of RE projects and may also be helpful in ensuring that opportunities for sustainable industrialisation arising from renewable electrification processes are realised because such capabilities are possible to reuse in other large infrastructure projects.

Finally, there are also a range of international attempts to create new and more interactive ways of learning about green solutions. Examples include e.g., C40 which involves a range of major cities in the world (www.c40.org) and the P4G[5] which has been established to help bring green growth policies into practice and may provide important opportunities for cross-country learning between public authorities, the private sector, and civil society through the establishment and interaction of national platforms.

Capabilities

A capability can be defined in the simplest form as 'having the capacity (resources, skills/competences, and knowledge) to carry out a task'. Capabilities can be locally defined as domestic as opposed to global but can also refer to capabilities at the sub-national (county, village) level. Capabilities – and in particular technological capabilities – are key to ensuring development of a country's industrial levels and also to its ability to learn, absorb, and make use of technologies developed elsewhere.

There are many different ways of understanding, categorising, and analysing technological capabilities. Figure 2.3 highlights that the stock of capabilities in service and manufacturing is particularly important for renewable energy and understanding the opportunities that projects in this field generate for sustainable industrialisation in low and lower-middle-income countries. The capabilities may relate to different areas of project functions and/or to different value chains.

With the rapid development of renewable energy technologies elsewhere in the world, development of capabilities and absorption capacity (Cohen and Levinthal, 1990) is key, if sub-Saharan Africa is to reap the full benefits from these which would mean achieving higher degrees of electrification through integration of renewable energy and successfully undergoing transformation processes leading to inclusive and sustainable industrialisation. Capabilities are important not just in the manufacturing part of the value chain – but also in the deployment and post-deployment steps.

The capabilities needed and available in the manufacturing, deployment, and user chains are embodied in people who have the skills needed to conduct different activities at different stages of manufacturing, deployment, and use chains. But they are also embodied in equipment and other physical technologies. They are also the result of combined activity at an organisational level. They are about practitioners in industry but also about policy skills, knowledge, and learning.

Our key point is that capabilities are not just about developing new technologies or new components in renewable energy. It is also (and maybe even more importantly) about how new technologies and new components in renewable energy systems are made use of in practice – through projects of different sizes and within different sub-sectors such as e.g., small vs. large wind and small vs. large solar.

This book is particularly interested in innovation capabilities and for this we draw on the work of Bell (Bell and Pavit, 1995; Bell, 2009; Bell and Figueiredo, 2012), Lall (1994, 1998), and others (c.f. Archibugi and Coco, 2005; Vidican, 2012; Watson et al., 2015; Baker and Sovacool, 2017; Hansen and Ockwell, 2014). A few of these have discussed innovation capabilities in the context of renewable energy technology manufacturing, deployment, and use. Based on their work, we can develop a typology of capabilities needed across the manufacturing, deployment, and use chains.

The different types of capabilities can be described as follows:

Physical technologies

Hardware that is bought for use by the firm, licensed or bought into the firm through a sub-contract, joint venture, or other partnership types. Examples of these technologies include cutting machines or furnaces during manufacturing; cranes or welding machines during deployment; and electricity meters or inverters during use.

Skills and knowledge/human capital

The skills and knowledge can be embodied in people i.e., through employing new staff or bringing into a project a consultant where the skills are not available within staff who are currently working on the project. It can also refer to sending staff on training to acquire the knowledge and skills needed. Related to this last point, this knowledge can also be codified in training manuals or instruction manuals that come with new equipment. However, skills and knowledge can also be learnt on the job over time because some skills and some types of knowledge are difficult to easily teach or codify into a training manual. Here we are not just referring to technical skills, i.e., in how to manufacture a turbine part or build a solar plant. We are also referring to a wider set of skills and knowledge that are needed throughout the renewable electrification project lifecycle: in how to finance and manage projects, in understanding and promoting conducive and supportive regulatory environments, etc.

Organisational change and linkages

Lall (1994, 1998) and Bell (2009) both recognised the importance of not only skills and knowledge embodied in individual people or technologies but also the importance of a broader, firm level understanding of how everything needs to fit together. This is very similar to the idea of 'core competences' (Hamel and Prahaled, 1990) or the idea that firms need to consider developing a unique set of abilities that gives them a competitive advantage over their competitors. However, Lall and Bell also focus on the importance of understanding the connections needed between different elements of the system of actors that are responsible for the successful completion of the various chains e.g., interfirm linkages and broader systems linkages. Lall (1992) notes the importance of having the right training institutions available to support local businesses and facilitatory regulatory systems – the right set of incentives and systems structures to support interfirm linkages.

Finally, as noted above, capabilities of different kinds are also very important in the public sector which solicits projects and regulates the policy domain. Many of the capabilities required in the public sector differ from those required in for instance firms producing key renewable energy technologies, because the role of the public sector in renewable electrification and sustainable industrialisation processes is different, but to ensure the restructuring of economic activities in this direction is not a small task and requires that all actors play their role. We return to the role of the public sector and policy makers in the last part of the book (Kingiri and Okemwa, 2022; this volume and Lema et al., 2022; this volume).

Outcomes

We define the outcomes of renewable electrification processes as benefits and/or resultant consequences of these processes including *employment (jobs)*, *local content*, *business opportunities*, and *new firms*. As such, they are closely linked to the capabilities that are developed or may be developed due to involvement in renewable electrification projects and processes. Emphasis is on outcomes in the form of capabilities generated through involvement in renewable energy projects and the possibilities these capabilities generate for supporting broader strategies for sustainable industrialisation and – in a broader perspective – also for issues related to questions of justice (cf. also discussions in Scoones et al., 2015).

It is an important hypothesis of this chapter (and the book as a whole), that renewable energy processes can be leveraged to result in these outcomes – but also that achieving these outcomes is not an automatic process. On the contrary, whether the outcomes are realised depends on a number of factors, such as the characteristics of the technologies introduced, the strategy of the supplying firm, the absorption capacity of the national-sectoral innovation system (including local firms), and the project specific features relating to different renewable energy projects. It may also depend on public policies such as policies promoting local contents in renewable energy projects and processes (see Gregersen and Gregersen, 2022; this volume and Kingiri and Okemwa, 2022; this volume).

Renewable electrification may result in increased employment and training opportunities in connection with e.g., manufacturing, installation, and maintenance of energy producing entities, such as wind- or solar-powered mini-grids or grid-connected wind- or solar-power plants. The amount of employment generated can be difficult to measure and may differ according to both the shape and the size of the technology in question (Hansen et al., 2018). The type of employment (and training) opportunities may also be influenced by introduction of renewable energy projects and hence attempts to document such changes are important.

New business opportunities and entirely new firms may emerge in connection with renewable electrification processes. New business opportunities (and firms) may for instance relate to the rapid development of solar pico-systems and solar home systems in recent years (Hansen et al., 2018). Existing and new firms may also include project development firms engaged in the preparation of renewable energy projects or firms involved in manufacturing, installation, and/or operation and maintenance. In the case of Kenya, there are also indications that some firms previously engaged in small wind energy have shifted their focus to solar energy (Hansen et al., 2018) or have become engaged in various types of hybrid systems.

Benefits and outcomes related to local content are related closely to government policies and whether government puts demands on foreign investors and companies to involve local companies or institutions when embarking on renewable energy projects in low and lower-middle income countries. Such demands may relate both to local engagement in project preparations (e.g., site preparation, access roads, and establishment of on-site services near large solar- or wind-power plants) and in the actual installation and operation and maintenance of the plants. Demands may be more or less specific and may have different levels of impact on local capabilities. If local companies succeed in getting involved in large-scale projects, they may obtain knowledge that can be referred to in future bids.

Measuring the outcomes of renewable electrification processes and the extent to which an additional number of jobs or new companies relate directly to these (and not to other factors) is a challenge. Statistical data, surveys, and in-depth case studies of critical projects can be used to gain insights into this, but often a combination of methods and a high level of triangulation between the various types of information is required.

Terrapon-Pfaff et al. (2014) also found that expectations related to productive uses and business development following on from renewable energy projects have been overestimated and that 'project design must explicitly incorporate activities that go beyond energy access in order for these to become an outcome of the project'. In this context, it is important to distinguish between direct outcomes in the form of employment, training, business opportunities, and local content directly related to the development and implementation/deployment of renewable energy projects on the one hand and indirect outcomes/effects of having access to electricity on the other. We are, in this chapter, primarily interested in the direct outcomes as explained above.

Finally, it should be mentioned that there are also potentially negative outcomes related to renewable electrification processes. Large-scale wind-projects such as the Lake Turkana Wind Power Project (LWTP) may for instance increase the demand for well-educated and experienced engineers to take over on operation and maintenance and thereby 'crowd out' other renewable energy projects or organisations unless mitigating efforts are made to increase number of engineers trained more generally and with the specific skills required. In some cases, large-scale projects also tend to overlook the requirements in the local communities (e.g., local communities may not get access to the electricity produced in Lake Turkana, but continue to rely on traditional sources of energy).

Concluding remarks

This chapter set out to investigate two overall research questions that shape this book: do renewable electrification efforts provide opportunities for sustainable industrialisation? If so, what are they and what are the conditions which may turn such opportunities into reality?

In the nested view (Figure 2.2) we illustrated how renewable energy projects (implemented and functioning in different localities) are embedded in both national-sectoral and technological innovation systems, the latter of which are often global in nature and forming the basis for global value chains focused e.g., on solar PV or wind energy. Learning spaces and opportunities for building capabilities are located at different levels – but they are also linked to different stages in the value chain: some relate to the production of core technologies such as solar PV and wind turbines, while others relate to the deployment and use of these (Figure 2.1).

We subsequently proposed that different types of interactive learning take place in firms, in sectoral and national innovation systems, and in the global value chains and complement intra-active learning within firms and organisations engaged in renewable energy activities. Both inter-active and intra-active learning are important potential opportunities for learning new skills and competences that may help build capabilities of different kinds. To capture the learning processes and development of capabilities, we find that renewable electrification processes must be analysed from different angles – including projects, national-sectoral innovation systems, and their links to global value chains. This is necessary due to the globalised nature of renewable energy, where production of e.g., solar panels and wind turbines takes place in one part of the world (often in China), whereas deployment and use takes place in other countries, including Kenya and other low and lower-middle-income countries.

Theoretically, such learning processes may help build up different capabilities that are not only key to the renewable energy agenda, but also include potential for underpinning sustainable industrialisation in low and lower-middle-income countries. Physical technologies, skills, and knowledge/human resources as well as capabilities related to organisational change and ensuring linkages with other actors and within the value chain are required, and developing these will also have likely spill-over to other sectors as some of the capabilities may be re-used in other sectors and processes. The conceptual framework for analysing the potential of renewable energy processes to contribute to sustainable industrialisation (Figure 2.3) breaks down the capability-related co-benefits in the renewable electrification process into three key elements: learning, development of capabilities, and the resulting (expected) outcomes.

The second research question asks what conditions may turn opportunities for sustainable industrialisation from renewable electrification processes into reality? In our view, important opportunities for renewable electrification processes to foster sustainable industrialisation arise when local actors are given the possibility to learn from getting involved in different parts of the renewable energy value chains and when they are enabled to use lessons learned in new projects or contexts. Since most of the manufacturing in the renewable energy value chains takes place outside Africa, local actors are mainly engaged in the deployment and use of the technologies. They may learn new skills and capabilities e.g., when acting as local suppliers on renewable energy projects providing services such as site preparation, transportation of equipment (e.g., wind turbines), or electrical wiring and connecting of new projects to the national grid or to mini-grids. Capabilities acquired may subsequently be used in other sectors and for other purposes such as transportation and storage of agricultural products, agro-business, or other products linked to new ways of industrialising.

In order to fully understand what inter-active and intra-active learning takes place at different levels, stages, and in different locations, empirical work is required. A number of the chapters in this book therefore explore the processes related to acquiring and developing capabilities from the different angles presented in the nested view and focusing on what outcomes are generated from the processes (cf. the conceptual framework). For instance, Gregersen and Gregersen (2022; this volume) focus on learning spaces developed more or less deliberately in large-scale wind projects in Ethiopia and Kenya and what type of learning and capabilities are built in these two cases. Wandera (2022; this volume) focuses on opportunities and barriers for deployment of small wind turbine technology, while Nzila and Korir (this volume) provide an overview of the capabilities present in ongoing renewable energy projects in Kenya. Hanlin and Okemwa (2022; this volume) focus on development of (project management) capabilities in a range of critical projects in solar and wind energy in Kenya, while Karjalainen and Byrne (2022; this volume) present a detailed categorisation of firms involved in solar PV in Tanzania and Kenya based on the level of innovativeness and what capabilities the firms possess. Bhamidipati et al. (2022; this volume) review the learning, capabilities, and outcomes generated (or not) in connection with the increasing amount of Chinese investments in renewable energy projects in sub-Saharan Africa. Finally, Kingiri and Okemwa (2022; this volume) investigate how the energy policies in Kenya have changed over time and the extent to which local content issues have been on the agenda as a means of building capabilities in renewable energy that are also useful for a broader range of projects and sustainable industrial development.

In the final chapter of this book (Lema et al., 2022; this volume) we will return to the nested view and the conceptual framework and draw out conclusions and lessons learned from the field as presented in the various chapters of the book. For now, it is important to reiterate that the considerable investments in renewable energy projects in sub-Saharan Africa expected in the coming years provide an open window to maximise opportunities for interactive learning. This also includes opportunities for learning from the import of technology that is taking place through these investments. Such learning is essential as renewable energy systems irrespective of size and type (shape) must be installed/deployed on the ground under what is often quite challenging conditions.

Acknowledgements

Support for research on which this chapter is based from the Danish Ministry of Foreign Affairs, Grant: DFC 14-09AAU is gratefully acknowledged.

Notes

1 A recent report from the Global Commission on the Geopolitics of Energy Transformation also points to the far-reaching geopolitical implications of an energy transformation driven by the rapid growth of renewable energy (Irena, 2016).
2 A recent special issue of The European Journal of Development Research edited by Lema et al. (2018) and in particular the article on 'Innovation Trajectories in Developing Countries: Co-evolution of Global Value Chains and Innovation systems', however, brings in new evidence to the field and is related closely to the approach suggested here (Lema, Rabellotti, and Sampath, 2018a).
3 Presentation by Prof. Abdelkader Djeflat, MAGHTECH, during the 2nd AfricaLics International Conference, Kigali 2015.
4 The role of the state in development and the extent to which the state should be involved in R&D and in production of e.g., electricity is a huge debate in itself. See e.g., Mazzucato (2015) and Mazzucato (2018).
5 http://um.dk/en/foreign-policy/p4g---partnering-for-green-growth-and-the-global-goals-2030/ (accessed 2 September 2020).

References

Archibugi, D. and Coco, A. (2005) 'Measuring technological capabilities at the country level: A survey and a menu for choice', *Research Policy*, 34(2), pp. 175–194. https://doi.org/10.1016/j.respol.2004.12.002

Baker, L. and Sovacool, B.K. (2017) 'The political economy of technological capabilities and global production networks in South Africa's wind and solar photovoltaic (PV) industries', *Political Geography*, 60, pp. 1–12. https://doi.org/10.1016/j.polgeo.2017.03.003

Bell, M. (2007) *Technological Learning and the Development of Production and Innovative Capacities in the Industry and Infrastructure Sectors of the Least Developed Countries: What Roles for ODA*. The Least Developed Countries Report Background Paper. [Online]. Available at: http://unctad.org/Sections/ldc_dir/docs/ldcr2007_Bell_en.pdf (Accessed: 30/08/2020).

Bell, M. (2009) 'Innovation capabilities and directions of development', STEPS Working Paper 33. [Online]. Available at: https://opendocs.ids.ac.uk/opendocs/handle/20.500.12413/2457 (Accessed: 30/08/2020).

Bell, M. (2012) 'Industrial technology transfer, innovation capabilities and sustainable directions of development', in Ockwell, D. and Mallett, A. (eds.) *Low-Carbon Technology Transfer: From Rhetoric to Reality*. London: Routledge. https://doi.org/10.4324/9780203121481

Bell, M. and Figueiredo, P.N. (2012) 'Building innovative capabilities in latecomer emerging market firms: Some key issues' in Amann, E. and Cantwell, C. (eds.) *Innovative Firms in Emerging Market Countries*. Oxford: Oxford University Press, pp. 24–110.

Bell, M. and Pavitt, K. (1995) 'The development of technological capabilities', *Trade, Technology and International Competitiveness*, 22(4831), pp. 69–101.

Bellini, E. (2017) 'CJIC starts construction of Kenya's first large-scale PV plant', *PV Magazine*. [Online]. Available at: https://www.pv-magazine.com/2017/04/21/cjic-starts-construction-of-kenyas-first-large-scale-pv-plant/ (Accessed: 30/08/2020).

Berkhout, F., Verbong, G., Wieczorek, A.J., Raven, R., Lebel, L. and Bai, X. (2010) 'Sustainability experiments in Asia: Innovations shaping alternative development pathways?', *Environmental Science and Policy*, 13(4), pp. 261–271. https://doi.org/10.1016/j.envsci.2010.03.010

Bolwig, S., Ponte, S., du Toit, A., Riisgaard, L. and Halberg, N. (2010) 'Integrating poverty and environmental concerns into value-chain analysis: A conceptual framework', *Development Policy Review*, 28(2), pp. 173–194. https://doi.org/10.1111/j.1467-7679.2010.00480.x

Brady, T. and Hobday, M. (2011) *Projects and Innovation: Innovation and Projects*. Oxford: Oxford University Press.

Cimoli, M., Dosi, G. and Stiglitz, J.E. (2009) 'The political economy of capabilities accumulation: The past and future of policies for industrial development' in Cimoli, M., Dosi, G. and Stiglitz, J.E. (eds.) *Industrial Policy and Development*. Oxford: Oxford University Press, pp. 1–16.

Coe, N.M., Dicken, P. and Hess, M. (2008) 'Global production networks: Realizing the potential', *Journal of Economic Geography*, 8(3), pp. 271–295. https://doi.org/10.1093/jeg/lbn002

Cohen, W. and Levinthal, D. (1990) 'Absorptive capacity: A new perspective on learning and innovation', *Administrative Science Quarterly*, 35(1), pp. 128–152. [Online]. Available at: https://www.jstor.org/stable/2393553 (Accessed: 03/09/2020).

Drinkwaard, W., Kirkels, A. and Romijn, H. (2010) 'A learning-based approach to understanding success in rural electrification: Insights from micro hydro projects in Bolivia', *Energy for Sustainable Development*, 14(3), pp. 232–237. https://doi.org/10.1016/j.esd.2010.07.006

Geels, F.W. (2002) 'Technological transitions as evolutionary reconfiguration processes: A multi-level perspective and a case-study', *Research Policy*, 31(8–9), pp. 1257–1274. https://doi.org/10.1016/S0048-7333(02)00062-8

Gereffi, G. (2014) 'Global value chains in a post-Washington consensus world', *Review of International Political Economy*, 21(1), pp. 9–37. https://doi.org/10.1080/09692290.2012.756414

Gereffi, G. and Korzeniewicz, M. (eds.) (1994) *Commodity Chains and Global Capitalism*. Westport: Greenwood Press.

Gregersen, C. and Gregersen, B. (2022) 'Interactive learning spaces: Insights from two wind power megaprojects', in *Building Innovation Capabilities for Sustainable Industrialisation: Renewable Electrification in Developing Economies*. New York: Routledge. https://doi.org/10.4324/9781003054665-8

Hamel, G. and Prahalad, C.K. (1990) 'The core competence of the corporation', *Harvard Business Review*, 68(3), pp.79–91.

Hanlin, R. and Hanlin, C. (2012) 'The view from below: Lock-in and local procurement in the African gold mining sector', *Resources Policy*, 37(4), pp. 468–74. https://doi.org/10.1016/j.resourpol.2012.06.005

Hanlin, R. and Okemwa, J. (2022) 'Interactive learning and capability-building in critical projects', in *Building Innovation Capabilities for Sustainable Industrialisation: Renewable Electrification in Developing Economies*. New York: Routledge. https://doi.org/10.4324/9781003054665-7

Hansen, U.E. (2019) 'China's involvement in the transition to large-scale renewable energy in Africa', IREK Working Paper no. 6, June 2019. [Online]. Available at: https://new.irekproject.net/wp-content/uploads/IREKPaper6.pdf (Accessed: 02/09/2020).

Hansen, U.E. and Ockwell, D. (2014) 'Learning and technological capability building in emerging economies: The case of the biomass power equipment industry in Malaysia', *Technovation*, 34, pp. 617–630. https://doi.org/10.1016/j.technovation.2014.07.003

Hansen, U.E., Gregersen, C., Lema, R., Samoita, D. and Wandera, F. (2018) 'Technological shape and size: A disaggregated perspective on sectoral innovation systems in renewable electrification pathways', *Energy Research and Social Science*, 42(2018), pp. 13–22. https://doi.org/10.1016/j.erss.2018.02.012

Henderson, J., Dicken, P., Hess, M., Coe, N. and Yeung, H.W.C. (2002) 'Global production networks and the analysis of economic development', *Review of International Political Economy*, 9(3), pp. 436–464. https://doi.org/10.1080/09692290210150842

Humphrey, J. and Schmitz, H. (2002) 'How does insertion in global value chains affect upgrading in industrial clusters?', *Regional Studies*, 36(9), pp. 1017–27. https://doi.org/10.1080/0034340022000022198

IRENA (2016) *Renewable Energy and Jobs: Annual Review 2016*. [Online] Available at: https://www.seforall.org/sites/default/files/IRENA_RE_Jobs_Annual_Review:2016.pdf (Accessed: 30/08/2020).

Jensen, M.B., Johnson, B., Lorentz, E. and Lundvall, B.Å. (2007) 'Forms of knowledge, modes of innovation and innovation', Lundvall, B.Å. (ed.) *The Learning Economy and the Economics of Hope*. London: Anthem Press.

Jurowetski, R., Lema, R. and Lundvall, B.-Å. (2018) 'Combining innovation systems and global value chains for development: Towards a research agenda', *The European Journal of Development Research*, 30(3), pp. 364–388. https://doi.org/10.1057/s41287-018-0137-4

Kaplinsky, R. and Morris, M. (2000) *A Handbook for Value Chain Research*. Brighton: University of Sussex, Institute of Development Studies.

Karjalainen, J. and Byrne, R. (2022) 'Moving forward? Building foundational capabilities in Kenyan and Tanzanian off-grid solar PV firms', in *Building Innovation Capabilities for Sustainable Industrialisation: Renewable Electrification in Developing Economies*. New York: Routledge. https://doi.org/10.4324/9781003054665-9

Kebede, K.Y. and Mitsufuji, T. (2017) 'Technological innovation system building for diffusion of renewable energy technology: A case of solar PV systems in Ethiopia', *Technological Forecasting & Social Change*, 114, pp. 242–253. https://doi.org/10.1016/j.techfore.2016.08.018

Kebede, K.Y., Mitsufuji, T. and Islam, M.T. (2015) 'Building innovation system for the diffusion of renewable energy technology: Practices in Ethiopia and Bangladesh', *Procedia Environmental Sciences*, 28, pp. 11–20. https://doi.org/10.1016/j.proenv.2015.07.003

Kingiri, A. and Okemwa, J. (2022) 'Local content and capabilities: Policy processes and stakeholders in Kenya', in *Building Innovation Capabilities for Sustainable Industrialisation: Renewable Electrification in Developing Economies*. New York: Routledge. https://doi.org/10.4324/9781003054665-11

Lall, S. (1992) 'Technological capabilities and industrialization', *World Development*, 20(2), pp. 165–186.

Lall, S. (1994) *Technological Capabilities*. Tokyo: United Nations University Press, pp. 264–301.

Lall, S. (1998) 'Technological capabilities in emerging Asia', *Oxford Development Studies*, 26(2), pp. 213–243.

Leach, M., Scoones, I. and Stirling, A. (2010) *Dynamic Sustainabilities: Technology, Environment, Social Justice*. London: Earthscan.

Lema, R., Iizuka, M. and Walz, R. (2015a) 'Introduction to low-carbon innovation and development: Insight', *Innovation and Development*, 5(2), pp. 173–187. https://doi.org/10.1080/2157930X.2015.1065096

Lema, R., Quadros, R. and Schmitz, H. (2015b). 'Reorganising global value chains and building innovation capabilities in Brazil and India', *Research Policy*, 44(7), 1376–1386. https://doi.org/10.1016/j.respol.2015.03.005

Lema, R., Sagar, A. and Zhou, Y (2016) 'Convergence or divergence? Wind power innovation paths in Europe and Aisa', *Science and Public Policy*, 43, pp. 400–413. https://doi.org/10.1093/scipol/scv049

Lema, R., Hanlin, R., Hansen, U.E. and Nzila, C. (2018a) 'Renewable electrification and local capability formation: Linkages and interactive learning', *Energy Policy*, 117, pp. 326–39. https://doi.org/10.1016/j.enpol.2018.02.011

Lema, R., Rabellotti, R. and Sampath, P.G. (2018b) 'Innovation trajectories in developing countries: Co-evolution of global value chains and innovation systems', *The European Journal of Development Research*, 30(3), pp. 345–63. https://doi.org/10.1057/s41287-018-0149-0

Lema, R. et al. (2022) 'Renewable electrification pathways and sustainable industrialization: Lessons learned and their implications', in *Building Innovation Capabilities for Sustainable Industrialisation: Renewable Electrification in Developing Economies*. New York: Routledge. https://doi.org/10.4324/9781003054665-12

Lundvall, B.-Å. (1985a) *Product Innovation and User–Producer Interaction*. Industrial Development Research Series, 31. Aalborg University Press. [Online]. Available at: http://vbn.aau.dk/ws/files/7556474/user-producer.pdf (Accessed: 30/08/2020).

Lundvall, B.-Å. (1985b) *Product Innovation and User–Producer Interaction*. Aalborg: Aalborg University Press.

Lundvall, B.-Å. (1992) *National Systems of Innovation: Towards a Theory of Innovation and Interactive Learning*, London: Pinter.

Lundvall, B.-Å. (2007) 'National innovation systems: Analytical concept and development tool', *Industry & Innovation*, 14(1), pp. 95–119. https://doi.org/10.1080/13662710601130863

Lundvall, B.-Å. and Lema, R. (2016) 'Growth and structural change in Africa: Development strategies for the learning economy' in Lundvall, B.-Å. (ed.) *The Learning Economy and the Economics of Hope*. London: Anthem Press, pp. 327–350.

Lundvall, B.-Å., Johnson, B., Andersen, E.S. and Dalum, B. (2002) 'National systems of production, innovation and competence building', *Research Policy*, 31(2), pp. 213–231. https://doi.org/10.1016/S0048-7333(01)00137-8

Lundvall, B.-Å., Chaminade, C., Joseph, K.J. and Vang Lauridsen, J. (eds.) (2009) *Handbook on Innovation systems in developing countries*. Cheltenham: Edward Elgar.

Malerba, F. and Nelson, R. (2011) 'Learning and catching up in different sectoral systems: Evidence from six industries', *Industrial and Corporate Change*, 20(6), pp. 1645–1675. https://doi.org/10.1093/icc/dtr062

Mathews, J. (2018) 'Schumpeter in the twenty-first century: Creative destruction and the global green shift' in Burlamaqui, L. and Kattel, R. (eds.) *Schumpeter's Capitalism, Socialism and Democracy: A Twenty-First Century Agenda*. London: Routledge, pp. 233–254.

Mathews, J.A. and Tan, H. (2013) 'The transformation of the electric power sector in China', *Energy Policy*, 52, pp. 170–180. https://doi.org/10.1016/j.enpol.2012.10.010

Mazzucato, M. (2018) *Mission Oriented Research and Innovation in the European Union*. [Online]. Available at: https://ec.europa.eu/info/sites/info/files/mazzucato_report_2018.pdf (Accessed: 30/08/2020).

Morris, M., Kaplinsky, R. and Kaplan, D. (2012) 'One thing leads to another: Commodities, linkages and industrial development', *Resources Policy*, 37(4), pp. 08–16. https://doi.org/10.1016/j.resourpol.2012.06.008

Ockwell, D. and Byrne, R. (2016): *Sustainable Energy for All: Technology, Innovation and Pro-poor Green Transformations*. Abingdon: Routledge. https://doi.org/10.4324/9781315621623

Ockwell, D. and Mallett, A. (2013) 'Low carbon innovation and technology transfer' in Urban, F. and Nordensvärd, J. (eds.) *Low Carbon Development: Key Issues*. Abingdon: Earthscan, Routledge, pp. 109–128.

Pietrobelli, C. and Rabellotti, R. (2011) 'Global value chains meet innovation systems: Are there learning opportunities for developing countries?', *World Development*, 39(7), pp. 1261–1269. https://doi.org/10.1016/j.worlddev.2010.05.013

Ponte, S. (2019) *Business, Power and Sustainability in a World of Global Value Chains*. London: Zed Press.

Ponte, S. and Ewert, J. (2009) 'Which way is "Up" in upgrading? Trajectories of change in the value chain for South African wine', *World Development*, 37(10), pp. 1637–1650. https://doi.org/10.1016/j.worlddev.2009.03.008

Romijn, H., Raven, R. and de Visser, I. (2010) 'Biomass energy experiments in rural India: Insights from learning-based development approaches and lessons for strategic Niche management', *Environmental Science and Policy*, 13(4), pp. 326–338. https://doi.org/10.1016/j.envsci.2010.03.006

Scoones, I., Leach, M. and P. Newell (eds) (2015) *The Politics of Green Transformation*. London: Routledge.

STEPS Center (2010) *Innovation, Sustainability, Development: A New Manifesto*. Brighton. https://steps-centre.org/wp-content/uploads/steps-manifesto_small-file.pdf (Accessed: 24/03/2021).

Stirling, A. (2009) *Direction, Distribution and Diversity! Pluralising Progress in Innovation, Sustainability and Development*, STEPS Working Paper 32, Brighton: STEPS Centre.

Sturgeon, T., Van Biesebroeck, J. and Gereffi, G. (2008) 'Value chains, networks and clusters: Reframing the global automotive industry', *Journal of Economic Geography*, 8(3), pp. 297–321. https://doi.org/10.1093/jeg/lbn007

Terrapon-Pfaff, J., Dienst, C., König, J. and Ortiz, W. (2014) 'A cross-sectional review: Impacts and sustainability of small-scale renewable energy projects in developing countries', *Renewable and Sustainable Energy Reviews*, 40, pp. 1–10. https://doi.org/10.1016/j.rser.2014.07.161

Vidican, G. (2012) 'Building domestic capabilities in renewable energy. A case study of Egypt', *German Development Institute, Studies Series*, 66. [Online]. Available at: http://hdl.handle.net/10419/199191 (Accessed: 30/08/2020).

Wandera, F.H. (2018) 'Applying the innovation systems framework to the study of the small wind turbine sector in Kenya: A review and research agenda', *IREK Working Paper* no. 3. Copenhagen/Nairobi/Eldoret: AAU, ACTS and MU. [Online]. Available at: http://irekproject.net/files/2015/09/IREKpaper3.pdf (Accessed: 30/08/2020).

Wandera, F. (2022) 'Understanding the diffusion of small wind turbines in Kenya. A technological innovation systems approach', in *Building Innovation Capabilities for Sustainable Industrialisation: Renewable Electrification in Developing Economies*. New York: Routledge. https://doi.org/10.4324/9781003054665-5

Watson, J Byrne, R., Ockwell, D. and Stuam, M. (2015) 'Lessons from China: Building technological capabilities for low carbon technology transfer and development', *Climatic Change*, 131(3), pp. 387–399. https://doi.org/10.1007/s10584-014-1124-1

3

CHALLENGES AND OPPORTUNITIES FOR THE EXPANSION OF RENEWABLE ELECTRIFICATION IN KENYA

Mbeo Calvince Ogeya, Philip Osano, Ann Kingiri, and Josephat Mongare Okemwa

Abstract

This chapter investigates the development of renewable grid electricity in Kenya from 1997 to 2019 using four elements of the multi-level perspective framework of socio-technical and economic transition. Opportunities and constraints for the development and growth of the wind and solar electricity subsectors are analysed. The outcome over the 22-year period has been mixed but remains promising in terms of the potential for increased investments in solar and wind power in rural areas. Five legal statutory and policy milestones facilitated the transformation that was witnessed in the renewable electricity subsector. Notably, the share of non-fossil electricity in the overall mix increased substantially reaching 90% in 2019, however contribution from wind and solar did not record any remarkable change between 2010 and 2018 – accounting for a high of 3% share. The coming to grid of the Garissa Solar Plant (55 MW) and the Lake Turkana Wind Power Project (310 MW) in late 2018 pushed the share to a high of 14.6% in 2020. Even so, factors including power politics, institutional inertia, and societal and cultural constraints have created a lock-in that is likely to limit the full realisation of Kenya's wind and solar energy.

Introduction

Access to modern energy is critical for economic growth, and poverty reduction. Sustainable Development Goal 7 seeks to ensure access to affordable, reliable, sustainable, and modern energy for all by 2030 (United Nations, 2015). Although Africa is endowed with rich energy resources, access to modern energy remains a challenge in the continent whose energy profile is characterised by low production, low consumption, and high dependence on traditional biomass energy. In

DOI: 10.4324/9781003054665-3

sub-Saharan Africa, an estimated 83% of households rely on traditional fuel such as firewood and charcoal for cooking and kerosene lanterns for lighting (Njiru and Letema, 2018). The African Agenda 2063 places a high priority on renewable energy in fostering economic growth and eradication of energy poverty (African Union & Commission, 2015). Although an estimated 600 million people are without access to electricity in sub-Saharan Africa, there has been marked progress recorded in electricity access. The World Energy Outlook Report for Africa notes that the number of people annually gaining access to electricity doubled from 9 million between 2000 and 2013, to 20 million between 2014 and 2018, with three countries – Kenya, Ethiopia, and Tanzania – accounting for 50% of those gaining access (IEA, 2019).

Ensuring energy for all is a key priority for African governments, many of which are increasingly investing in clean energy to transition their economies to renewable electricity that is dependent on clean sources such as hydropower, wind, solar, and geothermal. Sustainable transition to renewables is a complex and dynamic process with different policy, economic, social, and technological implications at the local, national, and regional levels.

This chapter presents an analysis of the opportunities and constraints for expansion of renewable electricity in Kenya. Our analysis draws upon the insights from four elements of the multi-level perspective (MLP) framework (Geels, 2011): (i) landscape pressure; (ii) institutional context and legitimisation; (iii) technology, development, and market function; and (iv) agency and power, to provide deeper analysis and evidence focused on the development of solar and wind energy in Kenya.

The chapter is structured as follows: first, the evolution of renewable electrification in Kenya is discussed, providing a context for the chapter through a detailed analysis of how renewable electrification has evolved since 1997 to date. This section exposes the country's energy demand and supply dynamics, and the policies and institutional reforms emanating from and in response to these dynamics. Next, we present the methodology and conceptual framework used. Then the opportunities and constraints for future expansion of the renewable energy sector in Kenya are reviewed using selected elements of the MLP framework followed by a conclusion.

The evolution of renewable electrification in Kenya

The Kenya Vision 2030, which is the country's development blueprint, aims at transforming the nation into a middle-income country by 2030 and acknowledges energy as one of the infrastructural enablers for this economic growth. As the industrial and manufacturing sectors expand as part of the Vision 2030 implementation, it is expected that energy consumption in these sectors will increase correspondingly, necessitating additional investment and financing to expand electricity to stimulate economic growth in the rural areas (GoK, 2007). The recent technological advancement, donor support, and energy system planning witnessed in Kenya have led to the rapid growth of electricity systems and energy access (Byrne et al., 2014). In Kenya, abundant wind, solar, biomass, and geothermal resources have led the government to seek the expansion of renewable energy generation to rural areas (Kiplagat, Wang, and Li, 2011).

Kenya electricity demand and supply (2008–2018)

The demand for electricity in Kenya has been increasing rapidly due to, among other factors, population growth, economic growth, and increased infrastructure investments in various sectors (Kiprop, Matsui, and Maundu, 2020). A linear steady growth in peak demand has been observed between 2008 to 2018 with an average growth rate of 5.6% (Kenya Power, 2019). The observed increase in energy demand has also been attributed to the World Bank-financed Last Mile Connectivity Project, and the introduction of the discounted night time electricity tariffs that may have created incentives for manufacturers to upscale the demand for electricity from the grid (Kenya Power, 2018). The Last Mile Connectivity Project has spurred the rapid increase in household connections from about 2.5 million in 2014 to 6.5 million in 2018. Even so, electricity peak demand has continued with a steady growth rate of about 5.6% annually, and is projected to rise to 4,244 MW in 2030 in the business as usual scenario, assuming historic demand growth trends (Ministry of Energy and Petroleum, 2018; Kenya Power, 2019; KNBS, 2020b). However, in a high growth scenario assuming full implementation of the Kenya Vision 2030 projects, the demand is estimated to reach 5,780 MW in 2030 (Ministry of Energy and Petroleum, 2018).

On the supply side, the effective capacity has increased by about 79% from 1,267 MW in 2008 to 2,265 MW in 2018. The trend shows a steady average growth rate of 6.1% between 2009 and 2014. Two spikes were observed in 2015 and 2019 resulting in higher growth in supply (average 7.5%) compared to the demand growth rate. These spikes were occasioned by the addition of 280 MW geothermal power in 2015 and 365 MW wind and solar power by 2019. The effective capacity thus increased to 2,631 MW in 2019. Thus, the updated capacity mix as of 2019 was: 745 MW fossil, 786 MW hydro, 649 MW geothermal, 21.5 MW biomass, and 411 MW other renewables – solar and wind accounting for 95% (Kenya Power, 2019; KNBS, 2020b).

The first on-grid solar photovoltaic (PV) system was realised in 2016 with the commissioning of the 0.25 MW grid-tie solar PV by Strathmore University, followed in 2019 by the 55 MW Garissa solar plant. The first grid-tie wind was realised in early 2000 – a capacity of 0.4 MW piloted in Ngong' Hills Kenya by the Kenya Electricity Generating Company (KenGen). It took about 12 years (to 2011) for KenGen to increase the grid installed capacity of wind power to 5.3 MW and another four years to grow to 25.5 MW in 2015 (Kenya Power, 2004, 2019). Private sector-led Lake Turkana Wind Power (LTWP) dispatched 310 MW wind capacity to grid in 2018, about 13 years after inception of the project in 2005 (Lake Turkana Wind Power Project, 2011). This development explains

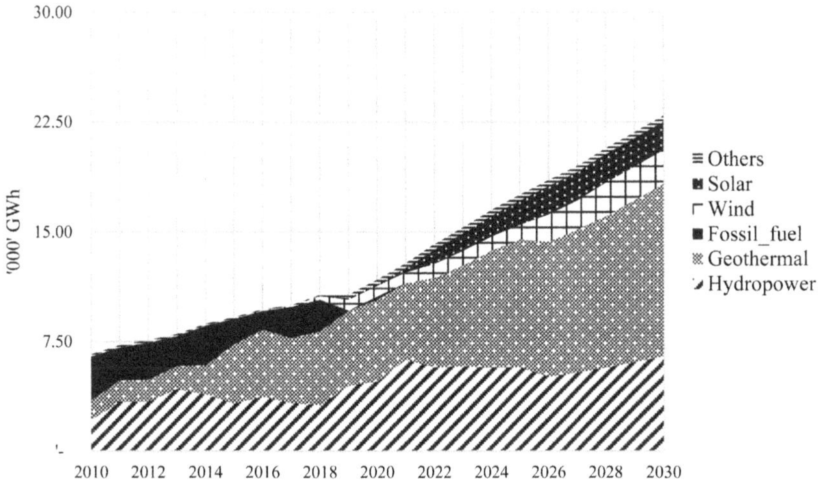

FIGURE 3.1 Projected generation in business as usual scenario. Data Source: authors.

the emergence of solar and wind energy in 2018 as illustrated in Figure 3.1, which presents a forecast of grid electricity supply by different electricity sources until 2030 based on existing and committed generation projects. There is a projected increase in renewables (solar and wind) contribution to about 18.5% of the total national grid mix by 2030.

Technically, the off-grid solar PV market is highly adopted; however, there is scanty verifiable data sources on the actual installed and effective capacities and proportion of off-grid and grid electricity supply. The national utility off-grid projects were implemented through the Rural Electrification Fund before the establishment of Rural Electrification Authority (REA) in 2006, later converted to Rural Electrification and Renewable Energy Corporation (REREC) in 2019. Until 2011 when 0.6 MW and 0.1 MW solar and wind capacity were installed for institutional electrification mainly in schools, health facilities, and trading centres, there was no data on off-grid solar or wind power generation managed by national utility. In 2018, there was no significant change in the REA off-grid solar and wind accounting for 0.69 MW and 0.55 MW respectively (Kenya Power, 2013, 2018). Other off-grid solutions including solar home systems, micro-solar home system, mini-grids, and solar lanterns have mainly been distributed by the private sector and often are targeted to residential, small, and medium scale businesses. According to the 2019 National Census Report, the share of households with standalone solar PV increased from 140 thousand households to 2.3 million households between 2009 and 2019 (KNBS, 2009, 2019); and another 5.2% of the total households using rechargeable solar lanterns. Other reports show that about 2.6 million solar PV modules were cumulatively imported into the country by 2017 (RENCON Consulting, 2018).

Policies and institutional transformations for electrification in Kenya

Although the history of the electrification process in Kenya dates to the pre- and post-independence period, we take as our reference point the enactment of the Kenya Electric Power Act 1997 to describe the socio-technical transition in the renewable electricity sector in Kenya and end the assessment with the enactment of the Energy Act of 2019 and implementation actions that have since followed the enactment. In the 22-year period from 1997 to 2019, five major milestones were realised in the policy and institutional reforms of electrification in Kenya. These include the Electric Power Act of 1997, the Sessional Paper No. 4 on Energy of 2004, the Energy Act of 2006, the National Energy Policy of 2018, and the Energy Act of 2019. Added to these was the enactment of Constitution of Kenya in 2010 which devolved planning and development of electricity and gas reticulation. The reforms brought about through these successive policies have been directed towards increasing energy demand, electricity access, especially renewable sources in the rural areas, and the expansion of the electricity capacity needed for industrialisation (Abdallah, Bressers, and Clancy, 2015).

The Electric Power Act of 1997 provided for regulation, generation, transmission and distribution, supply, and use of electricity. Amongst other reforms, the Act provided for the licensing and conflict resolution for Independent Power Producers (IPPs). Godinho and Eberhard (2019) note that the structure of Kenya's vibrant energy sector has been transformed by two waves of reforms. The first wave of reform was largely donor driven policy commencing in 1996. As part of this wave, the generation of electricity was separated from distribution and transmission. The second wave of reform commenced in 2002 led by the domestic champions whereby the trust of the first wave continued and independent regulations were strengthened. It included partial privatisation of the generation company, the Kenya Electricity Generating Company (KenGen) Limited (Godinho and Eberhard, 2019).

The Sessional Paper No. 4 on Energy of 2004 laid down the policy framework upon which the quality and affordability of energy services are provided to the domestic sector in the 2004–2023 period. The Sessional Paper also provided a framework for the Energy Act of 2006. The Sessional Paper recognised the challenges in the power sector including weak power transmission and distribution network, high losses, voltage fluctuation and intermittent power outages, high tariffs due to the high costs charged by Independent Power Producers (IPPs), and electricity demand (MoE, 2004). In 2006, the Energy Act 2006 was enacted. It provided for the establishment of the Geothermal Development Company (GDC), Energy Regulatory Commission (ERC), and the Rural Electrification Authority (REA). Moreover, several other new regulations, standards, and procedures were enacted. These included the Grid Code 2008, the Energy (Energy Management) Regulation 2012, Solar PV Regulation 2012, and the Revised Feed in Tariff (FiT) of 2012 (ERC, 2008, 2012; Ministry of Energy, 2012) that provided room for increased participation of private actors linked to small renewable energy generation projects.

The 2010 Constitution of Kenya (GoK, 2010) established a devolved system of governance, with energy considered as a component of devolution. The adoption of the new constitution led to the need for a review of energy policy, hence the development of the Energy Policy 2018 (MoE, 2018). The Policy, as well as the Energy Act 2019, stipulates the functions and roles of the ministries responsible for energy at the national and county levels. The national government provides overall guidance on policy, planning, and public investment. The county is responsible for amongst other things county planning, development including electricity and gas reticulation, implementation of county electrification programmes, and establishment of energy centres for promotion of renewable energy technologies, energy efficiency, and conservation. A number of policies, including the Energy Act of 2019, thus provide for a more consumer-oriented approach to energy generation and supply (GoK, 2010, 2011; The Energy Act, 2019, 2019).

The structural changes brought about by the Energy Act 2019 include the establishment of the Energy and Petroleum Regulatory Authority (EPRA), previously referred to as the Energy Regulatory Commission (ERC). The authority takes up all responsibility of the commission including enforcing local content requirements and regulation, issuing licences, setting tariffs, and approving power purchases and network contracts. It is also mandated to negotiate the power purchase agreements between the Kenya Power and Independent Power Producers. The Energy and Petroleum Tribunal was created to take over the functions of the Energy Tribunal. It is mandated to hear and determine energy and petroleum related disputes and appeals. The establishment of the Rural Electrification and Renewable Energy Corporation (REREC) took over the functions of the Rural Electrification Authority (REA). The Corporation is mandated to inter alia source additional funds for renewable energy; develop renewable energy master plans; and develop, promote, and collaborate with other agencies in the use of renewable energy technologies. Figure 3.2 shows a visualised summary of the transformation in the electricity sector policy.

Drivers of policy change and conditions for transformation

Energy has been identified as a strong pillar in Kenya's national development. It is critical in supporting Kenya's Vision 2030 in the area of manufacturing, agriculture, and commercial sectors (economic pillars) and education, health, gender equity, and environment (social pillar) (GoK, 2007). In addition, the focus of the Vision 2030 Medium Term Plan for the 2018–2022 period is on the Big 4 Agenda of Manufacturing, Improved Health, Affordable Housing, and Food Security.

Informed by the national policy agenda outlined above, we argue that the drivers of policy changes at the macro level are primarily economic growth,

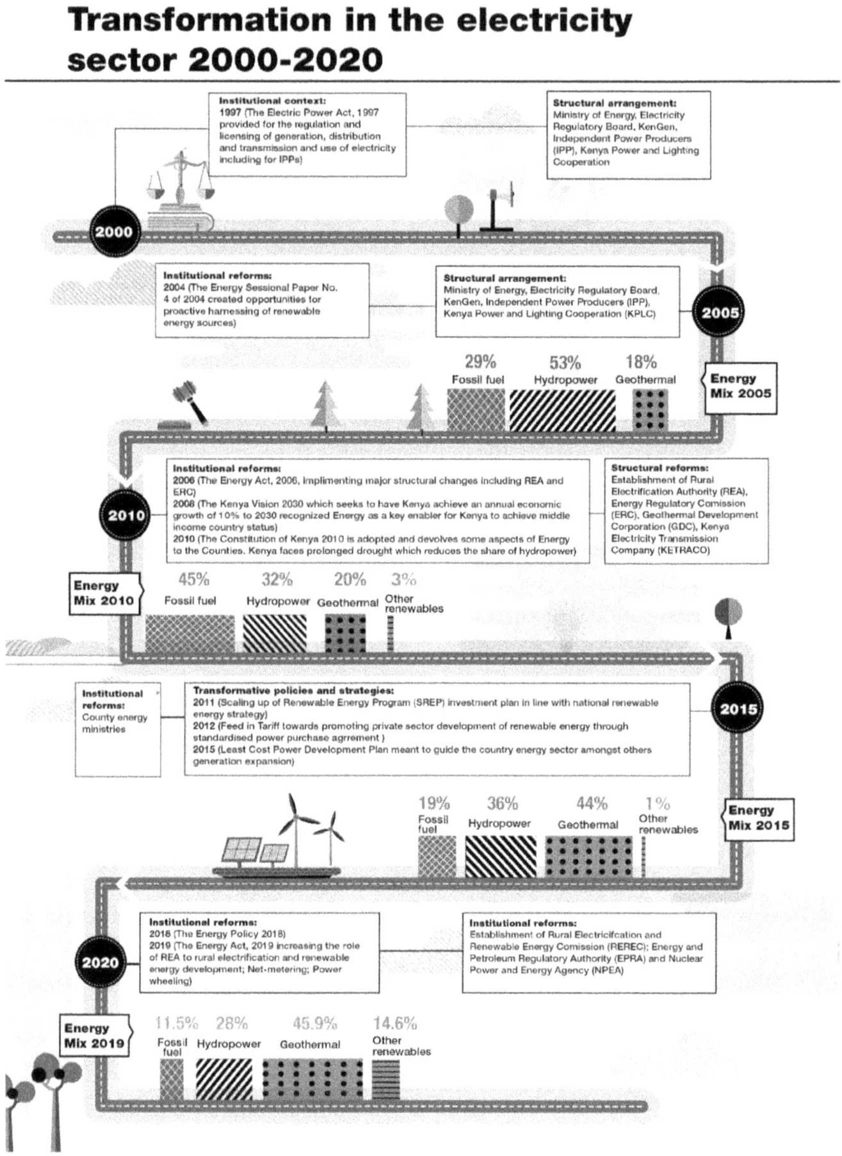

FIGURE 3.2 Policy transformation in the electricity sector. *Acronyms:* REA = Rural Electrification Authority; REREC = Rural Electrification and Renewable Energy Corporation; EPRA = Energy and Petroleum Regulatory Authority; LCPDP = Least Cost Power Development Plan. Source: authors.

industrial development, demographic change, and climate change. This comes with the need for promotion of energy efficiency, fostering international cooperation, capacity building in the energy sector, diversification of energy supply sources in ensuring security of supply, protection of investors, and promotion of cost effective and equitable pricing of energy products. Kenya has committed to the Sustainable Development Goal and put in place measures aimed at achieving Goal 7. Currently an estimated 11 million people do not have access to electricity (IEA, 2019; Kenya Power, 2018). Arguably, electricity access is attributed to increased economic growth, better provision of public service, and quality of life, as well as technological advancement in education, health, and agriculture (Blimpo and Malcolm, 2019).

Efforts towards economic growth and industrial development in the renewable energy sector have been perpetuated through international technology transfer and deployment. Ockwell et al. (2009) observe that proactive technology transfer can be equated to underpinned long term low carbon transfer which requires development of innovative capabilities, especially in developing countries. More important for the energy sector – which is our motivation for this chapter – is the policy change at the micro level that relates to building requisite local capabilities to enhance indigenous or local technological solutions. Kenya's universal access to electricity can only be achieved based on, in part, concerted efforts to enhance local capabilities that would promote both grid and off-grid technological solutions. Several publications expound on the nature of capabilities that are needed in this regard. These include organisational capabilities synonymous with 'innovation management', human capabilities, technological capacity, and policies (Lawson and Samson, 2001; Nussbaum, 2001; Ockwell et al., 2009). In acquiring innovation capabilities, developing countries governments, firms, and organisations learn through the process of technology transfer and deployment. This learning process is an accumulation of relevant capabilities whereby an integrated approach in technology transfer and local innovation in renewable electrification are complementary (Lema et al., 2018).

Methodology and conceptual framework

Data collection and analysis

The information and data used for this chapter were collected using several methods. We used secondary data on electricity demand and supply which was collated from institutional databases and reports. We carried out desk review of peer reviewed and grey literature looking at solar and wind subsectors and the entire renewable sector in Kenya. We also conducted document analysis of the policies that we reviewed. The systemic analysis of policy documents served to generate empirical data that informed an in-depth interrogation of the dynamics that constrain or support the sector's growth. The authors also participated in several policy processes and workshops, which allowed us to access useful materials such as presentations. Useful information was also obtained through both structured and unstructured interviews and informal conversations with key informants in the energy sector. The use of different sources enabled us to triangulate the data, thereby ensuring greater reliability and validity of the information and data that we used in our analysis.

The overall analysis was conducted using qualitative and quantitative methods. The qualitative information and data were first documented, and themes categorised according to the challenges and opportunities in the renewable energy sectors for both solar and wind. The recurring themes were highlighted and interpreted as critical observations by the authors. The quantitative data on the demand and supply was analysed using excel spreadsheet and the LEAP (Long Range Energy Alternative Planning) tool to obtain the energy demand projections for 2030.

Analytical framework: the multi-level perspective (MLP) in articulating challenges and opportunities for renewable electricity

Sustainable transition is viewed as multi-dimensional in nature because it involves a broad range of interacting actors and processes, including technologies, policy or power politics, economics or market space, and cultural expositions and the dynamics of structural changes (Geels, 2011; Smith et al., 2010). It is noted that the transition dynamics include mass (objects, actors, and infrastructure involved), speed (the pace at which socio-technical and their alignment develops), and direction (overall performance of system changes as a result of innovation), and in the process new products, services, business models, and organisations emerge, partly complementing and partly substituting existing regime (Farla et al., 2012; Smith et al., 2010).

Although the environmental push has significant influence on the socio-technical transitions, there are inherent systemic factors that seek to ensure that the norm continues (Geels, 2011). Many of such existing structural systems are stabilised through various lock-in mechanisms, for example economy of scale, sunk capital investment, infrastructure, and competencies (Geels, 2011). These lock-in systems make it difficult to extricate existing systems (also known as regime) or create tension. The multi-level perspective therefore conceptualises overall dynamics of patterns in socio-technical transitions. It views transition as a non-linear process that results from the interplay of development at three analytical levels: (1) niches (entrepreneurs, spinoffs, and start-ups innovating against regime); (2) socio-technical regimes (structures accounting for stability in a system); and (3) exogenous social–technical landscape (wider context which influence niche and regime) (Geels, 2011; Smith, Voß, and Grin, 2010).

The socio-technical regime is the structure that accounts for the stability of the existing socio-technical system. It is the set of rules that orient and coordinate the activities of social groups that reproduce the various elements of socio-technical system (Geels, 2014). Such rules include institutional arrangements, national policies and regulations, binding contracts, as well as social economic aspects of routines and shared believes, capabilities and competencies, lifestyle, and practices. Niche is described as a protected space, demonstrator projects or small market niches of special interest to demand actors who are willing to support the innovation (Geels, 2011). Niches have potential to disrupt regimes. Landscape is the wider context which influences niche and regime dynamics. This would include demographic trends, political ideologies, societal values, and micro-economic drivers such as the household income base. These are exogenous factors that niche and regime cannot influence in short term and they take a long time to make significant change.

This paper draws on four elements of the MLP as an analytical framework in building the discourse of energy transition in Kenya with a focus on wind and solar. These are: (1) the landscape pressure, referring to the demographic trends, macro-economic patterns, and societal values that influence regime and niche; (2) institutional context and legitimisation, referring to the legal frameworks within which a socio-technical transition occurs; (3) technology, development, and market function, as a sub-regime where innovations are tested and developed incrementally that could accumulate to stable trajectory; and (4) agency and power, which refer to the governance and the role of power politics in social–technical transition (Geels, 2014). We apply these elements in the analysis of both the opportunities and constraints.

Opportunities for future expansion of renewable electricity in Kenya

This section attempts to unearth the underlying opportunities for future grid expansion and the implications for renewables, particularly solar and wind.

The landscape pressure

The landscape pressure issue is mainly linked to economic, demographic, and environmental trends in Kenya. A stable intercensal population growth rate of 3.4% was observed between the 1962–1969 period and the 1979–1989 period, which then dropped to 2.9% for the following two decadal periods between 1989 and 2009. A further drop to 2.2% was marked in the 2009–2019 period (KNBS, 2020a). The country registered an increased economic growth rate from 1.8% in 2008 to a peak of 8.4% in 2010, followed by a decline to 4.6% in 2012. This was followed by a generally stable growth rate between 2012 (5.7%) and 2018 (6.3%) (KNBS, 2019).

The high growth trajectories and demographic trends have compelled the government to take different approaches in energy growth focusing on least cost power development plans (Ministry of Energy, 2011; Ministry of Energy and Petroleum, 2018). This planning process has also been influenced by the impacts of climate change and international obligations. As a signatory to the United Nations Framework Convention on Climate Change (UNFCCC) and having submitted the Nationally Determined Contribution (NDC) to the UNFCCC, Kenya is obligated to reduce emissions of greenhouse gases, including investing in clean and renewable energy with the consequent impact of the grid energy mix.

Institutional context and legitimisation

The energy reforms in Kenya were re-ignited in 2004 upon the enactment of the Sessional Paper No. 4 on Energy, which created multiple opportunities within renewable energy and rural electrification. It laid a policy framework upon which cost effective, affordable, and adequate quality energy services are to be made accessible to the national economy over the period between 2004–2023 through policy and structural transformation (MoE, 2004). Since then, several policies and regulations have enabled changes within the energy sector resulting in the deployment of renewable energy technologies, learning, and local capacity improvements for rural development.

A more detailed analysis of the transition of energy policies and regulation in Kenya and how this relates to opportunities to build requisite capabilities for renewable technologies diffusion is presented in Kingiri and Okemwa (this volume). The five legal and policy instruments that have been implemented to facilitate the transition to a more sustainable energy generation, transmission, and consumption ensuring reliability, price competitiveness for universal electrification, and economic growth, are highlighted below.

The Energy Act 2006 and Energy Act 2019

The Act provided for the creation of the Rural Electrification Authority with the mandate to plan and implement electrification programmes in rural areas. This highlights the importance of rural electrification at the time of developing the Act, hence the need for a dedicated agency. The Act liberalised electricity generation – especially the micro and mini power plants below one megawatt – a provision that has been maintained in the Energy Act 2019 under Section 117 (The Energy Act, 2019). As per the Act, there is no statutory obligation for own generation and use of electricity below 1 MW. Several private companies have invested in their own renewable electricity generation and community power plants, especially from small and micro-hydropower. Examples of such power plants include Brooke Bond, Tenwek Hospital, James Finley Tea Factory, Thima, Tungu Kabiru, Kathamba, Imenti, Diguna, and Munjwa (Mbaka and Mwaniki, 2016). This is a great opportunity for small scale producers and processors, especially in the rural areas whose electricity demand is below 1 MW.

Feed in Tariff

The Feed in Tariff (FiT) first issue was released in 2008. This was followed by two more revisions to the latest revised FiT policy of 2012. It stems from the Sessional Paper No. 4 on Energy 2004 and Section 103 of the Energy Act of 2006, both seeking to promote renewable energy generation. It presents a standardised power purchase policy for small scale renewable energy generation less than 10 MW and provides for review every three years. The standardised Power Purchase Agreement (PPA) features included sales at distributor level, first come first serve basis, and PPA offered to projects that demonstrated technical and economic viability and met all technical requirements for grid connection. This policy presented an opportunity for private sector development of renewable electricity to grid, especially solar and wind.

Experts in the sector observe that the potential of the FiT has not been realised, especially for solar and wind technologies. This is in part due to many non-committal applicants whose expression of interest totalled to over 3,418 MW (wind and solar) by the end of 2018 (MoE, 2018), but only about 11% had been realised by 2020. Since the establishment of the FiT policy in 2008, the first grid-tie solar within FiT framework (0.25 MW capacity) was reported a decade later in 2018 (Kenya Power, 2019). Some experts in the Ministry of Energy (MoE) attributed this delay to slow demand, utility unpreparedness, and lack of grid stability. The FiT policy has been upheld in the Energy Act of 2019 for the special purposes of catalysing renewable energy uptake, building local networks, stimulating innovation, and reducing GHG emissions. There are, however, ongoing debates for an auction system for variable renewable electricity generation for grid to be based on cost competitiveness. A similar programme – the South Africa Renewable Energy Procurement Program – provides a model for a successful power auctioning. In the past four competitive tendering processes, a total investment of USD 19 billion has been made, and the price of wind and solar fell by 46% and 71% in nominal terms between 2011 and 2016 (Eberhard and Kåberger, 2016).

Renewable energy regulations

There are five existing regulations and two proposed in the renewable energy sector. Table 3.1 lists these regulations with a description of their provisions. These regulations seek to ensure conformance to standards and codes of practice and have promoted formal capacity building, especially in the energy efficiency and photovoltaic (PV) sector.

A requisite certification to engage in the solar PV business and installation and as a certified energy manager requires tailored training. Both formal and professional associations have participated in this certification process and are conducting regular trainings. Before 2012, there were no certified solar PV technicians and energy managers in Kenya. At the time of writing this chapter, there

TABLE 3.1 Summary description of existing and proposed Renewable Energy Regulations

	Regulation	*Description*
1	The Energy (Solar Photovoltaic Systems) Regulation 2012	The regulation is made under Section 110 of the Energy Act 2006. Requires amongst other things that all persons designing and installing solar PV shall be licensed by ERC. All manufacturers, vendors, distributers, and contractors shall be licensed by ERC.
2	The Energy (Energy Management) Regulation 2012	All designated energy consuming facilities shall carry out energy audits once in every three years and all energy audit reports, implementation plans, and energy policies shall be submitted to ERC.
3	The Energy (Appliances Energy Performance and Labelling) Regulation 2016	These regulations are made under section 104 (Energy Efficiency and Conservation) of the Energy Act 2006. It requires laboratory testing and labelling of all electricity consuming appliances as per the Kenya standards and approved by the Kenya accreditation services.
4	Appliances (Energy Performance and Labelling) Regulation 2018	
5	Designation of Industrial, Commercial, and Institution Energy Users in Kenya	Requires all industrial, commercial, and institutional users of energy consuming a minimum of 180 MWh annually to comply with the Energy (Energy Management) Regulation of 2012.

Source: authors.

were about 485 registered certified solar photovoltaic contractors in classes of design and installation (class C1), design, installation and sales (class V1), and manufacture and/or importation of solar PV (class V2), and 376 registered technicians (EPRA, 2020). This capability has an opportunity for quality design and installation works, thereby enhancing rural electrification (EPRA, 2020). Fundamental questions raised during the discussion with stakeholders in the sector relate to effectiveness and implementation of the regulations. Stakeholders cited the lack of personnel capacity that has undermined the implementation of these regulations. The issues of capabilities have been cited as very instrumental in the sustainable industrialisation of the renewable energy sector in Kenya (see Kingiri and Okemwa, 2022; this volume).

Net metering

Net metering is a new provision under the Energy Act of 2019 and provides for any generator to supply the distributor through the net metering system. Section 162 of the Energy Act 2019 provides for consumers with electricity generators not exceeding 1 MW to enter into net metering agreement with electricity distributors or retailers. However, the implementation framework has not been developed for the same. Roux and Shanker (2018) make a comparative study on net metering policies in emerging and developing countries (two in Asia and six in Africa including Kenya). The comparative study elaborates key insights to effectively implementing net metering policies in emerging and developing economies. Such include compensation analysis that comes with net metering, political will, and implementation phases, e.g., commercial users first, then both residential and commercial users (Roux and Shanker, 2018). Even so, this is an opportunity for support of industrial production of solar PV to support adoption.

Power wheeling

'Power wheeling' is a term often used to describe a consumer transport of electricity from a generation source to a user point, paying service charges to transmission and distribution companies. It is also a new provision under the Energy Act 2019 to enhance power transmission between generator and user through paying service charges. This is a way to enhance private sector engagement in competitive tariffing through own power generation and use.

Technology, development, and market function

There are numerous niches at different levels within the energy system. The Kenyan renewables market has attracted numerous innovations in technology, business models, and supply chain, particularly in off-grid systems (Basu and Marett, 2016). These innovations have contributed greatly to electricity access. They are supported by the transformation in the policy regime and are gaining traction rapidly with evident disruption in the policy regime. Whereas the grid-tie solar PV is yet to develop and attract broader attention apart from e.g., the Garissa solar system in Kenya (see Hanlin and Okemwa, 2022; this volume), the off-grid systems have particularly matured in East Africa with remarkable innovative approaches (see Karjalainen and Byrne, this volume).

M-kopa is one such niche innovation, which was reported to be ranked amongst the world's top ten most innovative companies by Fast Company News (Fast Company, 2020). M-kopa implements a pay-as-you-go business model supported by mobile money transfer. In 2018, about 40% of Kenya's population had access to a mobile phone and were registered to enable them to do mobile money transfers (M-Pesa) (M-Kopa Solar, 2015). This provided an opportunity for a wider market reach and establishment of the innovative business models. The company's support is based on consumer asset financing through flexible credit facility. Its technology platform combines embedded global systems for mobile communication (GSM) and mobile payments. In only eight years of establishment between 2012 and 2020, about 750,000 homes were connected to M-Kopa-based micro-solar home systems with an average of 500 new connections made daily (M-Kopa Solar, 2020). BBOXX limited solar home system is another example of innovation within the off-grid solar home systems.

The remotely managed system has a flexible three-year payment plan increasing its reach in the market. These innovative approaches present opportunity for increased demand in off-grid solar, creating opportunities for local capacity development and expansion of solar and wind energy.

Agency and power

Urban and rural electrification advancement has been made possible as a result of the structural changes undertaken in the power sector. Rural electrification increased from 4% in 2004 to 23% in 2018 (Kenya Power, 2019; MoE, 2004). Figure 3.3 provides an illustration of the structural transformations that took place from 1997 to 2019. These were made possible because of the implementation of the Electric Power Act (1997), the Energy Act (2006), and the Energy Act (2019), named as first, second, and third waves of restructuring, respectively. It also includes information on how these three waves of restructuring presented opportunities for expansion of renewable energy in Kenya.

In the first wave of restructuring occasioned by the Electric Power Act 1997, the management of Kenya Power Company (KPC) – the electricity-generating arm of Kenya Power and Lighting Company (KPLC) – was separated and renamed KenGen in 1998. Kenya Power and Lighting Company (KPLC) remained with a dedicated mandate for electricity transmission and distribution, and KenGen as the State Corporation, remained with the mandate for electricity generation (KenGen, 2020). Under the Electric Power Act 1997, the power sector had three main actors; the Ministry of Energy, Kenya Power and Lighting Company, and KenGen (The Electric Power Act, 1997).

The second wave sought to perpetuate privatisation of electricity generation and distribution to enhance opportunity for private sector involvement in

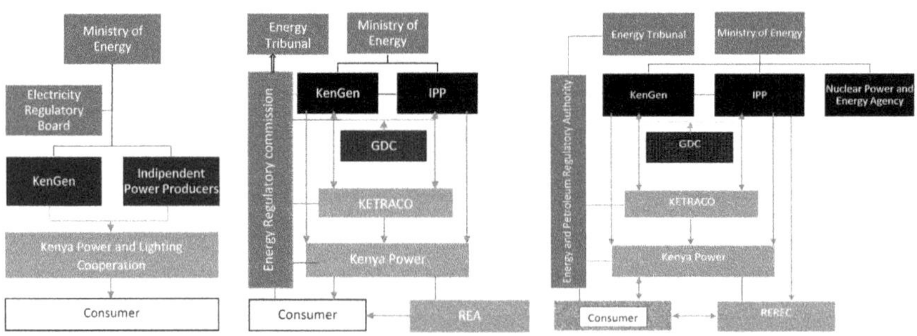

FIGURE 3.3 An illustration of the structural changes in electrification (1997–2019). Source: authors; based on the Electric Power Act 1997; Energy Act 2006; and Energy Act 2019.

renewable electricity generation and distribution as was recommended in the Sessional Paper No. 4 on Energy of 2004 (MoE, 2004). The privatisation started with KenGen selling 30% of its shares through public offers and the unbundling of Kenya Power and Lighting Corporation into three entities; Kenya Power (49.9% privately owned), Kenya Electricity Transmission Company (KETRACO) (100% government owned), and the Rural Electrification Authority (REA) (also 100% government owned). This thus called for an independent regulatory authority and a tribunal as confirmed in the Energy Act of 2006 (The Energy Act 2006, 2006). Kenya Power Company received the mandate to continue with distribution and management of old transmission lines and KETRACO was mandated to build new high voltage transmission lines. The common transmission infrastructure could support private sector investment and promote private distributors to enhance efficiency. The Rural Electrification Authority (REA) would venture an aggressive rural electrification programme with a target of reaching 40% rural electrification by 2020 (MoE, 2004).

The third wave creates opportunity for consumers to enhance adoption of rooftop solar under net metering and strengthen REA by transforming it into a state corporation, increasing its ability to promote renewable electricity in rural areas. The corporation adopts all functions of REA and much more, including sourcing additional funds for renewable electrification and developing and updating the Renewable Energy and Rural Electrification Master Plan in consultation with the county governments (The Electric Power Act, 1997; The Energy Act 2006, 2006; The Energy Act, 2019, 2019).

Constraints to future expansion of renewable electricity in Kenya

This section presents various constraints that would affect further expansion of renewable electricity in Kenya. Applying the MLP framework, we look at the constraints based on four elements; landscape, institutional context and legitimisation, technology development, and agencies and power.

The landscape pressure

In a rapid industrialisation landscape, a stable generator such as hydropower and other fossil-based sources would be preferred. According to the Kenya Vision 2030, economic growth outlook is projected at an annual growth rate of 10% higher than the current average of 5.7%. Upon implementation of the flagship projects, the national electricity demand pressure would exceed the rate of progression of renewable electricity (GoK, 2007; Ministry of Energy and Petroleum, 2018; KNBS, 2019). For instance, the development of Lamu Coal Power Plant is based on the projected expansion of infrastructure in the northern transport corridor, including the Lamu Port, South Sudan, Ethiopia Transport Corridor (LAPSSET) infrastructure. However, there is uncertainty around the government commitment to replace completely fossil-based energies. Arguably, the political economy of the energy sector would have a large influence on the choices and decisions to invest in renewables, including wind and solar.

Institutional context and legitimisation

In 2020 there were no known regulations enforcing power purchase agreement, yet it is a major determinant to electricity pricing and development of renewable energy. In an open market, the agreement lies between the buyer and the seller. Often, the contracts are defined with a take or pay system rather than energy delivered. As such, even with the rapid decline in fossil fuel-based generation, electricity tariffs remain high resulting in cost related bottlenecks to new customers and electricity consumption, and consequently constrained demand and reduced requirements for new generation. Moreover, the persistent high electricity prices can potentially be a factor contributing to unintended consequences of other policies. For instance, the high prices of electricity from the grid in combination with the net metering policy is likely to catalyse high levels of own generation, especially from solar PV, depriving Kenya Power of revenue.

Stakeholders consulted for this study emphasised that contractual obligations of the utility with power vendors have affected electricity tariffing. Future demand and supply forecast demonstrate a possibility of a grid free from fossil fuels except for grid stabilisation, yet the average cost of fuel charge is USD 4.5 cents/kWh which is about 20% and 30% overall electricity cost per kilowatt for domestic tariff one and domestic tariff two respectively (Regulus Limited, 2020). The statutory levies and charges account for about 44% of the total electricity bill. Institution-based levies include Energy and Petroleum Regulatory Authority levy, Water Resources Management Authority Level and Rural Electrification Program levy and other operational related levies and taxes include Value Added Tax, Fuel Cost Adjustment and Foreign Exchange. The variable charges are dependent on mainly global economic factors such as fossil fuel levy, inflation adjustment, and foreign exchange rate fluctuation adjustment related to fluctuation of hard currency against the Kenya shilling. Water resources levy is charged per kWh generated from hydro sources above 1 MW and charged at five cents per kilowatt hour. The four-variable electricity costs, including inflation adjustment, fuel costs, water resource management authority fees, and foreign exchange rate fluctuations, are published in the Kenya gazette on a monthly basis along with the trends in the past ten years (ERC, 2018). These and other charges have constantly maintained high tariffs with barely observed reductions irrespective of changes in the sector.

Technology, development, and market function

Most of the renewable energy technologies used locally are either sourced or imported from abroad. This presents several challenges including potential loss of foreign exchange and operations. This presents a bigger challenge related to technology transfer and adoption, and the extent to which this enhances local capabilities. We find this critical because the Energy Act 2019 to some extent provides for local content requirement which is intended to support the growth of local manufacturing.

This study shows that both wind and solar technologies are externally sourced, which has implications for technology development and deployment. Kenya over-relies on external support in four critical elements in technology development and transfer, namely equipment manufacturing, project development, construction and installation, and operations and maintenance (see Hanlin and Okemwa, this volume). Overall, Kenya imports almost all its equipment/technologies including solar and wind accessories. Most grid-tie project developments have been led by external consultants, and construction and installation is often sub-contracted to equipment suppliers. For instance, Vestas Limited was tendered to supply and install wind turbines in the LTWP project. Mostly local engineers are trained to operate the plant, but there are still high levels of dependency on external capacity for maintenance.

From a collaboration perspective, this highlights potential for foreign firms to contribute to local capabilities, but currently the policies are not clear on how the local content requirements should be implemented to ensure support for local manufacturing and industrialisation more generally.

Agency and power

Institutional monopoly and lack of capacity

Kenya Power and KenGen are the dominant agencies in the power sector based on the market shares they hold. Kenya Power is currently a monopoly, being the only legally recognised distributor of grid electricity in Kenya. KenGen is also a dominant producer because it currently controls a 75% share of the total electricity generation portfolio. This is despite the fact that both the generation and distribution of energy has been liberalised through the Energy Act (2006, 2019). However, the levels of control of the market by these two state agencies have created an artificial monopoly that is entrenching a system lock-in. The newly created agencies largely rely on Kenya Power's existing infrastructure, such as networks, transmission, and distribution lines, and financial collection systems, and KenGen's technical artefacts, such as exploration tools, data collection equipment, and historic surveys. Both REREC and KETRACO depend heavily on this established system and this dependence can create serious obstacles to innovation. The dominance by Kenya Power in the distribution sector creates a strong resistance to change and limits the potential emergence of new distribution companies and/or distribution models. The Energy Act 2019 provides for electricity retail through existing distribution lines at a cost determined by the licensee of the distribution line. Hence, whereas the institutional adjustments intended to be transformative, the knowledge, technology, and governance base is likely to remain unchanged for the near future.

Societal and cultural constraint

Key constraints for investments in solar and wind in Kenya are land policies and poorly developed land markets. The land tenure system in Kenya classifies land into four categories: public land, private land, community land, and forest land. While state corporations and agencies can compulsorily acquire land for infrastructure development, the high cost of land is sometimes very prohibitive. The challenge is more acute for private sector developers who have to negotiate directly with landowners for investments that target private and community land. As an example, the development of an evacuation high voltage line from Lake Turkana Wind Power was delayed by more than a year occasioned by challenges in obtaining wayleave access. Other renewable energy projects have struggled to get financial approval due to lack of necessary land rights documents. In other cases, rent-seeking behaviour, which has led to inflation of land purchase and lease prices, is becoming a major hindrance to investments in wind and solar plants. The environmental safeguards, such as environmental and social impact assessments, have in some cases also been faulted.

Resistant to change

It was noted during the stakeholder's engagement and mapping that some actors can be very influential in the value chain, even though they actually may be less influential in the contribution to systemic changes in policy processes. For instance, even though community members are mainly consulted during the environmental and social impact assessments, in which case they can provide information on their perceived positive or negative impact of the project to the community, they remain critical actors because they own the land on which the renewable electricity projects are developed. Virtually all renewable energy resources that are not on public land are located on land that is owned by a community or private entity. Land in Kenya has strong socioeconomic ties and is considered important as a source and means of wealth, and for agriculture, especially by the rural poor who form the majority of the population (Holde, Otsuka, and Place, 2010). Already, this valuable resource is shrinking with the swelling population, climate change impact, and migration. In some cases, it contributes to one's social status in a society, and in other cases it is important for cultural practices such as shrines. Any policy touching on land is in itself very sensitive. It has remained status quo, and the responsibility of investors to negotiate their way with the communities, including seeking wayleaves, makes it very difficult for investment. Kenya Power (2019) reports that land tenure has become one of the major bottlenecks in extending electricity in rural areas. Aggravating the land-based challenges is tied to the political angle of land ownership. Boone reiterates that land politics is a redistributive game that creates winners and losers (Boone, 2012).

Political will

Politics and power struggle have affected the investment (public and private) in the renewable energy sector, especially grid-tied investments. The nature of the sector as a business opportunity and power influence has made it critically political. In a focus group discussion, it was observed that the business angle has often blurred decision making, often resulting in partisan interests superseding objective analysis and decisions based on evidence. With the exploration and extraction of oil in Turkana and coal in Kitui, the country is at a crossroads of defining its energy future. A just transition to a low carbon economy that delivers poverty reduction and resilience and a more development-oriented growth presents a political decision dilemma. Who sets the terms of transitions and for whom? This raises political questions about the role of actors and whose interest takes preference (Newell et al., 2014). Political economy will therefore continue to play a key role in defining technological and social outcomes.

Conclusion

The renewable electricity subsector in Kenya has indeed changed from 1997 to 2019. Based on four elements of the multi-level perspective framework of socio-technical and economic transition, in this chapter we have analysed and discussed the opportunities and constraints for the development and growth of the wind and solar energy subsectors. The analysis showed a mix of opportunities and constraints but largely remains positive towards increased investment in renewable electricity (solar and wind) for rural electrification. During the period of the analysis, five major statutory and policy milestones facilitating the transformation in the energy sector were realised. Besides the Constitution of Kenya (2010), which devolved planning and development of electricity and gas reticulation, and energy regulation, the Electric Power Act (1997), the Sessional Paper No. 4 on Energy (2004), the Energy Act (2006), the Energy Policy (2018), and the Energy Act (2019) all presented different provisions for the creation of institutions and processes that promote investment in rural electrification and renewable energy, including solar and wind. These policy and structural changes in the electricity sector present opportunities and constraints in future renewable electricity development in Kenya as summarised below.

The opportunities in future deployment of renewable electricity include, inter alia, the high economic growth rate (an average of 5.6% annually) and a steady demographic growth (2.2% annually), significantly increasing energy demand. Renewable electricity is perceived to enhance job creation and employment, deriving more stakeholders' interest and private sector investment. The various legislation and policies, including the Feed in Tariff (FiT), net metering, and power wheeling, enhance increased harnessing of renewable electricity, especially by facilitating increased investments by the private sector. They also enhance flexibility of own generation and use of electricity. Finally, the unbundling and privatisation in the energy sector has broken institutional monopolies and could potentially increase private sector investment in renewable electricity generation.

Some of the glaring constraints in expansion of renewable electricity in Kenya could also be associated with high demand. A high double digit annual economic growth rate of 10%, as envisaged in the Kenya Vision 2030, could create more pressure on existing supply which may necessitate immediate investments in new power plants. Moreover, high dependency on foreign support in development of these renewable energy technologies limits the opportunity to develop and promote home-grown technologies in future renewable energy development. As reported in other chapters (see for instance Hanlin and Okemwa, 2022; this volume), besides manufacturing, Kenya relies heavily on foreign support in project development, construction and installation, and operation and maintenance.

The energy sector like other sectors also suffers political economy issues and this influences how things unfold, especially in the grid electricity subsector. Lastly, difficulties in negotiating purchase and leasing agreements on private and community land due to the sensitivity of land issues further limit investments in wind and solar PV projects.

Acknowledgements

Support for research on which this chapter is based from the Danish Ministry of Foreign Affairs, Grant: DFC 14-09AAU is gratefully acknowledged. We also acknowledge the IREK team especially the editors for insights, information, and valued critique.

References

Abdallah, S.M., Bressers, H. and Clancy, J.S. (2015) 'Energy reforms in the developing world: Sustainable development compromised?', *International Journal of Sustainable Energy Planning and Management*, 5, pp. 41–56. https://doi.org/10.5278/ijsepm.2015.5.5

African Union & Commission (2015) *Agenda 2063: The Africa We Want*. https://www.un.org/en/africa/osaa/pdf/au/agenda2063.pdf

Basu, A. and Marett, D.J. (2016) *Nationally Appropriate Mitigation Action: Access to Clean Energy in Rural Kenya through Innovative Market based Solutions*. Available at: https://www.undp.org/content/dam/LECB/docs/pubs-namas/undp-lecb-Kenya_Clean-Energy-NAMA-2016.pdf

Blimpo, M.P. and Malcolm, C.-D. (2019) *Electricity Access in Sub-Saharan Africa: Uptake, Reliability, and Complementary Factors for Economic Impact*. Africa Development Forum Series. Washington, DC: World Bank. doi:10.1596/978-1-4648-1361-0. License: Creative Commons Attribution CC BY 3.0 IGO

Boone, C. (2012) 'Land conflict and distributive politics in Kenya', *African Studies Review*, 55(1), pp. 75–103. https://doi.org/10.1353/arw.2012.0010

Byrne, R., Ockwell, D., Urama, K., Ozor, N., Kirumba, E., Ely, A., Becker, S. and Gollwitzer, L. (2014) *Sustainable Energy for Whom? Governing Pro-poor, Low Carbon Pathways to Development: Lessons from Solar PV in Kenya*. Brighton: STEPS Centre. Available at: https://core.ac.uk/download/pdf/30610839.pdf

Eberhard, A. and Kåberger, T. (2016) 'Renewable energy auctions in South Africa outshine feed-in tariffs', *Energy Science & Engineering*, 4. https://doi.org/10.1002/ese3.118

The Electric Power Act, no. 11 of 1997 (1997) https://www.scribd.com/doc/140005439/Electrical-Power-Act-1997-kenyan-laws

The Energy Act 2006, no. 12 of 2006, The Government of Kenya (2006) http://kenyalaw.org/kl/fileadmin/pdfdownloads/Acts/EnergyAct_No12of2006.pdf

The Energy Act, 2019, no. 1 of 2019, The Government of Kenya (2019) https://kplc.co.ke/img/full/ec4cPzz9sNDQ_1.%20ENERGY%20ACT%20No.%201%20of%2002019.PDF

Energy Regulatory Commission (ERC) (2018) 'Approval of the schedule of tariffs set by the energy regulatory commission for supply of electrical energy by Kenya power and lighting company limited pursuant to section 45 of the energy act, 2006', *The Kenya Gazette*. https://www.epra.go.ke/download/applicable-electricity-tariffs-june-2019/

EPRA (2020) 'Renewable energy registers' in *Energy and Petroleum Regulatory Authority*. Available at: https://www.epra.go.ke/services/renewable-energy-2/energy-audit-firm-register/

ERC (2008). *Kenya Electricity Grid Code*. https://eregulations.invest.go.ke/media/kenya%20grid%20code.pdf

ERC (2012). *The Energy (Solar Water Heating) Regulation, 2012*. Energy Regulatory Comission of Kenya (ERC).

Farla, J., Markard, J., Raven, R. and Coenen, L. (2012) 'Sustainability transitions in the making: A closer look at actors, strategies and resources', *Technological Forecasting & Social Change*, 79. https://doi.org/10.1016/j.techfore.2012.02.001

Fast Company (2020) *The World's Top 10 Most Innovative Companies of 2015 in Africa*, June 24. Available at: https://www.fastcompany.com/3041821/the-worlds-top-10-most-innovative-companies-of-2015-in-africa

Geels, F.W. (2011) 'The multi-level perspective on sustainability transitions: Responses to seven criticisms', *Environmental Innovation and Societal Transitions*, 1(1), 24–40. https://doi.org/10.1016/j.eist.2011.02.002

Geels, F.W. (2014) 'Regime resistance against low-carbon transitions: Introducing politics and power into the multi-level perspective', *Theory, Culture & Society*, 31(5), pp. 21–40. https://doi.org/10.1177/0263276414531627

Godinho, C. and Eberhard, A.A. (2019) *Learning from Power Sector Reform: The Case of Kenya*. Policy research working paper series 8819, pp.53. The World Bank. https://doi.org/10.1596/1813-9450-8819

GoK (2007) *Kenya Vision 2030*. Kenya: Government Press. https://www.researchictafrica.net/countries/kenya/Kenya_Vision_2030_-_2007.pdf

GoK (2010) *The Constitution of Kenya, 2010* (Revised Edition, 2010). National Council for Law Reporting with the Authority of the Attorney General. http://vision2030.go.ke/publication/the-constitution-of-kenya-2010/

GoK (2011) *Scaling-up Renewable Energy Program (SREP): Investment plan for Kenya*. https://www.climateinvestmentfunds.org/sites/cif_enc/files/Kenya_post_mission_report_March_10_2011.pdf

Hanlin, R. and Okemwa, J. (2022) 'Interactive learning and capability-building in critical projects', in *Building Innovation Capabilities for Sustainable Industrialisation: Renewable Electrification in Developing Economies*. New York: Routledge. https://doi.org/10.4324/9781003054665-7

Holden, S.T., Otsuka, K. and Place, F.M. (2010) *The Emergence of Land Markets in Africa: Impacts on Poverty, Equity, and Efficiency*. New York: Routledge. https://www.routledge.com/The-Emergence-of-Land-Markets-in-Africa-Impacts-on-Poverty-Equity-and/Holden-Otsuka-Place/p/book/9781936331666

IEA (2019) *Africa Energy Outlook 2019*. Available at: www.iea.org/africa2019

KenGen (2020) *Who We Are*, Accessed on April 30, 2020. Available at: https://kengen.co.ke/index.php/our-company/who-we-are.html

Kenya Power (2004) *Annual Financial Statement and Report for the Financial Year Ending June, 2003* [Annual Financial Statements]. https://www.kplc.co.ke/content/item/40/annual-reports-archives. https://www.kplc.co.ke/AR2004/Annual%20Reports%20&%20Accounts%202004-05.pdf

Kenya Power (2013) *Annual Report and Financial Statement, 2012* [Annual Financial Statements]. Kenya Power. https://www.kplc.co.ke/content/item/40/annual-reports-archives. https://kplc.co.ke/img/full/0XLOM2rTP95g_KENYA%20POWER%20ANNUAL%20REPORT%2020122013%20FA%20127,128.pdf

Kenya Power (2018) *Annual Report and Financial Statements for the year ended 30 June 2017*, p. 200. https://www.kplc.co.ke/content/item/40/annual-reports-archives. https://kplc.co.ke/AR2017/KPLC%202016%20-%202017%20Annual%20Report-.pdf

Kenya Power (2019) *Annual Report and Financial Statements for the Year Ended 30th June 2018*. https://www.kplc.co.ke/AR2018/KPLC%20Annual%20Report%2017_12_2018_Wed.pdf

Kenya National Bureau of Statistics (KNBS) (2009) *2009 Kenya Population and Housing Census: Counting our people for implementation of vision 2030* [National Census Report]. Vol II, Kenya National Bureau of Statistics.

Kingiri, A. and Okemwa, J. (2022) 'Local content and capabilities: Policy processes and stakeholders in Kenya', in *Building Innovation Capabilities for Sustainable Industrialisation: Renewable Electrification in Developing Economies*. New York: Routledge. https://doi.org/10.4324/9781003054665-11

KNBS (2019) *Economic Survey, 2019*. Government Press. Available at: https://www.knbs.or.ke/?wpdmpro=economic-survey-2019-popular-version

KNBS (2020a) *2019 Kenya Population and Housing Census Volume I: Population by County and Sub-County* [Census Report]. Available at. https://www.knbs.or.ke/?wpdmpro=2019-kenya-population-and-housing-census-volume-i-population-by-county-and-sub-county

KNBS (2020b) *Economic Survey 2020*. Available at: https://www.theelephant.info/documents/kenya-national-bureau-of-statistics-economic-survey-2020/

Kiplagat, J., Wang, R.Z. and Li, T.X. (2011) 'Renewable energy in Kenya: Resource potential and status of exploitation', *Renewable and Sustainable Energy Reviews*, 15, pp. 2960–2973. https://doi.org/10.1016/j.rser.2011.03.023

Kiprop, E., Matsui, K. and Maundu, N. (2020) 'The future contribution of demand side management to solving Kenya's energy insecurity problems', *International Journal of Environmental Science and Development*, 11, pp. 111–115. https://doi.org/10.18178/ijesd.2020.11.3.1235

Gregersen, C. and Gregersen, B. (2022) 'Interactive learning spaces: Insights from two wind power megaprojects', in *Building Innovation Capabilities for Sustainable*

Industrialisation: Renewable Electrification in Developing Economies. New York: Routledge. https://doi.org/10.4324/9781003054665-8

Lake Turkana Wind Power Project (2011) *Updated Environmental and Social Impact Assessment Summary* [Summary Report]. Available at: https://old.danwatch.dk/wp-content/uploads/2016/05/LTWP_ESIA_Executive_Summary_01.pdf

Lawson, B. and Samson, D. (2001) 'Developing innovation capability in organisations: A dynamic capabilities approach', *International Journal of Innovation Management*, 5(3), pp. 377–400. https://doi.org/10.1142/S1363919601000427

Lema, R., Hanlin, R., Hansen, U.E. and Nzila, C. (2018) 'Renewable electrification and local capability formation: Linkages and interactive learning', *Journal of Energy Policy*, 117, pp 326–339. Doi: https://doi.org/10.1016/j.enpol.2018.02.011

Mbaka, J. and Mwaniki, M. (2016) 'Small hydro-power plants in Kenya: A review of status, challenges and future prospects', *Journal of Renewable Energy and Environment*, 3, pp. 20–26. Doi: https://doi.org/10.30501/jree.2016.70096

Ministry of Energy (2011) *Updated Least Cost Power Development Plan 2011*—2031. http://kerea.org/wp-content/uploads/2016/08/Least-Cost-Power-Development-Plan-2011-2031.pdf

Ministry of Energy (2012). *Feed-in-Tariffs policy for wind, biomass, small hydros, geothermal, biogas and solar.* http://admin.theiguides.org/Media/Documents/FiT%20Policy%202012.pdf

Ministry of Energy and Petroleum (2018) *Updated Least Cost Power Development Plan 2017—2037.* https://www.decoalonize.org/2017-2037-least-cost-power-development-plans-lcpdp/

M-KOPA Solar (2015) *M-KOPA: The World's Top 10 Most Innovative Companies of 2015 in Africa. M-KOPA SOLAR*, Accessed on February 9, 2020. Available at: http://www.m-kopa.com/m-kopa-featured-as-the-worlds-top-10-most-innovative-companies-of-2015-in-africa/

M-KOPA Solar (2020) *Our Impact, M-KOPA Corporate*. Accessed in April 16, 2020, Available at: http://solar.m-kopa.com/about/our-impact/

MoE (2004) *Sessional Paper No. 4 on Energy*. Available at: https://renewableenergy.go.ke/downloads/policy-docs/sessional_paper_4_on_energy_2004.pdf

MoE (2018) *National Energy Policy*. Available at: https://kplc.co.ke/img/full/BL4PdOqKtxFT_National%20Energy%20Policy%20October%20%202018.pdf

Newell, P., Phillips, J., Pueyo, A., Kirumba, E., Ozor, N. and Urama, K. (2014) 'The political economy of low carbon energy in Kenya', *IDS Working Papers*, 2014(445), pp. 1–38. https://doi.org/10.1111/j.2040-0209.2014.00445.x

Njiru, C.W. and Letema, S.C. (2018) 'Energy poverty and its implication on standard of living in Kirinyaga, Kenya', *Journal of Energy*, 2018. Accessed 8 October 2020. Doi: https://doi.org/10.1155/2018/3196567

Nussbaum, M.C. (2001) *Women and Human Development: The Capabilities Approach*. University of Chicago, Cambridge University Press. Doi: https://doi.org/10.1017/CBO9780511841286

Ockwell, D., Ely, A., Mallett, A., Johnson, O. and Watson, J. (2009) *Low Carbon Development: The Role of Local Innovative Capabilities*. STEPS Centre. Available at: https://opendocs.ids.ac.uk/opendocs/handle/20.500.12413/2459

Regulus Limited (2020) *Electricity cost in Kenya*. Accessed on 2020, April 30, Available at: https://www.stimatracker.com/

RENCON Consulting (2018) *Study on Solar Photovoltaic Industry in Kenya* [Draft Report]. Available at: https://www.epra.go.ke/downloads/

Roux, A. and Shanker, A. (2018) *Net Metering and PV Self-consumption in Emerging Countries* [Report IEA-PVPS T9-18:2018; IEA PVPS Task 9, Subtask 4], pp. 56. International Energy Agency (IEA). Available at: https://iea-pvps.org/wp-content/uploads/2020/01/T9_NetMeteringAndPVDevelopmentInEmergingCountries_EN_Report.pdf

Smith, A., Voß, J.-P. and Grin, J. (2010) 'Innovation studies and sustainability transitions: The allure of the multi-level perspective and its challenges', *Research Policy*, 39(4), pp. 435–448. https://doi.org/10.1016/j.respol.2010.01.023

United Nations (2015) 'About the sustainable development goals', *United Nations Sustainable Development*. Available at: https://www.un.org/sustainabledevelopment/sustainable-development-goals/

4

CENTRALISED AND DECENTRALISED DEPLOYMENT MODELS

Is small beautiful?

Ulrich Elmer Hansen, Cecilia Gregersen, Faith H. Wandera, Nina Kotschenreuther, and Rebecca Hanlin

Abstract

The sectoral innovation system perspective has been developed as an analytical framework to analyse and understand innovation dynamics within and across various sectors. Most of the research conducted on sectoral innovation systems has focused on an aggregate-level analysis of entire sectors. This chapter argues that a disaggregated (sub-sectoral) focus is more suited to policy-oriented work on the development and diffusion of renewable energy, particularly in countries with rapidly developing energy systems and open technology choices. Based on preliminary insights from research carried out in 2016 and 2017, including mapping, interviews, and policy framework analysis, it focuses on size, distinguishing between small-scale (mini-grids) and large-scale (grid-connected) deployment paths in renewable energy. We explore how wind and solar markets in Kenya differ in terms of development and organisation, both across and within sectors, by examining the development and diffusion of solar photovoltaic (PV) and wind technology and how they evolve in these sub-sectoral systems. We find that innovation and diffusion dynamics differ more between small and large than between wind and solar. This has important analytical implications because the disaggregated perspective allows us to identify trajectories that cut across conventionally defined core technologies. This is valuable for ongoing discussions of electrification pathways in developing countries. We conclude the chapter by distilling the policy implications of these findings in terms of the requirements and incentive mechanisms that shape different pathways.

Introduction

Kenya, like many other countries around the globe, is currently facing momentous energy decisions. With a low rural electrification rate and a large proportion

DOI: 10.4324/9781003054665-4

of the population currently lacking access to electricity, increasing generating capacity and achieving 100% energy access is a key priority for the Kenyan government. While the current electricity system relies mainly on hydropower, the expansion of renewable energy (RE) sources, especially wind and solar power, has been given a high priority in national policies such as the national development strategy Vision 2030 and the rural electrification master plan (GoK, 2007; REA, 2009). In this context, Kenya faces a number of important technological choices in terms not only of which technologies to prioritise, but also how to deploy them. The current policy frameworks have enabled a combination of government and private-sector developments in the energy sector (See Ogeya et al., 2022; this volume, and Kingiri and Okemwa, 2022; this volume).

The concept of sectoral innovation systems (SIS) has been used to illuminate the factors affecting innovation dynamics within and across sectors. The SIS perspective is particularly concerned with highlighting sector-specific characteristics of industrial evolution (Malerba and Nelson, 2011). From the sectoral perspective, increasing attention is paid to RE sectors and their development. In this chapter, we argue that it is crucial to take a closer look at the RE sector and what constitutes such a sector in order to push further the disaggregation of trends in the sub-sectors of wind and solar PV. In examining differences in terms of size and shape across and between these sub-sectors, we raise questions regarding the definitions and boundaries of these renewable energy 'sectors'.

Thus, the key research question of this chapter is: how do wind and solar markets in Kenya differ in terms of development and organisation, both across and within sectors? We answer this question by analysing current status and developments across the mini-grid and large-scale market segments for wind and solar PV technologies, respectively. Then we use the SIS perspective to describe the characteristics of each sub-sector, their drivers, and barriers, and discuss the similarities and differences between them. As detailed and up to date information on the development and dynamics of the solar and wind markets in Kenya were found to be lacking, this chapter seeks to bring together insights from research conducted in 2015–2016.

The chapter is structured as follows: we start by briefly introducing the disaggregated sectoral innovation systems approach and the methodology. Then we present an overview of the current status and trends across the mini-grid and large-scale market segments for wind and solar PV in Kenya.[1] Next, we describe each of the four disaggregated sectoral innovation systems and their characteristics followed by a discussion of the similarities and differences across these sectors using the three main dimensions of the SIS approach as vectors. Finally, we offer concluding remarks on sub-sectoral pathways and policy implications based on the key findings of the research.

The disaggregated SIS perspective and methodology

Innovation systems approaches are increasingly used for the analysis of development challenges in Africa (Adebowale et al., 2014; Lundvall and Lema, 2014).

The sectoral systems perspective ascribes importance to learning, knowledge, and capability accumulation in the innovation process (Malerba, 2005). The SIS perspective is based on the underlying assumption that innovation dynamics are closely related to the specific characteristics of a given sector or industry. Innovation within a sector is a dynamic process, which constantly transforms the structure and boundaries of a given industry.

In this chapter, the focus is on analysing two low carbon technologies, namely solar PV, and wind technologies in Kenya. While there are profound differences between low carbon technologies (Lema et al., 2015), the differences within solar PV and wind energy as overarching technological categories are equally profound. To give an example, the notion of a 'solar technology' may be used as an umbrella term to describe solar-powered LED lamps, solar home systems, and utility-scale solar power plants. Common to these systems is the fact that they make use of solar panels as the underlying source of electricity generation. However, it is clear that there are significant differences between the respective users, producers, investors, actors, prices, scales, R&D intensities, value chains, technical characteristics, and competing technologies of these systems (Adebowale et al., 2014). As noted by Stephan et al. (2017), understanding such differences in sectoral configurations helps identify dynamics that otherwise go unnoticed. As a result, each of the sub-categories of these systems of technology may more appropriately be considered units of analysis in their own right. In the delineation of specific sectors, a key question therefore concerns the selection of an appropriate level of aggregation in the analysis.

Initially, Malerba defined SISs broadly as 'a set of new and established products for specific uses and the set of agents carrying out market and non-market interactions for the creation, production and sale of those products' (Malerba, 2002, p. 250). While this broad definition was developed with the intention to be able to cover research conducted at various levels of aggregation, most empirical studies in this field focus on a highly aggregated level of analysis covering the entire pharmaceutical, chemical, telecommunications, or biotechnology sectors (Malerba, 2005). In this chapter, we adopt a more disaggregated level of analysis in order to uncover in further detail the innovation dynamics within such overarching and broadly defined sectors.

Accordingly, the cases of solar and wind technologies examined in this chapter are understood as sub-sectors of the wider renewable energy sector, which in turn is considered a subset of the broader energy sector, and so forth. Based on this understanding, we distinguish between small-scale mini-grids and large-scale power plants using solar and wind technologies to generate electricity. Mini-grids are understood as decentralised (off-grid) systems consisting of power-generating assets and distribution with power capacities of between 0.2 kW and 2 MW connecting two or more individual households (Pedersen, 2016). Large-scale power plants are understood as grid-connected plants owned by utilities and/or private operators with installed capacities above 15 MW.

The above description translates into the conceptualisation of four different SISs in Kenya with distinctive sector-specific innovation features, which are explored in the chapter: (i) wind-powered mini-grids; (ii) large-scale, grid-connected wind-power plants; (iii) solar-powered mini-grids; and (iv) large-scale, grid-connected solar power plants. Following the SIS perspective, three main dimensions are used to guide the analysis of these four sectors (Malerba and Nelson, 2011):

- Knowledge and technologies
- Actors and networks
- Institutions

Overview of current status and trends across the mini-grid and large-scale market segments for wind and solar PV in Kenya

The analysis is based on research presented in Hansen et al. (2018). This article was based on insights from research conducted in 2016 and 2017 as part of the wider project on renewable electrification in Kenya entitled Innovation and Renewable Electrification in Kenya (IREK), which examines the implementation of wind and solar technologies in Kenya's renewable electrification process (IREK, 2018).

Table 4.1 presents a summary of the findings from Hansen et al. (2018). The main source of information was semi-structured interviews carried out in Nairobi in 2016 and 2017 with key actors involved in the sectoral systems. Actors and organisations interviewed include project developers, regulators, investors, plant operators, technology suppliers, donor agencies, and government agencies. To gain an overview of the market status and trends and to triangulate information, desk research reviewed and consulted a large variety of documents, including papers from the peer-reviewed literature, media reports, presentations, company press releases, and industry and other reports.

Looking at the overall wind sector, there is clear variation in the dynamics of small- and large-scale wind. The market for small-scale wind-based mini-grids appears to have stalled: very few hybrids exist or are planned, and private suppliers of wind-powered mini-grids have shifted focus (see also Wandera, this volume). In contrast, the market for large-scale wind projects is moving forward, with the flagship Lake Turkana project (see Gregersen and Gregersen, 2022; this volume) drawing massive attention, together with a number of other large-scale projects.

In the overall solar sector, the market for small-scale solar-based mini-grids is currently experiencing a period of significant momentum, with both private mini-grid operators and many donors involved with existing and planned hybrid greenfield mini-grids (Duby and Engelmeier, 2017). On the other hand, the market for large-scale solar projects has only moved to a very limited extent on

TABLE 4.1 Summary of current status and trends across sectors

Small	**Wind-powered mini-grids** • State-owned mini-grids include two operational wind hybrids owned by REA • Five KPLC diesel mini-grids are being retrofitted to include wind power • Scarce information regarding commercial wind- and solar-powered mini-grids indicate 80–100 small wind turbines installed by telecom players, NGO's, and commercial and household clients • Local manufacturer active since late 1990s, three foreign suppliers active since 2010, however focus and activities have shifted towards the emerging market for solar mini-grids • Approx. 20 companies currently offer imported turbines, but mainly as complement to their main product, solar PV • 19 wind-diesel hybrids planned as part of the implementation of the Kenyan government's rural electrification master plan 2009 • One donor-funded project (UNIDO) implemented in 2009	**Solar-powered mini-grids** • State-owned mini-grids include seven solar-diesel hybrids and one wind-solar-diesel hybrid, owned by REA, operated by KPLC • A further 15 state-owned mini-grids including solar power currently under construction • One local solar PV assembly plant and a number of local battery producers/ suppliers, but they mainly serve the domestic solar market • Nine hybrid solar-diesel stations being developed (in existing diesel-fired plants) and an additional 25 in initial proposal stage • Donor organisations actively promote and financially support a number of specific projects • Since 2012 a number of private foreign-owned companies have installed 20–30 grids (two of these with a formal licence to operate) and are in the process of significantly upscaling their activities • Most core components are sourced from renowned suppliers from Europe or the United States
Large	**Grid-connected wind-power plants** • Two operational plants, Ngong Power Station, owned by KenGen (establ. 1993), and the independently owned Lake Turkana Wind Power project • Two projects under development, (while one other project was cancelled at a late stage), all being developed in connection with the Kenyan Feed-in Tariff for wind-power project (FIT, first introduced in 2008, revised in 2012) by different consortiums with international actors	**Grid-connected solar power plants** • Five plants (<575 kWp) currently in operation, three financed mainly by international donors, two by private firms. Seven plants above 15 MW are planned/ under development • Delivered on turnkey basis from suppliers abroad • Involvement of local companies is limited to the construction stage and maintenance services during operation

(Continued)

TABLE 4.1 *(Continued)*

	• Private developers, donors, and development banks have provided financial support and advisory services to move the Kenyan Feed-in Tariff for wind-power project to financial closure • By 2013 a total of 236 applications were submitted under the FIT project, of which 20 were approved. However, a freeze on new power purchase agreements for wind and solar projects means that their development is uncertain	• One large-scale project (Garissa, owned by REA) is in operation, while a number of other larger scale projects developed by foreign suppliers seem to be under development (since 2012), most of which are supported by donors and development banks, but are struggling to secure funding and reach financial closure, hence most have not yet reached construction

Source: Hansen et al. (2018).

the ground, as existing projects are small in scale, and many large-scale projects remain at the planning stage. In the next section, these trends will be compared to the characteristics of the four disaggregated SISs.

The size and shape of wind and solar sectoral innovation systems

In the following section, the characteristics of the four SISs are explored and disentangled. The SIS perspective is used to describe the three dimensions – knowledge and technologies; actors and networks; and institutions – of the wind and solar sectors across the size and shape of the projects.

Sectoral innovation system characteristics of wind-powered mini-grids

The existing knowledge and technological base in the domestic industry for wind powered mini-grids in Kenya is characterised by relatively simple and small-scale technologies manufactured locally. Such small-scale systems can be tailored to different local contexts and manufactured from a range of locally available materials while still being relatively robust. As the turbines are typically produced by smaller manufacturers, universities, or NGOs involved in community projects, they do not require advanced engineering knowledge or skills. Thus, as opposed to formalised R&D, the domestic industry for small-scale wind turbines is generally characterised by a high level of informal knowledge and learning in the way that local artisans and blacksmiths tinker with various designs based on the available equipment and materials. While the wind turbines are produced and diffused at relatively low cost, final performance and standards tend to vary greatly. The locally produced systems are contrasted with the imported turbines used in the existing wind-diesel hybrid mini-grids, which are generally higher in performance and price levels (Vanheule, 2012). Due to the lack of experimentation with wind-powered mini-grids, related technical concepts, and commercial applications, limited specialisation and experience has been accumulated in this area.

The main supportive institutional conditions promoting the development of wind-powered mini-grids are related to initiatives adopted as part of the rural electrification master plan to hybridise the existing diesel-fired mini-grids with wind and solar (REA, 2009). These initiatives are supported and complemented by various donor programmes but are also driven by the increasing operational costs of the existing diesel-fired mini-grids. The main actors involved in the domestic industry are local wind-turbine manufacturers, NGOs, and local community entrepreneurs involved in various small-scale projects typically implemented by donors in rural villages (Harries, 1997; Bergès, 2009). A number of these projects include individual engineers and NGOs from abroad, involved in testing a specific technical design for rural applications, however with limited levels of commercialisation (Ferrer-Martí et al., 2012). The local manufacturers rely on local supply chains and distribution networks and typically make use of connections in the local environment for sourcing materials and related know-how. Government agencies promoting rural electrification in off-grid areas are typically also involved in specific projects either directly or indirectly via technical support. The Ministry of Energy and Petroleum is involved in the installation of wind speed data loggers at 20 m and 40 m. Local universities sometimes provide highly applied research input to specific projects such as a collaboration between Jomo Kenyatta University of Agriculture and Technology and the Japanese Government on small wind technology, but formalised R&D activities at universities focusing specifically on small scale wind is largely absent in Kenya.

Sectoral innovation system characteristics of large-scale, grid-connected wind power projects

The knowledge and technology base underlying the development of advanced large-scale wind turbines has evolved into a highly researched and capital-intensive process involving the continuous development of new materials, designs, and production methods. Thus, the development of utility-scale wind turbines involves both internal R&D carried out within industry lead firms and formalised R&D undertaken by research centres at universities or public research organisations. These R&D activities mainly draw on technical disciplines and engineering-based knowledge. The ongoing development efforts focus on reducing the price and improving the performance of wind turbines in order to increase the competitiveness of wind power compared to conventional sources of energy for power generation. As economic feasibility generally increases with the size of the wind turbines, the general trend in the industry has been towards the gradually increasing scale of wind turbines. The development of large-scale wind-power projects also draws on a broader set of organisational and administrative competences, including the skills and systems for turbine component manufacturing (e.g., supply chain management) and the knowledge required for engineering, procurement, and construction (EPC) contracting and the incorporation of third-party consultants (legal advice and engineering consultancy). In the projects under development in Kenya, the main contractors and wind-turbine suppliers have drawn upon a range of such knowledge bases and areas of expertise during project development.

International actors, such as pension funds, development banks, donors, and other types of financial institutions, play an important role in providing finance for the development of the projects. Due to the high national relevance of the projects as large infrastructure investments, national policy makers, regulatory bodies, and government agencies are also involved in developing them.

The government support for large-scale wind (and solar) is part of a broader objective to attract foreign investment in Kenya by making possible the inclusion of private, independent power producers (IPPs) in the energy sector. While direct involvement includes bilateral negotiations between project developers and the relevant authorities, indirect involvement includes political advocacy influencing the projects. While not being directly involved, local community and actor groups exert a strong indirect influence on project development, mainly due to disagreements over land rights issues. The main supporting instrument promoting the development of large-scale wind-power plants in Kenya is the feed-in tariff, which has undergone several revisions and currently has 50 MW cap for projects.

Sectoral innovation system characteristics of solar-powered mini-grids

The knowledge base underlying the development of solar-powered mini-grids in Kenya draws on a variety of disciplines and relies particularly on foreign expertise. In the case of the state-owned solar-diesel hybrids, the main expertise needed is in the area of turnkey contracting. The necessary technological skills of the total system suppliers relate mainly to the capacity to design the plants, manage the sourcing of key components, and undertake the construction and final commissioning of the plants. Since this expertise is not currently available from domestic suppliers in Kenya, European companies with significant experience in turnkey contracting and related engineering tasks dominate the development of these plants. Despite the technical capacity and knowledge accumulated in the domestic industry for solar home systems (Byrne, 2011; Karjalainen and Byrne, 2022; this volume), the local suppliers of core components (such as panels and batteries) seem disconnected from the development of solar-powered mini-grids. The private companies from abroad supplying solar-powered mini-grids on a commercial basis in Kenya draw mainly on engineering-based knowledge in the ongoing technical experimentation efforts to optimise their mini-grid systems. Experience from the telecommunications industry has also provided input into the development of business models based on pay-as-you-go (PAYG) systems specifically developed to target poor customers in rural, off-grid areas. This business model draws on knowledge about IT and software solutions and related data analysis and optimisation systems, as well as the use of smart metering and monitoring technologies. Some of these companies are engaged in client relations with (private) investors in solar-powered mini-grids, some of which are philanthropic foreign investors (Harrington, 2016). Collaborative networks have been established across a number of these companies, as well as linkages to foreign investors, headquarters, and component suppliers in Europe and the United States. A number of state and donor-funded programmes to hybridise the existing diesel-fired mini-grids are greatly influencing the enabling environment for the development of solar-powered mini-grids in Kenya.

However, the existing regulatory framework for rural electrification, which focuses on conventional grid-extension programmes, continues to play an important role in the development of commercial solar-powered mini-grids, resulting in lengthy approval and negotiating processes for project developers.[2] Challenges faced by many solar mini-grid developers still often include access to finance or ensuring affordability of the projects, as the higher cost of such small-scale energy production is borne by the consumers. The lack of focus on such new models for producing and distributing energy is also visible in the policy frameworks, where grid-owners and operators have called for stronger and clearer regulation regarding tariffs, integration, standards, and licensing, as well as the possibility for subsidy schemes (Duby and Engelmeier, 2017).

Sectoral innovation system characteristics of large-scale, grid-connected solar power projects

A key driver for the development of large-scale solar power plants in Kenya is the rapidly decreasing costs of solar panels. The experience of plants under development in Kenya indicates that designs for large-scale solar power plants are generally well proven globally, requiring only minor design and construction modifications to adapt them to local conditions. The knowledge and technological base underlying the development of large-scale solar power plants in Kenya thus draws greatly on foreign expertise in the delivery of plants on a turnkey basis. European companies with substantial experience in turnkey plant engineering, component sourcing, and commissioning have thus delivered the existing plants in cooperation with locally based consultancy companies. Due to the larger scale of the solar power plants currently under development in Kenya, their development draws on additional knowledge of EPC contracting and the related organisational expertise to manage the development of large infrastructure projects. Consequently, international contractors and technology suppliers with the technical expertise and management skills to develop an integrated plant design and to install and operate the system effectively have been involved in planning and developing the projects, as well as providing additional competences in the area of PPA contract negotiations, the legal aspects and detailed engineering tasks.

While development of the existing solar power plants has included industrial users and donors as the project owners, the larger scale solar power plants under development incorporate direct involvement from international investors, including development banks and donor organisations. However, the development of large-scale solar is generally being prevented by the difficulties project developers face in attracting finance from foreign investors, and concerns have been raised that the feed-in tariff system may be too low to incentivise foreign investments significantly (Hansen et al., 2015).

Discussion: sub-sectoral dynamics across size and shape

Distinguishing sectoral innovation system features across market segments and technologies has shown that it is worth considering the similarities and differences between the size and shape of the different sub-sectors of solar PV and wind energy in Kenya. In the following sections the three dimensions of Malerba's (2005) SIS framework are examined across the four sub-sectors (see Table 4.2).

Differences and similarities between knowledge and technology

Regarding the knowledge dimension, it is clear that both within and across the four SISs, each system is characterised by individually distinct knowledge bases. In fact, as noted by Malerba (2005), it is knowledge and technology that place the issue of sectoral boundaries at the centre of analysis. These differences therefore support the argument that a disaggregated sectoral analysis is necessary, perhaps particularly in respect of SIS size (Stephan et al., 2017). This is evident in that both large-scale wind and large-scale solar share some characteristics related to the size of the project, where EPC contractors and turnkey suppliers are present across the technologies. Many of the enabling aspects of this dimension are found in the intersections with the global sectoral characteristics where international actors have established themselves in the Kenyan market. This is notable because domestic actors seem disconnected, despite the technical capacity and knowledge that have been accumulated particularly in the domestic industry for solar home systems. There is little information on the involvement of local suppliers of either solar or wind components in any project. It is noteworthy, however, that across the solar and wind mini-grid sectors the knowledge base dimensions differ in terms of which actors with which knowledge bases are involved. While informal learning and knowledge characterise the wind mini-grid sector, the solar mini grid sector features rather engineering-based knowledge, with more involvement from both private actors and international donors. The solar-powered mini-grid sector is also highly

TABLE 4.2 Summary of sectoral innovation system dimensions across sectors

	Wind mini-grids	*Large-scale wind*	*Solar mini-grids*	*Large-scale solar*
Knowledge and technologies	• Small-scale and simple wind turbines • Informal learning and knowledge • Local craftsmen and engineers • Limited knowledge of wind-powered mini-grids • Absence of formalised R&D activities carried out at universities in small-scale wind turbines • Import of higher standard wind turbines	• Formalised R&D in large-scale wind turbines • Technical and engineering-based disciplines • Complex and capital-intensive capital goods • Experience in EPC contracting and planning of large-scale plants • Expertise in PPA contract negotiation and legal aspects • Design of project tailored to local conditions	• Engineering-based knowledge • Telecom expertise (mobile payment schemes, PAYG models) • Smart metering and monitoring systems • Data management and software optimisation tools • Consultancy and donor experience	• Engineering-based knowledge • Experience in turnkey contracting • Experience in EPC contracting and planning of large-scale plants • Knowledge system design integration and operation
Actors and networks	• Donors, NGOs, local manufacturers involved in small-scale development projects • Actors embedded in local and regional supply chains and distribution networks • Universities involved in practical and hands-on applied research in specific projects • Absence of private suppliers of wind-powered mini-grids • Importers of foreign wind turbines	• Industry lead firms, such as Vestas and General Electric • International investors, including development banks, donors, and pension funds • National policy makers and key government agencies (e.g., via direct negotiation with project developers) • Local community groups (opposing projects)	• European turnkey contractors • Local engineering and consultancy firms • Private suppliers of mini-grids owned by foreign expatriates • Foreign investors (direct plant investments and equity investments) • Foreign component suppliers • Examples of cooperatives and community-based solar mini-grids	• International EPC contractors • Technology suppliers • International investors, including development banks and donors • Industrial users

Institutions	• State and donor support for hybridisation of existing diesel-fired mini-grids • Apparent under-prioritisation compared to solar mini-grids	• Feed-in tariff for wind-power projects • Financial and advisory support from donors and development banks	• State and donor support for hybridisation of existing diesel-fired mini-grids • Significant funding from foreign investors	• Feed-in tariff for wind-power projects • Financial support from donors and development banks

Source: Hansen et al. (2018).

specialised, with business models and software catering to specific PAYG customer segments.

Differences and similarities between actors and networks

In the actor dimension, foreign industry actors play a role across large-scale wind, large-scale solar, and solar mini-grids. However, in wind mini-grids there is no significant presence of foreign industry actors; rather, small-scale domestic industry actors and foreign actors such as NGOs and donors focusing on small-scale development projects are dominant. While there are universities involved in practical and hands-on applied research in scientific projects, this does not translate into organised R&D in the domestic industry, and there is a notable absence of private suppliers of wind-powered mini-grids in the sector. In the solar mini-grid sector there are a number of private suppliers, foreign investors, and foreign component suppliers, as well as turnkey contractors. Across both large-scale wind and solar power projects, the role of lead firms in the global industry in the wind sector and international EPC contractors is clear.

The role of local community actors is visible in both large-scale wind projects and solar mini-grids, though there is not much evidence of community involvement in wind mini-grid projects, and in the case of large-scale solar, the users tend to be large industrial and/or government players. In large-scale wind projects, the role of national policy makers and governmental agencies has been notable through their direct negotiations with project developers over power purchasing agreements.

Differences and similarities between institutions

In terms of the institutional dimension of the SISs examined here, there are clear similarities in terms of the role of feed-in tariffs and power purchasing agreements in the large-scale solar and wind projects, while small-scale projects in both the wind and solar sectors are influenced most clearly by state and donor support for hybridisation of the existing diesel-fired mini-grids. What is noticeable, however, is that, despite the same overarching driver existing for the hybridisation of mini-grids because of the increasing operational costs of diesel-driven mini-grids, the solar mini-grid segment differs markedly in terms of actors and networks and has received more attention from international donors than wind mini-grids. A number of donor programmes and national plans also mainly support the development of hybrid wind-diesel mini-grids. However, compared to the support for solar-powered mini-grids, the development of wind-powered mini-grids seems to be somewhat under-prioritised in these initiatives. In a number of locations, especially in the eastern and northern parts of Kenya (such as the area surrounding Lake Turkana), which have particularly favourable wind resources, the development of wind-powered mini-grids can become economically viable, although optimising location also depends on local demand (GIZ, 2014).

Overall, the solar mini-grid market appears to have a more enabling environment that has led to the establishment of a commercial market for the sale of electricity services to rural communities. This private-sector approach to the provision of rural electrification via mini-grids seems to be unprecedented in Kenya and East Africa. Many of the active companies have been started by foreign expatriates with significant expertise in business start-ups, engineering, RE consultancy, telecommunications, and donor organisations. These companies have therefore brought a high level of technical and organisational expertise and management systems into Kenya, which has been combined with knowledge on energy use and needs in local communities collected by the companies over time (Rolffs et al., 2015).

However, across both wind- and solar-powered mini-grids, the challenge remains of the lack of a regulatory framework for the development of commercial mini-grids. Bilateral negotiations between the companies and key government agencies related to obtaining operational licences and approvals of end-user tariffs have shown to be challenging and lengthy (ESMAP, 2016). The prolonged negotiating process is partly related to the different objectives of government agencies and private operators. The commercial tariff proposed by the private companies is significantly higher than the universal tariff offered by the government through the conventional grid-extension programmes to support rural electrification. The regulatory authorities are generally hesitant in accepting the inclusion of private operators that are operating with business models based on low connection fees and high usage rates. In general, one aspect of the difficulties in attracting funding for RE projects is the unclear policy signals and ongoing discussions concerning the possible introduction of new incentive structures and regulatory models. Since the feed-in tariff system was revised in 2012 to its current form, a number of alternative models, such as an auction system, and a net metering system for smaller grid-connected projects, have been discussed. However, solar and wind power projects in various stages of development have been suspended in recent years due to a 'freeze' by Kenya Power and Lighting Company (KPLC) on signing new power purchase agreements.

Concluding remarks: sub-sectoral pathways and policy implications

In this chapter, we have aimed to analyse and understand innovation dynamics within and between various sub-sectors. Based on the SIS perspective adopted in this chapter there are not only profound differences between solar and wind technologies, but equally importantly differences within these technologies. Overall, the SIS perspective shows that, in terms of the key system dimensions, there is a greater similarity between large-scale wind and solar projects (size), than between projects within the same technologies (shape). The large-scale projects are characterised by scientific knowledge bases (R&D), with actors with EPC experience or turnkey contracting playing a large role. The projects are capital-intensive, involve management expertise and PPA negotiations, and generally involve foreign actors in terms of both technology and expertise, as well as investments. The large-scale sectors differ from small-scale wind and solar mini-grids, which are markedly characterised by decentralised electrification efforts and are highly dependent on tariff structures and cross-subsidies. The rural electrification domain is connected to discussions about grid extensions and sees many donor-driven hybridisation efforts (particularly in solar). However, it has also revealed that there are significant differences between the institutional conditions such as regulation and policy frameworks for wind and solar mini-grids, with the solar mini-grid SIS being strengthened by a range of drivers that have led to an unprecedented private-sector-driven approach. In contrast, the wind-power mini-grid projects seem to have suffered both from the comparative success of the solar mini-grid market and the apparent under-prioritisation of the sector by actors otherwise engaged in the mini-grid sector.

Our conclusions have important implications for ongoing policy discussions on the role of government in shaping electrification pathways. It supports the opposition to any 'one size fits all' policy incentive in the renewable energy sector – rather, policy -makers should think about how they want to shape electrification pathways across the sizes and shapes outlined here. Tailor-made policies can help shape the dynamics of each sub-sector, and stakeholders and decision-makers should ask themselves which aspects should be enhanced. The SIS perspective highlights how innovation systems are outcomes of interaction and co-evolution of both size and shape, but also across national borders and links to global industry trends. Yet the literature has also pointed out that knowledge created in specific sectors may not be easily acquired and transferred across sectors. Therefore, attention to nurturing each of these distinct sectors, how to set appropriate tariffs and incentives, but also how to establish a broader framework of technical and procedural regulations is required. The variations across sectors and the role of foreign expertise in driving certain sub-sectors also raise questions about building up the necessary capabilities and expertise within the local market. This call for future research to investigate further the 'structure' of sectoral systems and the kinds of policy mechanisms that may influence this. Furthermore, research into how interactions between and the co-evolution of such sub-sectoral innovation systems can help policy makers understand how regulations and incentive mechanisms may influence co-existing and complementary sub-sectoral systems.

Acknowledgements

The authors thank participants for feedback on the ideas presented in this article at the IREK workshop in Eldoret, Kenya in February 2017. They are also grateful for constructive and insightful comments from three anonymous reviewers. Support for research on which this chapter is based from the Danish Ministry of Foreign Affairs, Grant: DFC 14-09AAU is gratefully acknowledged.

Notes

1 Parts of this book chapter draw on Hansen et al. (2018).
2 An example of the continued focus of the grid operator and energy planning agencies in Kenya on grid extensions to promote enhanced access to electricity for the rural population is the so-called 'Last Mile Connectivity Project' (Gichungi, 2011).

References

Adebowale, B.A., Diyamett, B., Lema, R. and Oyelaran-Oyeyinka, O. (2014) 'Introduction', *African Journal of Science, Technology, Innovation and Development*, 6(5), pp. v–xi. https://doi.org/10.1080/20421338.2015.1010774

Bergès, B. (2009) 'Case study of the wind-based rural electrification project in Esilanke primary school, Kenya', *Wind Engineering*, 33(2), pp. 155–174. https://doi.org/10.1260/030952409789140982

Byrne, R. (2011) *Learning Drivers: Rural Electrification Regime Building in Kenya and Tanzania*. Doctoral Thesis (DPhil). East Sussex: University of Sussex.

Duby, S. and Engelmeier, T. (2017) *The World's Microgrid Lab*. Munich: TFE Consulting.

Energy Sector Management Assistance Program (ESMAP) (2016) *Current Activities and Challenges to Scaling Up Mini-grids in Kenya*. http://documents.worldbank.org/curated/en/173201489180103220/Energy-Sector-Management-Assistance-Program-ESMAP-annual-report-2016

Ferrer-Martí, L., Garwood, A., Chiroque, J., Ramirez, B., Marcelo, O., Garfí, M. and Velo, E. (2012) 'Evaluating and comparing three community small-scale wind electrification projects', *Renewable and Sustainable Energy Reviews*, 16(7), pp. 5379–5390. https://doi.org/10.1016/j.rser.2012.04.015

Gesellschaft für Internationale Zusammenarbeit (GIZ) (2014) *Where shall We Put It? Solar Mini-grid Site Selection Handbook*. Bonn, Germany: Deutsche Gesellschaft für Internationale Zusammenarbeit (GIZ).

Gichungi, H. (2011) *Progress Report on Use of Renewable Energy in Off-Grid Areas*, pp. 24. https://www.climateinvestmentfunds.org/sites/default/files/knowledge-documents/srep_learning_workshop_report_0.pdf

Government of the Republic of Kenya (2007) *The Kenya Vision 2030*. Nairobi: Government of Kenya.

Hansen, U.E., Pedersen, M.B. and Nygaard, I. (2015) 'Review of solar PV policies, interventions and diffusion in East Africa', *Renewable and Sustainable Energy Reviews*, 46, pp. 236–248. https://doi.org/10.1016/j.rser.2015.02.046

Hansen, U.E., Gregersen, C., Lema, R., Wandera, F. and Samoita, D. (2018) 'Technological shape and size: A disaggregated perspective on sectoral innovation systems in renewable electrification', *Energy Research and Social Science*, 42, pp. 13–22. https://doi.org/10.1016/j.erss.2018.02.012

Harries, M. (1997) 'Disseminating windpumps in rural Kenya: Meeting rural water needs using locally manufactured windpumps', *Energy Policy*, 30, pp. 1–18. https://doi.org/10.1016/S0301-4215(02)00060-5

Harrington, K. (2016) 'New smart solar microgrids speed up rural electrification in Kenya'. [Online]. Available at: http://www.aiche.org/chenected/2016/02/new-smart-solar-microgrids-speed-rural-electrification-kenya (Accessed: 06/072020).

IREK (2018) 'Innovation and renewable electrification in Kenya'. AAU/ACTS/Moi University [Online]. Available at: https://www.irekproject.net/ (Accessed: 12/01/2018).

Kingiri, A. and Okemwa, J. (2022) 'Local content and capabilities: Policy processes and stakeholders in Kenya', in *Building Innovation Capabilities for Sustainable Industrialisation: Renewable Electrification in Developing Economies*. New York: Routledge. https://doi.org/10.4324/9781003054665-11

Lema, R., Iizuka, M. and Walz, R. (2015) 'Introduction to low-carbon innovation and development: Insights and future challenges for research', *Innovation and Development*, 5, pp. 173–187. https://doi.org/10.1080/2157930X.2015.1065096

Lundvall, B.-Å. and Lema R. (2014) 'Growth and structural change in Africa: Development strategies for the learning economy', *Sustainable Industrialization in Africa*, 6(5), pp. 113–138. https://doi.org/10.1007/978-1-137-56112-1_6

Malerba, F. (2002) 'Sectoral systems of innovation and production', *Research Policy*, 31(2), pp. 247–264. https://doi.org/10.1016/S0048-7333(01)00139-1

Malerba, F. (2005) 'Sectoral systems of innovation: A framework for linking innovation to the knowledge base, structure and dynamics of sectors', *Economics of Innovation and New Technology*, 14(1–2), pp. 63–82. https://doi.org/10.1080/1043859042000228688

Malerba, F. and Nelson, R. (2011) 'Learning and catching up in different sectoral systems: Evidence from six industries', *Industrial and Corporate Change*, 20(6), pp. 1645–1675. https://doi.org/10.1093/icc/dtr062

Ogeya, M.C., Osano, P., Kingiri, A. and Okemwa, J. (2022) 'Challenges and opportunities for the expansion of renewable electrification in Kenya', in *Building Innovation Capabilities for Sustainable Industrialisation: Renewable Electrification in Developing Economies*. New York: Routledge. https://doi.org/10.4324/9781003054665-3

Pedersen, M.B. (2016) 'Deconstructing the concept of renewable energy-based mini-grids for rural electrification in East Africa', *Wiley Interdisciplinary Reviews: Energy and Environment*, 5, pp. 570–587. https://doi.org/10.1002/wene.205

Rolffs, P., Ockwell, D. and Byrne, R. (2015) 'Beyond technology and finance: Pay-as-you-go sustainable energy access and theories of social change', *Environment and Planning A*, 47, pp. 2609–2627. https://doi.org/10.1177/0308518X15615368

Rural Electrification Authority (REA) (2009) *Rural Electrification Master Plan: Electrification Action Plan 2009–2013*. Nairobi: REA.

Stephan, A., Schmidt, T.S., Bening, C.R. and Hoffmann, V.H. (2017) 'The sectoral configuration of technological innovation systems: Patterns of knowledge development and diffusion in the lithium-ion battery technology in Japan', *Research Policy*, 46, pp. 709–723. https://doi.org/10.1016/j.respol.2017.01.009

Vanheule, L. (2012) *Small Wind Turbines in Kenya: An Analysis with Strategic Niche Management*. Master Thesis. Delft, Netherlands: Delft University of Technology.

5

UNDERSTANDING THE DIFFUSION OF SMALL WIND TURBINES IN KENYA

A technological innovation systems approach

Faith H. Wandera

Abstract

Kenya has a long history of wind energy development, but the potential for stand-alone small scale wind for electricity generation remains relatively unexploited to date. This prompts the questions of whether a technological innovation system for small wind exists, how well the system functions are fulfilled, and how the diffusion of technology is either blocked or induced. Using an innovation systems perspective and a technological innovation systems approach, data was sourced from literature, small wind business firms, and other actors. A technological innovation system for small wind is established to exist as most of the key actors are present; however, the system functions are weakly fulfilled. Key blocking mechanisms are in the functions of market formation, knowledge development, resources mobilisation, and guidance of the search. The function of creation of legitimacy has a weak inducing effect. The need for government support to conduct site-specific assessments, set specific development goals, and enhance financial support for small wind development is implied.

Introduction

Kenya is estimated to have a decentralised market of about 6.7 million households (MOEP, 2016), and a possible paradigm shift to decentralised electricity generation exists (Pueyo, 2015). Small wind turbines (SWT) less than 100 kW (Pitteloud and Gsänger, 2017) have the potential to contribute to a decentralised, small-scale pathway of electricity supply to enhance access to clean energy sources, particularly in remote areas where extension of the grid is not financially and economically viable compared to a large-scale pathway that relies on grid supplied electricity (Hansen et al., 2018). With correct placement, SWTs offer

DOI: 10.4324/9781003054665-5

potential for ensuring a pollution-free environment (Berges, 2007) in remote areas over a shorter gestation period (Ashok, 2007) and could guarantee the creation of more jobs per dollar invested per kilowatt-hour generated than fossil generation through direct job creation over their lifetime (Lewis and Wiser, 2005).

But the decision to tap into SWT requires a stronger evidence base. A discussion with a renewable energy consultant in Kenya indicated the price of SWT to be 1.5 to 2 times that of solar photovoltaic (PV). For every watt of wind installed one needs 3–5 $ compared to 2–2.5 $ for solar for installations less than 100 kW. In decentralised electrification, PV neither competes nor substitutes SWT. There are over 200,000 solar PV systems installed in Kenya, and annual sales are estimated at 20,000 systems (MOEP, 2016) with 30–40 distributor firms and 600 trained technicians. By comparison, the installed number of SWT systems in Kenya is estimated to be 500, mostly of 1–10 kW serving the communications industry, with about 20 SWT distributor firms and limited availability of trained technicians. The trend in the solar PV industry has been made possible by the emergence of innovative business models and the rapid decline in global prices of solar PV, thus making it a dominant technology in decentralised electrification.

While many studies have been conducted on the diffusion of SWT in developed countries, relatively fewer studies on this topic have been conducted in developing countries, especially in East Africa. This chapter presents the Technological Innovation Systems (TIS) approach to studying the diffusion of SWT in Kenya and hence provides evidence that supports the contribution of SWT to electrifying communities who lack access to grid electricity. To the author's knowledge, this is the first time small wind turbine diffusion has been considered from an innovation systems perspective. Use of the TIS framework is based on its ability to enable the understanding of the adoption and impact of technologies, as well as to reflect the socio-economic context present in developing countries (Edsand, 2016). These are key areas of interest for policy makers in making choices on the types of technology to promote. The chapter contributes methodological insights on how to adjust indicators of the TIS framework when applied in a developing country context.

The research findings presented in this chapter argue for the improvement of energy policy frameworks by increasing attention to the role of SWT in decentralised energy access in Kenya, particularly the setting of long-term goals for renewable energy (RE) development that incorporate SWT, allocation of adequate resources, local manufacturing and market development, conducting site-specific assessments in potential areas, and support for the integration of SWT in hybrid mini-grids (Nema, Nema, and Rangnekar, 2008). The study applied a mix of methods to answer the following research questions: does a TIS for diffusing SWT exist in Kenya? If it does, how well does it function? What are the main inducing and blocking mechanisms within the TIS?

The chapter is structured as follows: first insights on studying technology diffusion in developing countries using the technological innovation systems framework are presented. Then the methodology is described covering the scope and boundaries of the research, details of indicators, and method employed in data collection. An empirical analysis of the SWT TIS in Kenya covering the description of actors, assessment of the performance of the system functions, and a highlight of the blocking and inducing mechanisms is presented followed by an evaluation of the performance of the system functions. Lastly, the chapter concludes on the study findings, key policy issues, usefulness of the TIS framework, and the prospects for technology diffusion.

Studying technology diffusion in developing countries using the technological innovation systems framework

The National, Regional, and Sectoral/Technological Innovation Systems concepts focus on explaining the nature and rate of technological change (Hekkert et al., 2011). The emergence of the TIS framework alongside other innovation system approaches points towards the rise in systemic approaches to the study of technology development. The systemic approach contrasts with the 'linear model of innovation', where knowledge flows are modelled simply as research that leads to product or technology development which enter the market (OECD, 2008) without attention to feedback mechanisms.

The TIS framework focuses on technology specific factors while taking into account interactions with other systems at sectoral and national levels, thus providing a methodological tool for addressing complex dynamics through the aggregation of influences. It takes the technology as the focal point around which to consider system interactions and the influence of internal and external pushes and pulls rather than a sectoral or national focus. The benefits of the TIS framework are the capability to analyse the complex nature of the emergence and growth of new industries as well as the facilitating factors, termed 'inducing mechanisms', and obstacles, termed 'blocking mechanisms' (Jensen et al., 2007). Inducing or blocking mechanisms are likely to vary with the specific circumstances and context being examined, thereby enabling the translation of obstacles to intervention measures in the form of systemic instruments and policy mixes and enhancement of the inducing mechanisms for better performance of the system (Bergek et al., 2015).

Many studies have applied the TIS framework to emerging clean technology sectors, which signifies its importance as a major building block of sustainability transitions research (Bergek et al., 2015). The application of the TIS framework in the understanding of diffusion processes in developing economies is still a subject for debate. While some argue that the TIS framework was initially designed to understand diffusion of innovations in developed countries, which makes the possibilities for its application to developing or emerging economies limited (Kebede and Mitsufuji, 2016), others support its application. This study draws inspiration from various studies that have used the TIS framework to analyse RE sectors in developing country contexts (van Alphen, Hekkert, and van Sark, 2008; Furtado and Perrot, 2015; Kebede and Mitsufuji, 2016; Kingiri and Fu,

2019; Tigabu, Berkhout, and Beukering, 2014). Based on the extensive review of literature conducted, until now no study has utilised a TIS framework to study small wind turbines in sub-Saharan Africa.

A technological innovation systems perspective of the system components

The structural components of a TIS include the actors responsible for generation of knowledge (e.g., knowledge institutes, educational organisations, industry, market actors, government agencies, and supportive organisations); formal or informal institutions for formulating the rules of the game that shape human interaction (e.g., hard legislation including standards and intellectual property rights and soft legislation including ethics, norms, and behaviour); and networks and technological factors which are interactions regulated by for example institutional practices, education, and supply and demand (Hekkert et al., 2011). The focus is specifically on the role of actors, institutions, and networks in the diffusion of SWT.

Actors function in networks which may be either localised or globalised in nature. Interactions between individuals, groups, and organisations are regulated by institutions (Edquist, 2001) which are relevant for reducing uncertainty by providing information, managing conflicts and cooperation, and providing incentives for innovation. Institutions relevant in technology diffusion include publicly funded research and development (R&D), regulations and policy instruments, and technical norms such as the materials and equipment (Markard and Truffer, 2008). Literature suggests that learning networks are a crucial determinant in a firm's ability to obtain success with a new technology (Lewis, 2007). Networks also facilitate changes in the social dimension such as the user practices, regulation (Hekkert et al., 2007), and resource market activation campaigns and partnerships as a potential means of creating the demand, pressure, policy, regulatory foundation and interaction with international markets (Bruton, Ahlstrom, and Obloj, 2008) that is necessary for sustaining distributed energy markets. Because institutions and networks of agents take part in the generation, diffusion, and utilisation of specific technologies, the focus of analysis is technological innovation systems (Hekkert and Negro, 2009).

Such analysis has been used to identify systemic problems that hamper the development and diffusion of technological innovations (Negro, Alkemade, and Hekkert, 2012). The systemic problems relate to market structure, infrastructure, institutions, interactions, and capabilities. Measuring how innovation systems function is considered as a big breakthrough in innovation systems research (Hekkert et al., 2011). It entails the use of specific diagnostic questions in assessing the performance of specific functions. Comparing the main channels for knowledge flows, identification of bottlenecks, and suggestion of policies and approaches helps improve fluidity (Hekkert et al., 2011). Institutional mapping focuses on formal rather than informal institutions because of the difficulties associated with systematic mapping of the latter despite their relevance in influencing the performance of systems. By tracing the links and relationships among industry, academia, government, private sector, non-governmental development actors, and other agencies, systemic failures that impede a transition towards renewables and obstruct the formation of powerful functions could be identified (al-Saleh, 2011). These could be in the form of insufficient linkages between the actors, lack of entrepreneurial spirit, weak science and technology infrastructure and capabilities, under-developed educational and research capabilities, technical hurdles associated with renewables (mainly intermittency and high costs), lack of democracy and transparency in terms of policy making, and weak advocacy coalitions (al-Saleh, 2011). A combination of a few or all these could result in poor performance of the system functions, thus affecting the diffusion of RE technologies.

New technologies usually require long periods of nurturing before they achieve price parity that attracts larger segments in the market (Jacobsson and Johnson, 2000). In a TIS, the innovation activity of each actor contributes to one or several functions of the TIS, thus determining the performance of the system (Markard and Truffer, 2008), and hence how well diffusion takes place. Market development may depend on the emergence of new and innovative alliances of actors to meet the service needs of the excluded poor while delivering profits for business firms (Kuratko, 2010). Low diffusion could result from poor articulation of demand occasioned by inadequate or incorrect information. Demand could be stimulated through creation of awareness on the functioning thereby stimulating the uptake of technology. Technology diffusion occurs through the interactions of actors within networks. The diffusion of technologies undergoes five distinct phases, namely pre-development, development, take-off, acceleration, and stabilisation.

The formation of markets is not spontaneous, and articulation of demand particularly in the pre-development and development phases depends on the potential of the innovation system to create markets (Jacobsson and Johnson, 2000). New technologies are likely to suffer competition from incumbent substitutes offering better returns, thus causing the new product to be associated with a high price and low utility value. In the absence of nurturing, it may be difficult for the new technology to overcome these disadvantages. Changing the technological base of a firm happens slowly, implying that their search is often restricted to local developments, thus rendering them ignorant of available opportunities for diffusing technology (Jacobsson and Johnson, 2000).

Diffusion of small wind turbines

Wind technology has existed in Kenya since 1986 for mechanical water pumping. Later, large-scale systems were developed for electricity generation, for example the Ngong Wind (25 MW) power project and the Lake Turkana Wind (310 MW) power project (Kenya Miniwind, 2018). Further, SWT have been

TABLE 5.1 Diffusion of SWT in selected countries, no. of systems installed and total capacity (kW)

Country	*No. of systems installed*	*Date of Data*	*Total capacity (kW)*	*Date of data*
China	732,000	2015	415,000	2015
USA	160,995	2015	230,400	2015
United Kingdom	28,917	2015	146,192	2015
Italy	1,725	2015	59,833	2015
Germany	17,000	2015	26,000	2015
Japan	10,500	2011	5,258	2011
South Korea	1,900	2011	1,200	2011
Denmark	1,036	2015	8,600	2015
Kenya	500	2018	Not available	2018
Morocco	200	2012	700	2012
India	Not available	2017	2400	2012

Source: Pitteloud and Gsänger, 2017 except for Denmark installed systems data which is based on http://www.windworks.org/cms/index.php?id=64&tx_ttnews%5Btt_news%5D=3772&cHash=a66b51c490535bfe2d4f98c7dffb3c14 and Kenya data which is sourced from Hansen et al. (2018).

documented to be viable for electrification in developing countries (Foster, 2011), but examples of these are limited in Kenya. A literature search did not identify examples of SWT diffusion in neighbouring regions, except Ethiopia where the industry is not well developed (Eales, 2014). Table 5.1 shows the diffusion of SWT in selected countries across the world. The upper capacity limit of small wind ranges between 15 kW and 100 kW for the five largest small wind countries (Pitteloud and Gsänger, 2017). The definition of SWT by IRENA (2015) is based on the International Electrical Commission (IEC) 61400-2 standard whose definition corresponds to 50 kW. This study was based on the (Pitteloud and Gsänger, 2017) definition which gives the upper limit of SWT as 100 kW.

The search by national governments for viable electrification solutions is often limited by the cumbersome nature of initiating interactions with actors who may not be within easy reach (Foster, 2011). Special measures may therefore be required to help promote technology diffusion, including diffusion of RE and SWT in particular. An in-depth understanding of the functioning of the TIS for SWT in Kenya is thus critical for defining policies and actions that may contribute to increased access to electricity services, particularly in wind resource rich areas where grid extension is not economically feasible. This section has discussed the structuring of the TIS, identification of systemic problems, the role of the various components, and technology diffusion with specific reference to SWT.

Methodology

The methodology is based on broader knowledge from a forthcoming PhD dissertation examining technology innovation systems, capabilities, barriers, and opportunities for diffusing SWT in Kenya, and a journal paper on 'The innovation system for diffusion of small wind in Kenya: strong, weak or absent? A technological innovation system analysis' (Wandera, 2020). The scope of this study was the SWT TIS in Kenya narrowed by excluding grid-connected systems larger than 100 kW and mechanical applications for water pumping. Hence the system boundary reference for the study was 'grid-connected or isolated SWT systems less than 100 kW' for energy service provision to communities lacking access to the grid because of geographical location or economic and feasibility reasons.

The seven functions of the TIS framework by Hekkert et al. (2007) were used as the organising principle to identify the blocking and inducing mechanisms and assess the performance of the SWT TIS in Kenya. Choice of the Hekkert et al. (2007) framework over the Edsand (2017) framework was informed by the technology being analysed, the concepts guiding the study, and the indicators adopted. The indicators of the TIS framework were adjusted as indicated in Table 5.2 because the TIS was originally intended for application in developed rather than developing countries. The two have different contexts of technology development which are likely to vary with respect to structure, actors, and processes. The adjustment may therefore imply that the indicators used in this study are only applicable to developing country contexts similar to Kenya.

The adjustments included: for Function 1 (entrepreneurial activities), new entrants and number of experiments were not considered, as only very few firms focus on SWT (entrepreneurial focus in Kenya is on solar PV). The number of experiments on SWT is low and would present data collection constraints. For Function 2 (knowledge development), patents were not considered, as literature review established that most of the business firms were engaged in importation rather than manufacturing and even less so in R&D. Other aspects of knowledge development elaborated in Table 5.2 were considered. Function 3 (knowledge diffusion through networks) focused on the nature and frequency of interactions rather than the number, owing to the possible inaccuracies posed by the difficulty of accurately recalling figures particularly since record keeping on interactions is not the norm. For Function 5 (market formation), the suggested indicator 'niche markets introduced' was not considered, as niches for SWT are not well defined and would present difficulties in data collection. Emphasis was on the creation of an enabling environment, a function performed by governments. Functions 4, 6, and 7 were utilised without change.

Multiple data collection methods were applied, including email questionnaires, face-to-face interviews, and literature search of refereed journals and published reports using the following search terms: technology diffusion; small wind turbines; technology innovation systems; and decentralised electrification. The different types of data were triangulated so as to strengthen the validity of the findings (Aguinis et al., 2011). Both qualitative and quantitative data were collected from business firms and actors in the SWT innovation system, including government agencies, networks, consultants, and development partners. Business firms were identified through an

TABLE 5.2 Recommended versus applied indicators (sourced from various literature) for analysing the performance of a technological innovation system

Function	*Recommended indicators in TIS literature (Hekkert et al., 2011)*	*Adjusted indicators applied in this study*
1 Entrepreneurial activities	New entrants; the number of diversification activities of incumbent actors; and the number of experiments with the new technology	Activities in SWT business firms; number of firms in the TIS; sales of SWT; local manufacturing; use of external expertise; access to financing
2 Knowledge Development	R&D projects; patents; and investments in R&D	Availability of up to date knowledge; intensity of higher level education for skills development; adequacy of the training curriculum; evidence base (learning by doing/using) (commercialisation); level of experimentation; venturing into wind hotspots; reverse engineering; conducting feasibility assessments; linkages with public research organisations; local worker skills development
3 Knowledge diffusion through networks	Number of workshops and conferences devoted to a specific technology topic; the network size; and intensity over time	Availability of networks; focus of networks; networking capabilities with local and foreign actors; interactions and knowledge sharing; knowledge of the Kenyan market; integration of new knowledge; awareness on benefits of SWT; public awareness of technology; no. of systems and demonstrations installed
4 Guidance of the search	Specific targets set by governments or industries regarding the use of a specific technology and the number of articles in professional journals that raise expectations about new technological developments	Government long-term goals for SWT; industry long-term goals for SWT; incorporation of SWT in ongoing projects; utilisation of lessons from successful and unsuccessful projects; coverage of positive SWT developments in professional journals

(*Continued*)

TABLE 5.2 (*Continued*)

Function	*Recommended indicators in TIS literature (Hekkert et al., 2011)*	*Adjusted indicators applied in this study*
5 Market formation	The number of niche markets introduced; specific tax regimes for new technologies; and new environmental standards that improve the chances for new environmental technologies	Development focus; support for developing SWT mini-grids; supply chain development; availability of leadership; institutional structure; policy planning and implementation; availability and adequacy of regulations, standards, incentives, and local content; clarity of policy signals on SWT
6 Resources mobilisation	Detection through interviews whether or not inner core actors perceive access to sufficient resources as problematic	Government investments/priorities; budgetary support for SWT; foreign investments/technical assistance; risk taking by financing institutions; availability of business models
7 Creation of legitimacy/ counteract resistance to change	Rise and growth of interest groups and their lobby actions	Availability of lobby groups; lobbying through formal networks; lobbying via informal networks

Source: Based on Hekkert et al. (2011).

internet search of websites of firms engaged in the manufacture, importation, and sale of SWT technology in Kenya. Additional actors were identified through literature and snowballing during interviews. Altogether, 14 representatives of business firms and 12 other actors were interviewed in Kenya in late 2017 and early 2018 bringing the total number of interviewed to 26, in addition to 11 emailed questionnaires. Two of the business firms were local manufacturers of SWT. Quantitative data was analysed using descriptive statistics and interpretations made from tables and histograms generated. Data from interviews, case studies and published literature was analysed by thematic coding, identification of common themes, and presentation in narrative form. Evaluation of the fulfilment of the TIS functions was done by assigning values (−1 representing a strong blocking effect and +1 representing a strong inducing effect).

SWT TIS Kenya: an empirical analysis

The Hekkert et al. (2011) framework recommends mapping out the structure of the SWT technological innovation system. It entails identifying the actors and

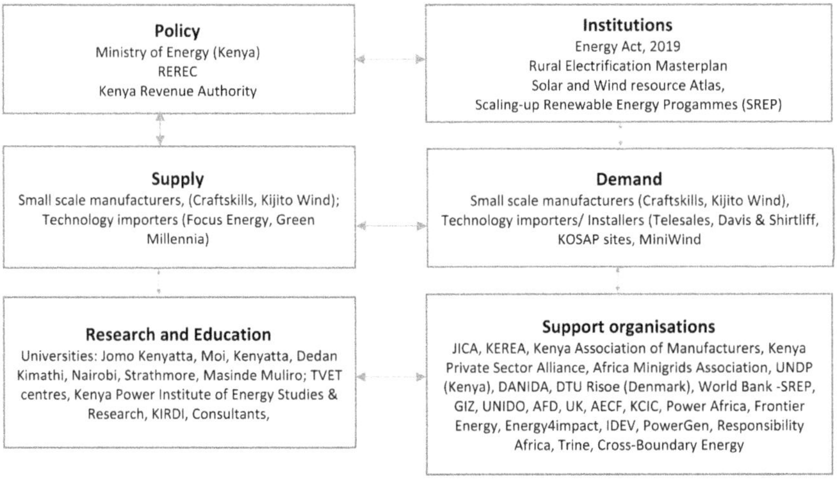

FIGURE 5.1 Actors in the small wind turbine Technological Innovation System in Kenya. Source: author; utilising actor groupings in Hekkert et al. (2011).

the relations between them and hence the structure within which all the identified actors operate. Figure 5.1 is a mapping of the actors in the SWT innovation system in Kenya.

Analysis of the performance of the SWT TIS functions

The performance of the SWT TIS was analysed based on the seven system functions, namely: (1) entrepreneurial activities; (2) knowledge development; (3) knowledge diffusion through networks; (4) guidance of the search/articulation of demand; (5) market formation; (6) resources mobilisation; and (7) creation of legitimacy/counteract resistance to change and the interactions.

Function 1: entrepreneurial activities

Literature sources established the existence of approximately 20 companies offering imported SWT in Kenya, predominantly installers of solar PV systems who complement their energy product portfolio with SWT products on a demand basis (Kamp and Vanheule, 2015). The bias to solar PV is explained by the low prices and greater availability of business models (reported by five out of six business firms). Among these are two local manufacturers of SWT whose operations in Kenya date back to the 1980s. One of them specialises in water pumping, while the other focuses on electricity generation. By comparison, interview with the Kenya Renewable Energy Association (KEREA) indicated that there are more solar PV importers (about 30–40) supported by about

300–400 distributors who target the rural areas of Kenya to grow their business. Most of the business firms are concentrated in the city of Nairobi which is relatively well served by the grid. Only one firm operates an outlet located in wind resource rich areas. These areas are difficult to access because of poor road infrastructure and insecurity (reported by five out of six business firms). Consequently, the solar PV products enjoy better market penetration compared to SWT whose sales by individual companies are in the range of one to five units over a four year period (reported by 13 out of 24 interview respondents). It is estimated that the installed number of systems is 500 in the whole country, mostly in the range of 1–10 kW (Hansen et al., 2018). Findings from literature, however, indicate SWT to be a better investment in the most appropriate sites when compared with solar PV on a technology versus technology basis (Fleck and Huot, 2009). Most project stakeholders including developers, promoters, and users indicate the lack of information to be a major obstacle (7 out of 11 interview respondents). The transfer of new knowledge from foreign firms with better experience in diffusing SWT is limited which therefore constrains the acquisition of technological innovations by firms (8 out of 11 interview respondents). Most firms were established to rely on advertising through their websites, an approach that limits public access to information and consequently contributes to the low awareness on SWT technology benefits (7 out of 11 respondents).

Function 2: knowledge development

The status of knowledge development and the technological base for SWT in Kenya is established from literature to be highly informal and comprises largely small-scale locally manufactured and imported systems (Hansen et al., 2018). The business firms indicated acquiring new knowledge from exhibitions and trade fairs despite the majority of these events focusing on solar PV which renders the acquisition of knowledge on SWT negligible. The training curriculum in tertiary agencies was indicated to be out of date with SWT developments in other parts of the world (8 out of 11 respondents), a factor that constrains the generation of new designs of SWT. Two prototypes developed at Jomo Kenyatta University of Agriculture and Technology in the 1990s are yet to be commercialised and only one local company is in commercial production. The focus of tertiary learning was more on PhD and Masters level rather than technical training which is needed more in the Kenyan SWT industry.

Site-specific assessments which are crucial to the installation of SWT are hardly conducted (8 out of 11 respondents) and available national resources such as the Solar and Wind Energy Resource Atlas are hardly consulted. Interaction with international actors, such as the Nordic Folkecenter in Denmark which has extensive experience in testing of SWT and has made valuable contribution to the diffusion of SWT in Denmark, was reported to be non-existent. A partnership between the Ministry of Energy (MOE) and the Danish Technical

University (DTU) which is highly experienced in site-specific assessments for SWT installation was initiated in 2018 through the Kenya Miniwind Project. The project was terminated in late 2019 because the SWT market in Kenya and sub-Saharan Africa was established to be small.

Function 3: knowledge diffusion through networks

The only network available for SWT business firms is KEREA but the membership is dominated by solar PV companies, and therefore the discussion on SWT is rather limited. Study tours organised by KEREA to the European region were reported to offer little benefit for learning by Kenyan business firms who preferred to venture into the Asian markets which offered cheaper RE products such as solar PV. The diffusion of knowledge through fairs and exhibitions is similarly inhibited by the inherent focus of these events on solar PV. This was confirmed from the programmes of the fairs and exhibitions held during the data collection period, despite the research finding that these events constituted important modes of knowledge exchange. Knowledge exchange with established centres of excellence, such as the Nordic Folkecenter which has a good reputation of diffusing research findings from the testing of small wind turbines in the Nordic countries, was established to be limited (4 out of 11 interview respondents). The sharing of knowledge between business firms, financial institutions, tertiary institutions, and government and R&D agencies was reported by survey respondents (6 out of 11) to be limited due to the fear of disclosing business secrets. This constrains learning experiences beneficial for developing new designs through modifications for improved performance and therefore it inhibits the diffusion of innovations (Lundvall, 2017). Available information on local markets for solar/wind hybrids was indicated to be out of date with the most recent developments (15 out of 24 interview respondents). This limits the ability of business firms to assess the actual demand for SWT technology even in areas of high potential such as the northern parts of the country.

Function 4: guidance of the search/articulation of demand

The Sustainable Energy for All Action Agenda recommends the development of clear goals and targets that could facilitate electricity access through the use of decentralised supply options, such as small wind in areas where the wind speeds permit off-grid integration, but so far, plans, such as the rural electrification master plan, do not clearly portray the demand from such sources (MOEP, 2016). Among the planned projects are the 26 mini-grids being developed by Rural Electrification and Renewable Energy Corporation (REREC), formerly known as Rural Electrification Agency (REA), and 13 planned mini-grids by the private sector – as outlined by the government in early 2019 (Kenya Miniwind, 2018), none of which incorporate small wind. Further, the experience of Safaricom Company which depends on small wind turbine technology to power the communication sector is yet to elicit lessons on the benefits of small wind to low-income households and business sectors (3 out of 11 respondents). China and India are documented to represent developing economies that have leapfrogged the diffusion of SWT to supply electricity in remote regions (Lewis, 2011). However, Kenya, which is documented to have good wind speeds in some regions, is yet to be listed as successfully diffusing small wind. Inadequate government support for decentralised electrification, and in particular mini-grids, limits their contribution to electricity access (Pueyo, 2015).

Function 5: market formation

The Kenyan Government's actions in the RE sphere demonstrate an inherent bias to solar PV installation in schools and health centres, while large wind systems such as Lake Turkana and Kipeto Wind projects are promoted for grid connection, a strategy interpreted as targeting foreign investment for large projects (Hansen et al., 2018) at the expense of SWT (6 out of 11 interview respondents). This is evident in the national electrification plans and donor programmes, the majority of which favour hybrid mini-grids largely based on solar PV and diesel and only a few of these incorporate SWT (Hansen et al., 2018). The energy sector operated without an official policy until 2004 but even then, for 13 years the policy was primarily focused on the development of geothermal, small hydropower, and solar PV. Later years have aroused interest in developing small hydropower which culminated in a small hydropower Atlas, but up to late 2019, government interest in SWT has been minimal. The incorporation of net metering among other provisions in the Government of Kenya (2019) Energy Act is expected to open up avenues for diversifying decentralised electricity generation. However, the extent to which these provisions will stimulate the diffusion of SWT remains unclear given that the obstacles to local manufacturing of SWT still remain pertinent in the form of taxation of batteries which are part and parcel of SWT systems; absence of Value Added Tax (VAT) waiver on imported SWT turbines; and the insufficiently developed supply chain for SWT countrywide (Berges, 2007).

Function 6: resources mobilisation

According to data collected from case studies, more financing is available for solar PV development than SWT (Pueyo, 2015). For example, the development support available from Power Africa (RES4Africa Foundation, 2015) is focused on solar PV and none of the projects supported includes SWT. The availability of angel investment as well as government financing for developing SWT projects is still lacking in Kenya, according to two consultants, six business firms and five other actors. The Danish funded Kenya Miniwind was a four year project aimed at supporting SWT development at a cost of 10.7 Million Danish Kroner[1] (Rambøll, 2017).

The key objective of this project was to demonstrate that a partially locally produced and operated wind turbine could contribute to affordable and reliable electricity in Kenya through mini-grids, stimulating employment and growth, and creating a market for smart mini-grids in Kenya and the region. The economic sustainability of the project was based on studies by Vestas which indicated the possibility of developing a new kW wind turbine which could compete with solar PV mini-grids in areas established to have medium inland wind regimes. This project was terminated in 2019 because of the small size of the SWT market in Kenya and sub-Saharan Africa.

Generally, financing institutions in Kenya have been reluctant to bear the risks associated with the promotion of SWT technology which is dominated by small companies of one to five employees. The lack of financing opportunities limits the growth potential of small firms and their ability to develop viable business models, hence reduces the dream of leapfrogging SWT technology within the sub-Saharan region to a mirage. The cost of installing SWT can be reduced by relying on local skills for constructing the tower whose price is about one-third of the total system cost and sourcing the electronic components from the Kenyan market since these are similar to those used for solar PV (interview with a manufacturer of SWT in Denmark). This implies the availability of a latent potential to reduce the cost of importing technology from foreign manufacturers since importation would be limited to highly specialised components, such as the blades, which require a certain level of expertise for the manufacturing process.

Function 7: creation of legitimacy and counteracting resistance to change

Some formal lobbying for the diffusion of SWT is confined to research groups such as the Innovation and Renewable Electrification in Kenya (IREK) project (www.irekproject.net) which successfully presented to the Kenyan Parliament a case on the need to incorporate the development of local capabilities for RE electrification in the Energy Act (Government of Kenya, 2019). The IREK project is coordinated by the African Centre for Technology Studies (ACTS) and hence reflects the role of Non-Governmental Organisations (NGOs) in creating legitimacy. Others include Delft University in the Netherlands which has conducted extensive research on SWT in Kenya using the multi-level perspective and strategic niche management approach (Kamp and Vanheule, 2015); and Jomo Kenyatta University of Agriculture and Technology which had previous collaboration on a capacity-building initiative with the Japanese International Cooperation Agency and later developed low cost prototypes (Haroub, Ochieng, and Kamau, 2015; Kasera, Ochieng, and Kinyua, 2015). Other groups such as the wind empowerment group through collaboration with the Low Carbon Energy for Development Network, Loughborough University, United Kingdom (Leary et al., 2018) have shown some interest in SWT development in Kenya, but their operations have been more at the international level. The United Nations

Development Programme initiated discussions with Ministry of Energy (MOE) to collaborate in developing the market for SWT in Kenya in 2015, but the initiative failed to materialise due to unavailability of financing, according to sources from MOE. The Kenyan Government's collaboration with the Danish government in the Kenya Miniwind Project and Vestas was expected to open up avenues for the diffusion of SWT, but with its termination, this is unlikely to be achieved. The two local manufacturers of SWT have unsuccessfully lobbied for recognition of SWT in government circles but so far they have not realised any significant policy support

Evaluation of the fulfilment of the seven functions of the SWT TIS in Kenya

This section presents an evaluation of the fulfilment of each of the seven functions of the SWT TIS in Kenya. The explanation for the evaluation of each function is presented in Table 5.3 (later in this chapter) which was generated by examining how each indicator contributes to the fulfilment of each function with respect to the diffusion of SWT.

According to Hekkert et al. (2011), understanding the performance of the TIS functions requires recognition of the elements of the functions that are facilitatory (inducing) and those that present obstacles to the fulfilment of the system functions (blocking). The assessment of the contribution of each function to the performance of the TIS is only feasible after constructing the narrative, through assignment of +1 for a positive contribution and -1 for a negative contribution (Table 5.3). The resulting sum of the data points thus gives insights into the overall effects of each function and whether they can be rated as either positive (inducing) or negative (blocking) (Hekkert and Negro, 2009). Based on this approach, the inducing or blocking effect for each function is discussed below.

Function 1: entrepreneurial activities is largely limited because of the primary focus of SWT firms on solar PV and the lower number of firms compared to solar PV, low sales of SWT, low levels of local manufacturing, limited use of external expertise in operations, and limited access to financing (Table 5.3). The cumulative effect of this function is -4.

Function 2: knowledge development is largely unfulfilled because of limited availability of up to date knowledge, sector doubts about the adequacy of the training curriculum at tertiary level, limited level of experimentation, absence of venturing into hotspots, absence of reverse engineering and commercialisation, unavailability of test facilities, and limited linkage between business firms and public research. Tertiary learning agencies exist but only one is active in SWT, and available local manufacturers have attempted to commercialise with limited success. The cumulative effect of this function is -7.

Function 3: knowledge diffusion through networks is only partially fulfilled by the availability of networks which mostly focus on solar PV, limited levels of networking with local and foreign actors, and low interactions and

TABLE 5.3 The established performance of the SWT Innovation System in Kenya (using modified indicators (+ =inducing, - = blocking)

Functions	*Indicators (as modified in the methodology)*	*Description of the indicators by the study respondents (qualitative and quantitative)*	*+/-*
1 Entrepreneurial activities	Activities focus	SWT firms focus on solar PV which has about 30–40 importers	-1
	Number of firms in the TIS	20 SWT firms	+1
	Sales of SWT	A total of one to five units over four years	-1
	Local manufacturing	Two local manufacturers	-1
	Use of external expertise	Negligible	-1
	Access to financing	Limited	-1
2 Knowledge development	Availability of up to date knowledge	Limited	-1
	Intensity of higher level education for skills development	One tertiary learning institution active	+1
	Adequacy of the training curriculum	Outdated (3 out of 11 respondents)	-1
	Evidence base (learning by doing/using) (commercialisation)	Only two manufacturers engaged in learning by doing and in commercial production	+1
	Level of experimentation	A 2018 project shelved, earlier experimentation did not result in diffusion	-1
	Venturing into wind hotspots	Limited (3 out of 11 respondents)	-1
	Reverse engineering	Two prototypes not commercialised	-1
	Conducting feasibility assessments	Non existent	-1
	Availability of test facilities for SWT	Zero reported	-1
	Linkages with public research organisations	Reported as low	-1
	Local worker skills development	Limited enhancement of skills of available engineers	-1
3 Knowledge diffusion through networks	Availability of networks	One functional network	+1
	Focus of networks	Solar PV	-1
	Networking capabilities with local and foreign actors	Reported to be low	-1
	Interactions and knowledge sharing	4 out of 11 respondents	-1
	Knowledge of the Kenyan market	Two local manufacturers	+1

(Continued)

TABLE 5.3 *(Continued)*

Functions	*Indicators (as modified in the methodology)*	*Description of the indicators by the study respondents (qualitative and quantitative)*	*+/-*
	Integration of new knowledge	Reported as low by business firms	-1
	Awareness on benefits of SWT	Reported as low by business firms	-1
	Public awareness of technology and awareness creation	Reported as low by business firms	-1
	No. of systems and demonstrations installed	About 500 systems installed	+1
4 Guidance of the search/ articulation of demand	Government long-term goals for SWT	Limited number of SWT projects identified in energy sector plans and budgets compared to solar PV	-1
	Industry long-term goals for SWT	None reported or found in literature	-1
	Incorporation of SWT in ongoing projects	0 out of 39 ongoing mini-grid projects by government and partners	-1
	Utilisation of lessons from successful and unsuccessful projects	3 out of 11 respondents	-1
	Coverage of positive SWT developments in Kenya in professional journals	None reported or found in literature	-1
5 Market formation	Development focus	6 out of 11 respondents indicate strong government focus on solar PV and wind	-1
	Support for developing SWT mini-grids	Regulatory framework absent	-1
	Supply chain development	Not evident from interviews	-1
	Availability of leadership	Unfavourable government leadership	-1
	Institutional structure	None established from literature	-1
	Policy planning, implementation	Focused on solar PV and large wind	-1
	Availability and adequacy of regulations, standards, incentives, and local content	None identified in literature	-1
	Clarity of policy signals on SWT	Unclear	-1

(Continued)

TABLE 5.3 (*Continued*)

Functions	*Indicators (as modified in the methodology)*	*Description of the indicators by the study respondents (qualitative and quantitative)*	*+/-*
6 Resources mobilisation	Government investments/ priorities	Strong bias to solar PV and large wind	-1
	Budgetary support for SWT	None identified	-1
	Foreign investments/ technical assistance	Very limited	-1
	Risk taking by financing institutions	Highly reluctant to take risks (6 out of 12 respondents)	-1
	Availability of business models	Not available when compared with solar PV	-1
7 Creation of legitimacy/ counteract resistance to change	Availability of lobby groups	Research groups e.g., IREK and Delft University, one tertiary learning institution	+1
	Lobbying through formal networks	Wind empowerment group United Nations Development Programme (stalled project), Kenya Miniwind Project shelved	+1
	Lobbying via informal networks	Two local manufacturers have lobbied but with no success	-1

Source: author; based on TIS functions by Hekkert et al. (2007).

knowledge sharing. The level of integration of new knowledge in firms and awareness of the benefits of SWT is also low. The low awareness is compounded by limited numbers of demonstrations. The local manufacturers have a good knowledge of the market unlike the distributor firms whose understanding of the market is limited. The cumulative effect of this function is -3.

Function 4: guidance of the search is weakened by the absence of long-term goals for SWT in both industry and government, non-incorporation of SWT in ongoing mini-grids, limited utilisation of lessons from successful and unsuccessful projects, and no coverage of positive SWT developments in Kenya in professional journals. The cumulative effect of this function is -5.

Function 5: market formation is constrained by the development and policy focus by the leadership in the energy sector on solar PV and large wind, absence of a regulatory framework for mini-grids and institutional structure for SWT development, inadequacy of the SWT supply chain, and lack of incentives and standards for SWT. The cumulative effect of this function is -8.

Function 6: resources mobilisation is blocked by low prioritisation of SWT in government investments and budgetary allocation, limited availability of foreign investments in SWT, low risk taking by available government financing institutions, and limited availability of business models compared to solar PV. The cumulative effect of this function is -5.

Function 7: creation of legitimacy is weakly fulfilled by the availability of lobby groups such as the Innovation and Renewable Electrification in Kenya research group among other formal organisations. Informal organisations have tried to lobby unsuccessfully. The cumulative effect of this function is +1.

The phase of development of the SWT TIS was determined by asking specific diagnostic questions relating to each phase. The pre-development phase of the TIS is characterised by the availability of a working prototype, the development phase by commercial application, the take-off phase by fast market growth, and the acceleration phase by market saturation (Hekkert et al., 2011). The preceding assessment of the performance of the SWT TIS in Kenya indicates that it is still in the pre-development phase because a working prototype exists but efforts to commercialise are limited. All the conditions in the remaining three phases are yet to be fulfilled.

Conclusion

On the existence of an SWT TIS in Kenya, it is concluded based on the mapping of actors, institutions, and networks that most of the actors required to constitute a TIS are present, except demand which was not analysed in this study. The research question on how well the Kenyan SWT functions are fulfilled was answered by analysing the TIS functions. Based on the scale defined in the methodology section, the four functions with the strongest blocking effect are market formation, knowledge development, resources mobilisation, and guidance of the search. The functions of entrepreneurial activities and knowledge diffusion are weakly blocking, while the function of creation of legitimacy is weakly inducing. The identified weaknesses in the Kenyan SWT TIS therefore render it unfavourable for the diffusion of SWT. This finding corroborates the observation that the SWT TIS is still in the pre-development phase despite the long history of SWT in Kenya. This finding is supported in literature by Negro, Alkemade, and Hekkert (2012) who observed that diffusion of RE technologies is sometimes a slow and tedious process.

Available literature suggests that focusing policy intervention on remedying the poor functionality of the TIS functions could strengthen the inducing mechanisms and minimise the effect of the blocking mechanisms (Hekkert et al., 2011). Key areas of intervention in view of the small market revolve around enhancing government support for SWT development by conducting site-specific assessments in potential areas, enhanced R&D, support for the integration of SWT in hybrid mini-grids (Nandi and Ghosh, 2010), setting specific goals for SWT development, allocation of government and development partner support for SWT development, and support for local manufacturing and market development.

The number of empirical studies that apply the TIS framework to the developing country context in published literature is established to be growing steadily. This could be a sign that the TIS framework is increasingly found beneficial for analysing the performance of technological innovation systems in different sectors of developing economies. Using the TIS framework to study the SWT innovation system in Kenya is thus considered a contribution to the growing interest in the use of this framework in the developing country context with specific reference to the African region. Using this framework in the developing country context entails adjustment of the indicators to the specific national and sectoral context under study, as well as weighting of the research findings using the same indicators applied to empirical studies. This facilitates objective interpretation and enhanced reliability of the recommendations.

The limitations faced in studying the diffusion of SWT using the TIS framework relate to the static nature of the analysis which makes it difficult to determine whether the SWT TIS is advancing or declining. Furthermore, different weighting mechanisms could produce variations in the results in which case the application to other contexts could be limited. Additional empirical studies that enhance the assessment of indicators for specific technologies are thus recommended. This study points towards the possibility of implementing solar PV/SWT hybrids (Carmago et al., 2019), provided site-specific assessments are conducted in specific areas. Accounts of the Kenya Miniwind Project (2018), however, indicate the Kenyan market for SWT to be small, a finding which led to termination of the project. Research studies, such as Johannesen (2019) on the integration of small wind turbines in mini-grids in Kenya, are considered important contributions towards attaining the inflexion point in the diffusion curve for SWT from the pre-development to the development phase. The blocking and inducing mechanisms identified in this chapter, if addressed, could lead to this transition.

Acknowledgements

The research on which this chapter is based was supported by the Danish Ministry of Foreign Affairs, Grant: DFC 14-09AAU. All input by the editorial team towards finalising this chapter is greatly appreciated. This chapter is based on an upcoming PhD research by the author and a journal paper (Wandera, 2020).

Note

1 1 Danish Kroner = United States Dollar 0.15 or Kenya Shillings 15.27 at time of writing.

References

Aguinis, H., Chaddad, F.R., Harrigan, K.R., Hoon, C. et al. (2011) 'Research methodology in strategy and management', *The International Journal of Health Planning and Management*, 26(4), pp. 488–490, https://onlinelibrary.wiley.com/doi/full/10.1002/hpm.1118

van Alphen, K., Hekkert, M.P. and van Sark, W.G.J.H.M. (2008) 'Renewable energy technologies in the Maldives-realising the potential', *Renewable and Sustainable Energy Reviews*, 12(1), pp. 162–180. https://doi.org/10.1016/j.rser.2006.07.006

Ashok, S. (2007) 'Optimised model for community-based hybrid energy system', *Renewable Energy*, 32(7), pp. 1155–1164. https://doi.org/10.1016/j.renene.2006.04.008

Bergek, A., Hekkert, M., Jacobsson, S., Markard, J., Sandén, B. and Truffer, B. (2015) 'Technological innovation systems in contexts: Conceptualising contextual structures and interaction dynamics', *Environmental Innovation and Societal Transitions*, 16, pp. 51–64. https://doi.org/10.1016/j.eist.2015.07.003

Berges, B. (2007) *Development of Small Wind Turbines*. Lyngby, Denmark: Technical University of Denmark, pp. 66–69.

Bruton, G.D., Ahlstrom, D. and Obloj, K. (2008) 'Entrepreneurship in emerging economies: the research go in the future', *Entrepreneurship Theory and Practice*, 32(January), pp. 1–14. https://doi.org/10.1111/j.1540-6520.2007.00213.x

Carmago, L.R., Gruber, K., Nitsch, F. and Dorner, W. (2019) 'Hybrid renewable energy systems to supply electricity self-sufficient residential buildings in central Europe', *Energy Procedia*, 158(2019), pp. 321–326. https://10.1016/j.egypro.2019.01.096

Eales, A. (2014) 'Locally manufactured small wind turbines in Ethiopia: Is the levelised cost of energy competitive with alternatives for rural electrification, and if so, where?', p. 149. Available at: https://www.academia.edu/7865944/ (Accessed: 20/06/2018).

Edquist, C. (2001) 'The systems of innovation approach and innovation policy: An account of the state of the art' in DRUID Conference, Aalborg, (December), pp. 12–15. Available at: https://www.researchgate.net/publication/228823918 (Accessed: 20/06/2018).

Edsand, H.E. (2016) 'Technological innovation systems and the wider context: A framework for developing countries', *UNU-MERIT Working Papers*, 17, p. 38. Available at: https://www.eadi.org/publications/publication_55276/ (Accessed: 25/06/2018).

Edsand, H.E. (2017) 'Identifying barriers to wind energy diffusion in Colombia: A function analysis of the technological innovation system and the wider context', *Technology in Society*, 49, pp. 1–15. https://doi.org/10.1016/j.techsoc.2017.01.002

Fleck, B. and Huot, M. (2009) 'Comparative life-cycle assessment of a small wind turbine for residential off-grid use', *Renewable Energy*, 34(12), pp. 2688–2696. https://doi.org/10.1016/j.renene.2009.06.016

Foster, R. (2011) 'Energy in developing countries', *Trade, Investment, and Sustainable Development*, 11(3), pp. 27–28. Available at: https://digitalcommons.wcl.american.edu/cgi/viewcontent.cgi?article=1487&context=sdlp (Accessed: 29/06/2018).

Furtado, A.T. and Perrot, R. (2015) 'Innovation dynamics of the wind energy industry in South Africa and Brazil: Technological and institutional lock-ins', *Innovation and Development*, 5(2). https://doi.org/10.1080/2157930X.2015.1057978

Government of Kenya (2019) 'The energy act no 1 of 2019', *Kenya Gazette Supplement Acts, 2019*, 29(29). Available at: https://www.epra.go.ke/download/the-energy-act-2019/ (Accessed: 01/12/2019).

Hansen, U.E., Gregersen, C., Lema, R., Samoita, D. and Wandera, F. (2018) 'Technological shape and size: A disaggregated perspective on sectoral innovation systems in renewable electrification pathways', *Energy Research and Social Science*, 42, pp. 13–22. https://doi.org/10.1016/j.erss.2018.02.012

Haroub, H.A., Ochieng, F.X. and Kamau, J.N. (2015) 'Development of a low cost rotor blade for a H: Darrieus wind turbine', 2(3), pp. 92–99. Available at: https://pdfs.semanticscholar.org/0341/1a1c7e31c2b7bbb97661a3c31ffc3cc697d5.pdf (Accessed: 10/03/2019).

Hekkert, M.P. and Negro, S.O. (2009) 'Functions of innovation systems as a framework to understand sustainable technological change: Empirical evidence for earlier claims', *Technological Forecasting and Social Change*, 76(4), pp. 584–594. https://doi.org/10.1016/j.techfore.2008.04.013

Hekkert, M.P., Suurs, R.A.A., Negro, S., Kuhlmann, S. and Smits, R.E.H.M. (2007) 'Functions of innovation systems: A new approach for analysing technological change', *Technological Forecasting and Social Change*, 74(4), pp. 413–432. https://doi.org/10.1016/j.techfore.2006.03.002

Hekkert, M.P., Negro, S., Heimeriks, G,, Harmsen, R.O. and Jong, S.D. (2011) 'Technological innovation system analysis: A manual for analysts', Utrecht University. Available at: https://www.semanticscholar.org/paper/Technological-Innovation-System-Analysis-A-manual-Hekkert-Negro/68e1abecbbe0da073c7e63d95dbb750f5d910024 (Accessed: 20/11/2018).

Jacobsson, S. and Johnson, S. (2000) 'The diffusion of renewable energy technology: An analytical framework and key issues for research', *Energy Policy*, 28(9), pp. 625–640. http://dx.doi.org/10.1016/S0301-4215(00)00041-0

Jensen, M.B., Johnson, B., Lorenz, E. and Lundvall, B.Å. (2007) 'Forms of knowledge and modes of innovation', *Research Policy*, 36(5), pp. 680–693. https://doi.org/10.1016/j.respol.2007.01.006

Johannsen, R.M. (2019) 'Hybrid PV and wind mini-grids in Kenya : Barriers and potential for diffusion Hybrid PV and wind mini-grids in Kenya : Barriers', 4. Available at: https://www.irekproject.net/portfolio/wp4/ (Accessed: 11/02/2020).

Kamp, L.M. and Vanheule, L.F.I. (2015) 'Review of the small wind turbine sector in Kenya: Status and bottlenecks for growth', *Renewable and Sustainable Energy Reviews*. https://doi.org/10.1016/j.rser.2015.04.082

Kasera, A.A., Ochieng, F.X. and Kinyua, R. (2015) 'Design and testing of a low cost and higher efficient savonius wind turbine's rotor blade for low wind speed applications', *Open Access Journal of Sustainable Research in Engineering*, 2(1), pp. 23–35. Available at: http://ir.jkuat.ac.ke/handle/123456789/2246 (Accessed: 20/03/2019).

Kebede, K.Y. and Mitsufuji, T. (2016) 'Technological innovation system building for diffusion of renewable energy technology: A case of solar PV systems in Ethiopia', 114(2017), pp. 242–253. https://doi.org/10.1016/j.techfore.2016.08.018

Kenya Miniwind (2018) *Market for the Integration of Smaller Wind Turbines in Mini-grids in Kenya*. Available at: https://backend.orbit.dtu.dk/ws/files/158704528/Kenya_mini_grid_market_study_FINAL.pdf (Accessed: 05/05/2019).

Kingiri, A.N. and Fu, X. (2019) 'Understanding the diffusion and adoption of digital finance innovation in emerging economies: M-Pesa money mobile transfer service in Kenya', *Innovation and Development*, 10(1), pp. 67–87. https://doi.org/10.1080/2157930x.2019.1570695

Kuratko, D.F. (2010) 'Corporate entrepreneurship: An introduction and research review' in Audretch D.B. and Zoltan, J.A. (eds.) *Handbook of Entrepreneurship Research*. Available at: https://link.springer.com/book/10.1007/978-1-4419-1191-9 (Accessed: 13/01/2020).

Leary, J., To, L.S. and Alsop A. (2018) 'Is there still a role for small wind in rural electrification programmes?', *LCEDN Briefing Paper* 2, pp. 1–17. Available at: www.lcedn.com (Accessed: 13/01/2020).

Lewis, J.I. (2007) 'Technology acquisition and innovation in the developing world: Wind turbine development in China and India', *Studies in Comparative International Development*, 42(3–4), pp. 208–232. https://doi.org/10.1007/s12116-007-9012-6

Lewis, J.I. (2011) 'Building a national wind turbine industry: Experiences from China, India and South Korea', *International Journal of Technology and Globalisation*, 5(3/4), p. 281. https://doi.org/10.1504/IJTG.2011.039768

Lewis, J. and Wiser, R. (2005) 'Fostering a renewable energy technology industry: An international comparison of wind industry policy support mechanisms', *eScholarship Publishing*. Available at: https://escholarship.org/uc/item/6cf1r3z5 (Accessed: 08/03/2018).

Lundvall, B.-Å. (2017) *The Learning Economy and the Economics of Hope*. London and New York: Anthem Press. https://doi.org/10.26530/oapen_626406

Markard, J. and Truffer, B. (2008) 'Actor-oriented analysis of innovation systems: Exploring micro-meso level linkages in the case of stationary fuel cells', *Technology Analysis and Strategic Management*, 20(4), pp. 443–464. https://doi.org/10.1080/09537320802141429

Ministry of Energy and Petroleum (MOEP) (2016) *Sustainable Energy for All Action Agenda*. Available at: https://www.renewableenergy.go.ke/asset_uplds/files/SE4All%20AA%20Report%20Final%2010%20March%202016.pdf (Accessed: 10/11/2018).

Nandi, S.K., and Ghosh, H. R., (2010). 'Prospect of windePV–battery hybrid power system as an alternative to grid extension in Bangladesh', *Energy*, 35(2010), pp. 3040–3047. https://doi.org/10.1016/j.energy.2010.03.044

Negro, S.O., Alkemade, F. and Hekkert, M.P. (2012) 'Why does renewable energy diffuse so slowly? A review of innovation system problems', *Renewable and Sustainable Energy Reviews*, 16(2012), pp. 3836– 3846. https://doi.org/10.1016/j.rser.2012.03.043

Nema, P., Nema, R. K., and Rangnekar, S. (2008) 'A current and future state of art development of hybrid energy system using wind and PV-solar: A review', *Renewable and Sustainable Energy Reviews*, 13(2009), pp. 2096–2103. https://doi:10.1016/j.rser.2008.10.006

Organisation for Economic Co-operation and Development (OECD) (2008) 'National innovation systems', *International Journal of Entrepreneurship and Innovation Management*, 8(1), p. 74. https://doi.org/10.1504/IJEIM.2008.018615

Pitteloud, J.-D. and Gsänger, S. (2017) *2017 Small Wind World Report Summary*. Available at: https://issuu.com/wwindea/docs/swwr2017-summary (Accessed: 10/01/2019).

Pueyo, A. (2015) 'Pro-poor access to green electricity in Kenya', *Pro-Poor Electricity Provision*, 135. Institute of Development Studies, pp. 29–31.

Rambøll, B. (2017) *Supporting Sustainable Mini-grid Development and Local Production of Wind Turbines Using the Case of Kenya*. Available at: https://orbit.dtu.dk/en/projects/supporting-sustainable-mini-grid-development-and-local-production (Accessed: 15/08/2018).

RES4Africa Foundation (2015). Available at: https://www.res4africa.org/wp-content/uploads/2016/05/Power-Africa-Fact-Sheet.pdf (Accessed: 11/06/2018).

al-Saleh, Y.M. (2011) 'LR1-An empirical insight into the functionality of emerging sustainable innovation systems: the case of renewable energy in oil-rich Saudi Arabia', *International Journal of Transitions and Innovation Systems*, 1(3), p. 302. https://doi.org/10.1504/IJTIS.2011.042662

Tigabu, A., Berkhout, F. and Beukering, P. (2014) 'Technology innovation systems and technology diffusion : Adoption of bio- digestion in an emerging innovation system in Rwanda', *Technological Forecasting and Social Change*, 90(January 2016), pp. 318–330. https://doi.org/10.1016/j.techfore.2013.10.011

Wandera, F.H. (2020) 'The innovation system for diffusion of small wind in Kenya: Strong, weak or absent? A technological innovation system analysis', *African Journal of Science, Technology, Innovation and Development* [Online]. https://doi.org/10.1080/20421338.2020.1771979

6

ARE THE CAPABILITIES FOR RENEWABLE ELECTRIFICATION IN PLACE?

A Kenyan firm-level survey

Charles Nzila and Michael Korir

Abstract

Improved and sustained access to cleaner electricity remains central on the global development agenda. In this regard, Kenya has set plans for deployment of renewables and a target of achieving universal electricity access by the year 2030. This process of renewable electrification depends not only on finance and technology, but also on the availability of requisite capabilities for deployment and use of the technologies. This chapter seeks to assess which capabilities are already in place and where the shortfalls are across five different renewable energy (RE) technologies and five different steps in the value chain. It focuses on capabilities related to deployment rather than manufacturing of renewable energy technologies. The chapter draws on one of the most comprehensive surveys undertaken in the sector in Kenya to date. It covers 71 firms and organisations involved in renewable electrification projects. The observed capability levels put the RE deployment related capabilities in Kenya as relatively high on average but with noticeable bottlenecks. The results also indicate that while management capabilities are generally high, there are a number of areas that need improvement, especially with respect to the ability to identify, assess, negotiate, and finalise terms of financing. The survey results further show that the overall capability levels are highest in the solar photovoltaic (PV) domain. The findings presented in this chapter can help to inform actors and interventions geared towards enhancing renewable electrification in Kenya including directing a new paradigm – from continued dependence on external actors in most steps of the RE value chain to the targeted development of local capabilities.

Introduction

Renewable electrification is widely regarded as one of the key solutions to the threat of energy poverty and slow actualisation of the sustainable development

DOI: 10.4324/9781003054665-6

goals (SDG's) and improved livelihoods in many developing economies. Hence clean energy and particularly access to clean electricity remain central in virtually every major developmental agenda in the world today. Consequently, many national/regional multi-faceted interventions have been fronted and continue to receive increased attention (Almeshqab and Ustun, 2019; Pedersen, 2016; Zhang and Gallagher, 2016). At the global level, one key intervention in this regard is the United Nations led SDG on ensuring access to affordable, reliable, sustainable and modern energy for all (www.un.org). However, the goal of providing universal energy access to all by 2030 requires new innovative initiatives that are economically feasible and sustainable (Nzila et al., 2012; Flüeler et al., 2012) and enhanced capabilities (Lema et al., 2018), which are rather quite a challenge (Oshiro, Kainuma, and Masui, 2016; Azzuni and Breyer, 2018). Nonetheless, the concept of renewable electrification (RE) capabilities has received little empirical consideration (Pedersen, 2016; Byrne, Mbeva and Ockwell, 2018) hence it is not surprising that many developing countries in general continue to consider different pathways for enhancing modern energy access.

Kenya has a promising potential for power generation from renewable energy sources owing to abundant renewable energy resources such as solar, hydro, wind, biomass, and geothermal. The main models for enhancing electricity access especially in the rural areas in Kenya as well as in most East Africa Community (EAC) states include the creation of an enabling environment for renewable energy markets (Byrne et al., 2012) as well as capability related interventions especially through rural electrification via the use of mini- and micro-grid infrastructure and the last mile connectivity among others (Almeshqab and Ustun, 2019). In the context of this chapter, capability is construed as the competitive capacity (in terms of resources, knowledge, and skills/competencies) to perform a task within a firm or project. In this regard capabilities are generally what makes renewable energy (RE) firms or projects (within a country) perform differently among other competing and partnering firms or projects. The variation amongst firms' or projects' performance could be explained by specific differences that arise from different strategic capabilities including deployment of strategies, competencies, and resources.

Research on capabilities in renewable energy has continued to evolve via multi-sectoral analysis (Lindbom et al., 2015) as well as the assessment of energy access and impact using a variety of models (Cole, 2018). Indeed, while capabilities are widely perceived to have a potentially greater impact on the sustainability of renewable electrification (Lema et al., 2018; Cole, 2018), there is little empirical consideration especially from the developing economies perspective. Consequently, there is an ongoing debate pertaining to the inadequacy of empirical data on the role of capabilities and their outcomes in renewable electrification projects as well as the approach of conducting and presenting these assessments (Kirchherr and Urban, 2018). In addition, there is an apparent lack of studies focused on the analysis of capabilities for renewable electrification in emerging economies and particularly the East African Region where access levels, the urban/rural distribution, high use of unclean fuel and abundance of fossil fuels (Nzila et al., 2012) continue to present further unique challenges including pricing out of renewables.

The assessment of capabilities in RE projects therefore serves as a basis for decisions regarding various enterprise/project activities such as business development measures and investments (technology, training, research and development, policy, etc.) as well as engagements with other enterprises or organisations. The business/project activities have mutual influence on the actual project outcomes. In addition, the overall capabilities, decisions, business/project activities, and the outcomes collectively and individually influence energy access and hence sustainable industrialisation. Consequently, firm-level capabilities are vital elements for anchorage of sustainable industrialisation. Hence, understanding the characteristics of firm-level capabilities is critical for sustainable industrialisation. In this regard, if renewable electrification in Kenya is analysed in terms of capabilities, there is a higher prospect of identifying the decisions and business/project activities that are necessary to enhance energy access in the country.

Towards this end, the goal of this chapter is to develop and implement a capability assessment framework to investigate the character of capabilities and to determine whether the distribution of capabilities differs among the renewable energy projects in Kenya. This goal was actualised through the development and execution of a survey that was built around the characteristics (activities) and outcomes of capabilities that are manifested in renewable energy firms/projects. The survey therefore sought to establish *which capabilities are in place and where are the shortfalls?* In addition, it also sought to establish *to what extent do technology sectors (solar, wind, small hydro, geothermal, and hybrids) influence the development and deployment of capabilities?* The survey explored two research-guiding hypotheses, based on a series of recent research contributions, which were:

H_0: Outcomes of the development of capabilities in renewable energy projects are independent of the type of technology.

H_1: Outcomes of the development of capabilities in renewable energy projects depend on the type of technology.

This chapter therefore seeks to empirically employ a capability assessment framework integrated into a survey setting to examine the renewable energy business and project activities in Kenya against the development of capabilities. The unit of analysis in the survey was renewable energy business firms and energy projects in Kenya. The scope of the survey entailed capabilities in the deployment chain of RE and did not include investigations of capabilities for manufacturing of core RE technologies. Finally, the survey focused on projects and firms located in Kenya.

The value premise of the chapter is based on both the empirical and results fronts. Firstly, the unique empirical approach employed by the chapter in the investigation of capabilities in Kenya's renewable energy sector is novel. Indeed, while many writers have underscored the importance of investigating RE capabilities, very few have articulated a strategy to measure them and actually proceeded to systematically measure them empirically as done in the current chapter. The strategy promoted in this chapter has further leveraged multiple indicators per capability category to provide an in-depth insight into the inherent capabilities of RE projects in Kenya. Secondly, the chapter demonstrates empirically that there is a strong association between the choice of RE technology and the development hence distribution of capabilities. Consequently, the chapter postulates that while the share of solar PV technology in Kenya's energy mix continues to trail other RE technologies such as wind and small hydro, the reported higher development of capabilities in solar PV RE projects implies that there is a likelihood of an upward trajectory in the share of solar PV in the country's energy mix. In addition, the results reported in the chapter have substantial ramifications especially on modalities for leveraging on the capabilities and the sustainability of RE hence industrialisation in developing economies.

The structure of the chapter is therefore organised as follows: the current section has introduced the chapter, presented an overview of capabilities in the context of Kenya's sustained drive towards renewable electrification, and spelt out the key research questions and hypothesis of the study. In the following section we present the theoretical framework and disaggregation of capabilities and their indicators as well as the assessment framework. Then the methodology for disaggregating capabilities and hence an empirical design for measuring the capabilities that was embedded in a survey to assess the capabilities in RE projects/firms in Kenya is presented. The proceeding section details and discusses the survey results in terms of general outlook of capabilities in RE projects in Kenya prior to presenting a detailed analysis of the five main strands of capabilities that were investigated. Finally, we wrap up with conclusions and recommendations.

Theoretical framing

The drive to increase the deployment of renewable energy in Kenya to increase energy access particularly in rural areas has seen the country deploy a variety of (free) market instruments such as regulatory policies and fiscal incentives. This approach has been supported by proponents of policies, public financing, and aid programmes in the East African Region (Kirchherr and Urban, 2018; EAC, 2016). However, critics have remained largely sceptical and called for much more active state intervention besides arguing that the widely acknowledged free markets alone cannot deliver the envisaged level of electrification (Byrne, Mbeva, and Ockwell, 2018). The disapproval could probably further be attributed to capability issues (both technical and non-technical) related to deployment of existing and transferred technologies. It is therefore apparent that there is need for enhanced capabilities (Bell, 1990; Bell and Pavitt, 1993), which include financial and management, and technical and supplementary/supporting capabilities, in addition to interventions such as innovations in business (model) development, production processes, distribution models, marketing, cost structures, and consumer financing. In the current section we therefore probe the capability assessment framework, which is promoted in the chapter, and provide an overview of the theoretical underpinnings of capabilities and their outcomes with respect to renewable electrification in Kenya. The section is specifically structured to address theoretical framing of capability accumulation and its assessment that is further pursued and elaborated upon in the following sections.

Theoretical framework on capabilities for renewable electrification

The status of renewable electrification capabilities and their outcomes within the energy sector are bound to shape developments in energy access. However, owing to a rather low projected level of electrification of 30% by 2030 in sub-Saharan Africa (Almeshqab and Ustun, 2019), if no changes in models, policies, and practices are made then it follows that the universal access objective will most likely not be achieved. While there has been some progress towards the goal of universal energy access as well as in the development of the requisite capabilities (Cole, 2018; Kirchherr and Urban, 2018), the underlying challenges remain countless. Meanwhile, the generally recognised success of the Kenyan photovoltaic market (Byrne et al., 2014; Rolffs, Ockwell, and Byrne, 2015; Ockwell and Byrne, 2016) has been linked to the extended period of active interventions (including public funded nurturing of capabilities, building of actor networks, and directed technology development) however, there is still little empirical evidence to draw from, especially from the perspective of capabilities in the RE sector.

Capability accumulation, disaggregation, and assessment

The literature on the concept of capability has advanced considerably over the last three decades since the early contribution of Bell (1984) hence leading to diversity in its nuances and taxonomy (Lall, 1992; Bell and Pavitt, 1993, 1995; Panda and Ramanathan, 1996; Archibugi and Coco, 2005; Augier and Teece, 2006; Garcia-Muiña and Navas-López, 2007; Freeman and Soete, 2009). The Capability Accumulation (CA) taxonomy was developed by, amongst others, Lall (1992) as well as Bell and Pavitt (1993, 1995), where they recognised capabilities as the resources needed to generate and manage change while the accumulation of these capabilities was acknowledged through learning processes. According to Panda and Ramanathan (1996), the concept can be regarded as a set of functional abilities, reflected in the firm's performance through various activities and whose ultimate purpose is firm-level value management and the development of organisational abilities that are difficult to duplicate. While this classical taxonomy (Table 6.1) has been extensively used in empirical studies to unravel firm-based capabilities especially in developing economies (Figueiredo, 2017; Bell and Figueiredo, 2012; Lundvall et al., 2009; Archibugi and Coco, 2005), its critics have been sceptical owing to a variety of issues including its static nature, inability to provide information about how the organisation got to such a level of maturity, its narrow focus on technological innovations, as well as heterogeneity of capability accumulation between and within firms (Figueiredo, 2001, 2017; Dutrénit, 2004).

The other main taxonomy that attempts to explain the creation of capabilities emanates from the Resource Based View (RBV) literature. The RBV taxonomy regards a firm as a bundle of resources, competences, and capabilities, and considers its heterogeneity and competitive advantage to be based on different combinations of these elements (Penrose and Penrose, 1995; Teece, Pisano, and Shuen, 1997). In view of the RBV perspective, Garcia-Muiña and Navas-López (2007) conceptualised strategic technological capability as the generic knowledge-intensive ability to jointly mobilise different scientific and technical resources that enable a firm to implement competitive strategy and create value in a given environment so as to successfully develop its innovative products and/or productive processes. Other previous approaches (e.g., Lall, 1992; Bell, 1984), presented rather different generic aspects such as acquisitive, operative, adaptive, innovative, supportive, and marketing capabilities. Similarly, Momeni, Nielsen, and Kafash (2015) presented capability in terms of structural, personnel, and operational dimensions. Nevertheless, one common drift in all the different viewpoints is that sustained development of any sector of an economy is largely dependent on the prevailing capabilities, their formation, and their outcomes. However, because resource-centred terms such as resources, assets, and capabilities are often used in different contexts, the RBV approach is deemed to be rather insufficient (Lindbom et al., 2015). Indeed, Eisenhardt and Martin (2000) had earlier explicitly challenged the RBV approach by arguing that there are identifiable processes, such as those domiciled within a project life cycle, that can explain the nature of competitiveness. In addition, the apparent lack of specificity in the disaggregation of capabilities has largely remained prominent.

The relationship between the capability accumulation and the resource-based view approach in the creation of capabilities therefore shows overlaps, variations in terminology, and general lack of consensus. Indeed, terms such as resources, assets, and capabilities are commonly used but most often in different contexts and meanings (De Bakker and Nijhof, 2002). Consequently, there is need to harmonise the different viewpoints, for instance by considering task-related parameters such as generation and diffusion of inventions and innovations, competitiveness, and research and development infrastructure. This alternative approach is thus regarded as the Task Based View (TBV) approach that is summarised in Table 6.1.

Drawing from the foregoing discourse, the key paradigm that is pursued in this chapter is to integrate the preceding three taxonomical approaches to analyse capabilities on renewable electrification. The framework used in the chapter draws from the preceding approaches and considers specific task based sub-constructs, that is, financial, structural, and operational, as well as their outcomes. The framework also considers the life cycle approach that covers all the firms' strategic capabilities including the time dimension as espoused by DeSarbo et al. (2005).

TABLE 6.1 Summary of taxonomical analysis of technological capabilities

Taxon	*Elements /types*	*Features*	*Indicators*
Classical Taxonomy	Embodied, disembodied, codified, and tacit	• Capital goods and equipment • Infrastructure • Human capital/skills • Scientific and technical expertise • Manuals, blueprints, patents, scientific publications • Qualifications of labour force	• Import of capital goods and equipment • Inward foreign direct investment • Technology licensing payments • Imported technology • Level of staff education • Absolute numbers per period
	Capability accumulation	• Learning processes • Organisational abilities	• Functional abilities • Firm-level value management
RBV Taxonomy	Acquisitive, operative, adaptive, innovative, supportive, and marketing	• Scientific and technical resources • Knowledge-intensive ability	• Innovative products or processes • Level of output • Absolute numbers per period
	Structural, personnel, operational	• Industrial performance	• Level of output
	Bundle of resources, competencies, and capabilities	• Heterogeneity and competitive advantage based on different combinations of the elements	• Absolute numbers per period
TBV Taxonomy	Generation and diffusion of inventions and innovations	• Inventions/innovations output • Application and dissemination of inventions/innovations	• Number of inventions/innovations per period • Absolute numbers per period
	Competitiveness	• Industrial performance	• Level of output
	Research &Development	• R&D resources/intensities	• R&D expenditure
	Infrastructure	• Technology infrastructure	• Infrastructure capital deployment

Source: authors; illustration adapted from Archibugi and Coco, 2005

Framework for capability indicators

Analysis of capabilities in a business or project can be done using a framework that disaggregates the main typologies of capabilities essential for renewable electrification. An analytical operationalisation (Reichert et al., 2011; Panda and Ramanathan, 1996; Lall, 1992) of such framework yields four main dimensions, namely financial, structural (management), operational (technical), and personnel and supplementary capabilities (Table 6.2).

Besides financial and engineering capabilities, the development of local technical capability to plan, execute (construct, commission, and operate), and

TABLE 6.2 Analytical operationalisation summary of the dimensions of capabilities

Dimension	*Activities*	*Elements*
Financial and management capability	• Financial acquisition and management • Strategic management and steering	• Identification, assessment, negotiation, and finalisation of terms • Development and implementation of strategic plans, organisational structures, monitoring, and evaluation
Operational (technical/design and engineering) capability	• Planning and execution activities (design and engineering, construction and technology acquisition)	• Project evaluation, procurement, support, process improvements, planning, monitoring, and control
Implementation capability	• Implementation activities (feasibility studies, performance of civil works, erection and commissioning), planning, monitoring and controlling processes • Quality assurance, inspection, and inventory control	• Planning, monitoring, and coordination of project implementation activities
Servicing and maintenance capability	• Project sustenance, servicing, and maintenance activities	• Carrying out maintenance, planning, monitoring, and coordinating service activities
Personnel and supplementary capability	• General acquisition (procurement, consumables, and human resource)	• Planning, monitoring, and coordination of acquisition processes (consumables, human resources, etc.)

Source: authors

sustain projects (maintain the generation and improve service quality) is the key for the realisation of sustainable and enhanced energy access. This chapter therefore employs the project lifecycle perspective and the dimensions of capabilities that are summarised in Table 6.2 to analyse capabilities in the renewable energy firms and projects. The effective and accurate analysis of the different aspects of capabilities from the project lifecycle viewpoint has been linked to monumental bearing in the assessment of renewable electrification projects (Ye et al., 2019). In this connection, pursuant to this generic discussion of dimensions of capabilities, subsequent sections of this chapter present the analysis of these dimensions as they relate to RE specific projects and the required capabilities.

Methodology and rationale for the assessment of capabilities

Measurement and indicators of capabilities

The procedure for the assessment of capabilities for deployment of renewable electrification in Kenya as covered in this study entailed:

- Analysis of renewable electrification related operations undertaken within a project or enterprise and the identification of the ability and knowhow necessary to perform the operations.
- Development of a set of indicators for assessing the operations with respect to capabilities, outcomes, and influencing factors.
- Analysis of extent to which the capabilities are deployed within the project/enterprise, their outcomes, and influencing factors.

The specific categories of capabilities that were investigated included financial and management, design and engineering, project implementation, servicing and maintenance, and supplementary capabilities.

Financial and management capability

The assessment of financial and management capability entailed the evaluation of financial acquisition and management as well as strategic management and steering. The specific capability elements that were assessed in this case included the capability to identify, assess, negotiate, and finalise terms of financing; deploy integrated financial systems; develop and implement strategic plan for the organisation; create new organisational structures; plan, monitor, and control research and development (R&D) projects; as well as make strategic decisions, and implement and integrate them in the organisation activities. The key indicators that were evaluated included the level of independence with respect to the respective capabilities.

Design and engineering capability

The evaluation of design and engineering capability involved the assessment of the capability to: undertake routine design and detail engineering of process; adapt both internally and externally acquired technologies; duplicate internally/externally acquired technologies; identify, assess, negotiate, and finalise the terms of the technology to be acquired; identify, assess, negotiate, and finalise the terms of acquiring raw materials, supporting facilities, spare parts, and consumables; undertake process/product improvements and development of new ones; and plan, monitor, and control design, engineering, and contract activities. The specific indicators that were evaluated included the total number of detail engineering studies completed; level of independence in carrying out detail engineering studies; cost and time overrun due to lapses in detail engineering; cost of technology adaptation as a percentage of the cost of technology acquired internally as well as externally; number of internally and externally acquired technologies duplicated; and the number of new or significantly improved services, process, technology, or design accomplished and the level of independence when carrying out the respective detail engineering activities.

Project implementation capability

The assessment of project implementation capability involved the evaluation of the capability to: support project feasibility studies, parameter estimation, value engineering, site engineering (in terms of technical, economic, financial, environmental, and social aspects); perform civil works; erect and commission equipment; plan, monitor, and control construction, erection, and commissioning activities; effectively utilise and control the conversion technologies and auxiliary processes; and undertake quality assurance, inspection, and inventory control. In this regard, the specific indicators that were employed for the evaluation of project implementation capability elements included total number of feasibility studies of projects completed; number of large civil construction works done by the firm; number of items of equipment erected and commissioned; cost and time overrun during civil construction and erection due to lapses in planning, monitoring, and control activities; successful start-up rate; actual unit availability as a percentage of planned unit availability; inventory turnover of major spare parts; level of independence when carrying out the respective project implementation activities including feasibility studies; and site selection, preparation of construction specifications, and cost estimation.

Servicing and maintenance capability

The assessment of servicing and maintenance capability involved the evaluation of the capability to: carry out maintenance (preventive, routine, corrective, improving, and predictive maintenance) and offer technical advice; diagnose problems, undertake corrective actions (repairs, maintenance, replacement and disposal); and plan, monitor, and coordinate service activities and capacity. The specific indicators that were employed included the total maintenance hours used as a percentage of gross available hours, average time taken to restore electricity supply after a failure or fault has occurred (system average restoration index), and average time taken to attend to interruptions.

Supplementary capabilities

The assessment of supplementary capabilities included the evaluation of the capability to: acquire competent human resources; acquire the necessary technology and raw materials; and plan, monitor, and coordinate resource acquisition processes. In this regard, the key indicators that were employed to assess the respective supplementary capabilities included level of independence during the procurement of human resources; the acquisition of technology and raw materials; and the development of procurement strategies.

A detailed elaboration of the different capability elements, their respective indicators, as well as the criteria for evaluating the level of deployment of the respective capabilities is espoused in the Innovation and Renewable Electrification in Kenya (IREK) working paper on capabilities, outcomes, interactive learning, and influencing factors (Nzila and Korir, 2020). The working paper further provides the questionnaire that was used to assess the various capability categories and their outcomes for renewable electrification in Kenya.

Empirical design, development, and execution of the survey

The survey sought to interview Kenyan firms dealing with solar, wind, geothermal, small hydro, and biomass/biogas. The design of the survey (Kasunic, 2005) was based on preliminary work from two IREK project workshops held in 2016 and 2017. During the preliminary exercise, a database was developed to map out the renewable energy projects in the country. The survey respondents were then selected from the database. To obtain a representative sample of the firms, a random recruitment process was undertaken from the sample frame (Table 6.3) generated from the prior-constructed database.

The survey questionnaire was in the form of a structured interview protocol, with close ended and a combination of dichotomous and multichotomous questions. The questionnaire consisted of items that were intended to collect data on five main sections, namely general information, business/project activities, learning and development of capabilities, benefits and outcomes, and influencing factors.

The data collection technique was based on a cross-sectional survey method owing to its nature of utilising a questionnaire to collect views and information from respondents at a single period in time as well as to identify the relationship between predefined variables. The choice of the target geographic area and

TABLE 6.3 The survey sample frame and response rate

Technology	*Sample frame*	*Sample size*	*Response*	*% Response*
Small hydro	15	14	10	71%
Wind	11	11	7	64%
Solar	57	50	41	82%
Geothermal	4	4	2	50%
Biogas	6	6	3	50%
Hybrid	9	9	8	89%
Total	102	94	71	76%

Source: authors

population was motivated by the need to obtain sufficient data. To this end, a clustering approach was employed whereby the population of the respondents was clustered in terms of solar PV, wind energy, small hydropower, geothermal, and biomass projects in Kenya. The details of the sample frame and sample size are provided in Table 6.3. The sample size was computed according to Cochran's formula (Taherdoost, 2017; Gill, Johnson and Clark, 2010) at 95% confidence interval and 5% margin of error.

Data analysis

An analysis of the questionnaires was conducted in multiple stages. A number of descriptive statistics were done, followed by distribution tests, multiple responses, and multiple dichotomy data analysis using numerical Python. The analysis involved the use of bar charts and line curves to visualise output for the different variables in each question. The graphs represent the total number of respondents who indicated their intention of affirming their responses to the respective questions. The survey data was analysed using the Python software. The survey results are presented both qualitatively and quantitatively through analysis consisting of summary statistics reported for the full sample frame.

Results and discussions

This section presents the descriptive statistics of the survey findings pertaining to business/project activities prior to delving into findings on the development of capabilities, their outcomes, and the linkages therein. An analysis of the mutual connection of the survey data and the evaluation of the responses is also presented.

Survey response and data inventory

The aggregate survey response rate of 76% (Table 6.3) implied an overall non-response bias of only 24%. According to the standards for survey research

(Draugalis, Coons, and Plaza, 2008; Converse et al., 2008), this reported response rate is quite acceptable since the norm for such mixed mode surveys is to target response rates close to 60%. Moreover, based on the respective response rates, it was evident that the respondents from the hybrid and solar PV sector could be generally presumed to be the most motivated groups.

The data inventory from the survey is summarised in Table 6.4 in terms of a summary-mapping matrix of the respondents while Table 6.5 presents selected characteristics of the private/industry energy projects. As shown in the matrix, the respondents were characterised by a variety of players of which private actors were the majority amongst all the categories. There were distinct sectoral differences with more actors being in the solar PV space. The role of the private/industry sector in the solar PV space is also evidently clear. These findings are in agreement with other researchers (Rolffs, Ockwell, and Byrne, 2015; Ockwell and Byrne, 2016; Karjalainen and Byrne, 2022; this volume) and they reaffirm that, unlike the other low carbon energy categories, solar PV technology remains a significant player in the Kenyan energy market segment. While these

TABLE 6.4 Mapping of the survey respondents

Categories	*Solar PV*	*Wind*	*Small Hydro*	*Geothermal*	*Biomass*	*Hybrid*	*Total*
Private/Industry	31	6	2	1	3	5	48
Quasi Private	-	-	1	-	-	-	1
University/Research	2	-	-	-	-	-	2
Consulting	1	-	-	-	-	-	1
Public Authority	3	-	4	1	-	2	10
NGO	1	1	-	-	-	-	2
Religious Mission	-	-	3	-	-	1	4
Cooperative Society	3	-	-	-	-	-	3
Total	41	7	10	2	3	8	71

Source: authors

TABLE 6.5 Key characteristics of the private/industry energy projects

Details	*Solar PV*	*Wind*	*Small Hydro*	*Geothermal*	*Biomass*	*Hybrid*
Installed nominal capacity (MWp)	0.01–55	0.01–310	0.095–21	3.7–110	0.01–0.15	0.25–2.7
Type	Grid/mini-grid	Grid/mini-grid	Mini-grid	Grid	Mini-grid	Mini-grid

Source: authors

findings do not resonate with the current energy mix in the country where the share of solar PV remains largely obscure, it is evident that, based on its prominence across all the respondents, solar PV in the country's energy mix is likely to increase considerably and might even leapfrog other low carbon resources such as wind and small hydro power.

Involvement in business/project activities

With a view to understanding the project activities and the inherent capabilities, the respondents were asked a group of questions focusing mainly on their involvement in their latest renewable energy project. Respondents were required to indicate the type of activities they were involved in from a project life cycle perspective, in terms of project initiation, planning, implementation, monitoring and evaluation, and follow-up and closure. The resultant responses are presented in Figure 6.1. Pertaining to project initiation, the specific activities where most respondents were involved included proposal development, feasibility studies, and business plan development (n = 51, 48, and 43, respectively) whereas there was least involvement in the delimitation of boundaries (n = 22) in the projects.

Concerning project planning, the results indicate that the activities where the majority of the firms were involved included the development of project plan, definition of operational requirements and development of financial plan (n = 48, 45 and 45, respectively) while the least involvement was reported in systems integration and construction of basic civil works (n = 31). In the case

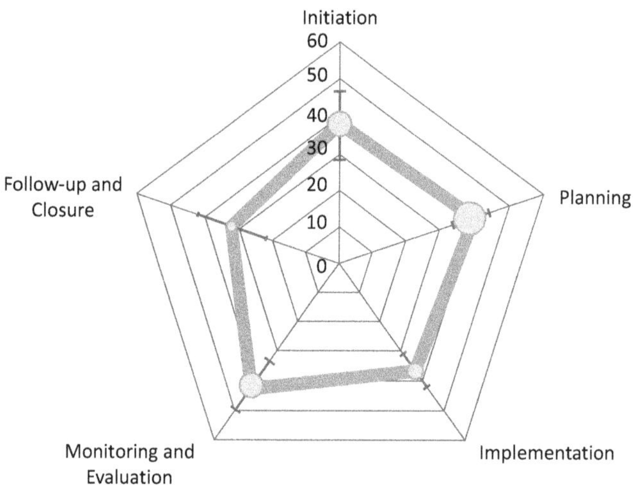

FIGURE 6.1 Level of involvement in business/project activities. *Source: authors. NB: the size of the point markers denotes the relative weighted level of involvement of private/industry sector respondents within a category while error bars represent standard deviation within the responses.*

of project implementation, most respondents had distinct involvement in capacity building and provision of project management services while there was least involvement in the provision of turnkey solutions (n = 43, 39, and 28, respectively). Pertaining to the involvement in project monitoring and evaluation, most respondents were involved in performance assessment and quality management while there was least involvement in change management (n = 54, 49, and 27, respectively). With respect to project follow-up and closure, the majority of the respondents were involved in the provision of instruction and training for staff as well as writing of project reports whereas there was least involvement in writing handbooks (n = 42, 41, and 20, respectively).

Traditionally, most firms found it essential to be actively involved in the entire project lifecycle, hence the need to thinly spread the available resources/capabilities. However, nowadays there is more merit in specialisation on niche activities while engaging other specialist consulting firms in activities where there are deficits. The reported distinctive involvement of the firms in the project activities, from project initiation to follow-up and closure, thus gives a clear indication that most of the surveyed firms still follow the classical approach in the deployment of the existing capabilities. The foregoing also indicates that the studied firms constitute suitable candidates for further analysis of capabilities from a project lifecycle perspective.

General outlook of capabilities

An overview of all the capabilities typified in the different firms pertaining to the present study is presented in Figure 6.2 and further clarified in the following sections. The capabilities are characterised in terms of the demonstrated magnitude with respect to capability category and levels (high, moderate, and low). About 43%, 35%, and 22% of the respondents are shown to have high, moderate, and low financial and management capabilities. In addition, an association between the category and distribution of capabilities was observed. Generally, the most predominant capabilities (where the highest association between the category and distribution of capabilities was observed) were the design and engineering capability (χ^2 (10) = 0.6, p = 1.0) as well as the financial and management capability (χ^2 (10) = 1.83, p = 0.997). Within the two categories of capabilities, the highest association was observed in solar PV and small hydro. In contrast, servicing and maintenance capability was the most lacking capability where the lowest association between the category and distribution of capabilities was observed (χ^2 (10) = 5.22, p = 0.88). In this category, the lowest association was observed in biogas, geothermal, and wind energy, respectively. The order of predominance of capabilities in low carbon electrification firms and projects in Kenya was therefore observed to be design and engineering, financial and management capabilities, planning and implementation, servicing and maintenance, and supplementary capabilities, respectively.

Among the respondents having high financial and management capabilities, 22%, 30%, and 13% were also responsible for the high proportion of design

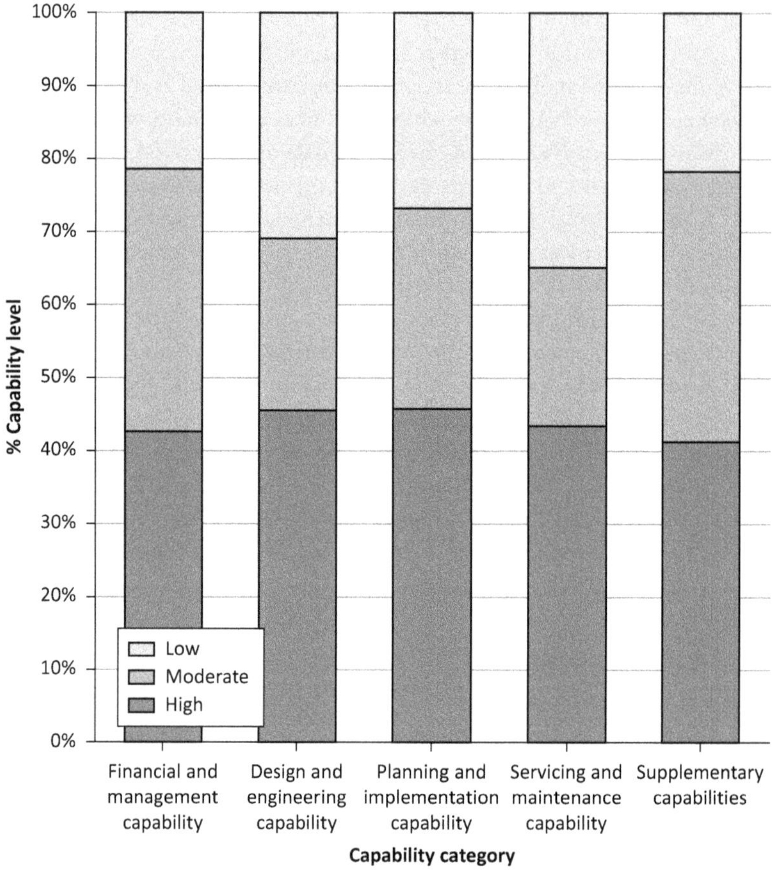

FIGURE 6.2 Summarised analysis of capabilities in RE firms/projects. *Source: authors.*

and engineering, planning and implementation, and servicing and maintenance capabilities, respectively. Similarly, among the respondents who are shown to exhibit moderate financial and management capability, about 31% were responsible for the overall high service and maintenance capabilities. On the other hand, among the respondents shown to have high design and engineering capability, about 66% and 50% were respectively responsible for the high capabilities in planning and implementation as well as servicing and maintenance capabilities.

Based on the categories of financial management capabilities that were evaluated, it can be affirmed that most of the firms that were interviewed have preponderance for financial acquisition, strategic management, and steering of RE projects. The specific financial management capability levels are further elucidated in Figure 6.3. In contrast, servicing and maintenance capability is observed to be the most lacking capability with over 35% of the firms interviewed

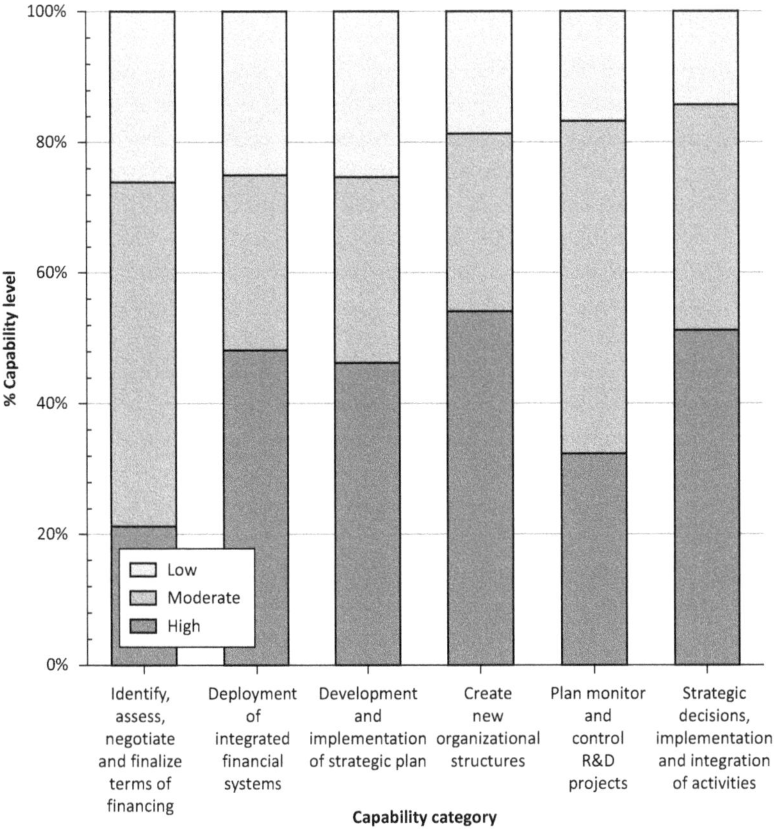

FIGURE 6.3 Analysis of financial and management capabilities. *Source: authors.*

reporting low capability levels. This finding has a direct bearing on the diagnostic and maintenance abilities of the firms hence adverse ramifications on both the short- and long-term sustainability of RE projects being undertaken by the interviewed firms. In addition, the propensity for potentially reduced quality of energy service is quite apparent. The servicing and maintenance capability levels are further elucidated below.

Generally, based on the indicators considered and the resultant responses, from the high and moderate capability levels (Figure 6.2) it can be concluded that the RE capabilities in Kenya are relatively high on average (above 65%) but with noticeable bottlenecks as elucidated in subsequent sections. Often it is not always clear what most scholars (Teece, 2007; Byrne et al., 2012, among others) mean when they state that capabilities are low. Conversely, the survey presented in this chapter takes all capabilities across the project cycle and compares across sectors thus bringing out more 'meat to the bones' by specifying types of capabilities and sub-capabilities and the extent to which respondents think they are manifested or not.

Financial and management capabilities in low carbon energy projects

A general overview of the different elements under financial and management capabilities in low carbon energy enterprises and projects is presented in Figure 6.3 relating the capability categories against the respective levels. Among the surveyed firms, the most predominant financial and management capability are the capabilities to make strategic decisions, and to implement and integrate the organisation activities along with the capability to create new organisational structures, both of which have a high capability level of 53% and about 80% when high and moderate capability levels are combined. Furthermore, when this result is viewed with respect to the assessment matrix, it is evident that the majority of the firms report having high financial management, strategic, and steering capabilities while the most lacking capabilities are manifested with respect to financial acquisition as well as planning, monitoring, and control of research and development projects.

Managerial capabilities are vital for the development of managerial and innovative initiatives, organisational improvements, and general competitiveness (Teece, 2006, 2007). These findings therefore imply that the surveyed firms have high potential for managerial innovations hence – theoretically at least – the capacity to put in place a competitive financial strategy for renewable electrification. Nevertheless, the survey results also indicate that between 70% and 80% of the firms are potentially prone to external exploitative manipulation and predatory strategies due to the noticeably missing capability to identify, negotiate, and finalise the terms of the necessary financing as well as the capability to plan, monitor, and control research and development projects. Hence there is an apparent dependence on external consultants to mitigate the missing capabilities. Meanwhile, it is noteworthy that generally the terms of implementation of most projects are routinely sealed during the finalisation of financing terms. In this connection, the respective missing financial and management capabilities are likely to predispose the surveyed firms to non-strategic undertakings as well as financing deals that are tailored to perpetuate foreign businesses to the detriment of local content and capability development.

Design and engineering capabilities in low carbon energy projects

Pertaining to design and engineering capabilities (Figure 6.4) it is shown that routine design and detail engineering, and process improvements are the most predominant capabilities. Conversely, adaption of acquired technologies and the capability to identify, assess, negotiate, and finalise the terms of technology acquisition were the most lacking capabilities. The cost of technology adaptation was reported as the main impediment behind the low capability in the adaption of the acquired technologies whereas increased reliance on external actors was

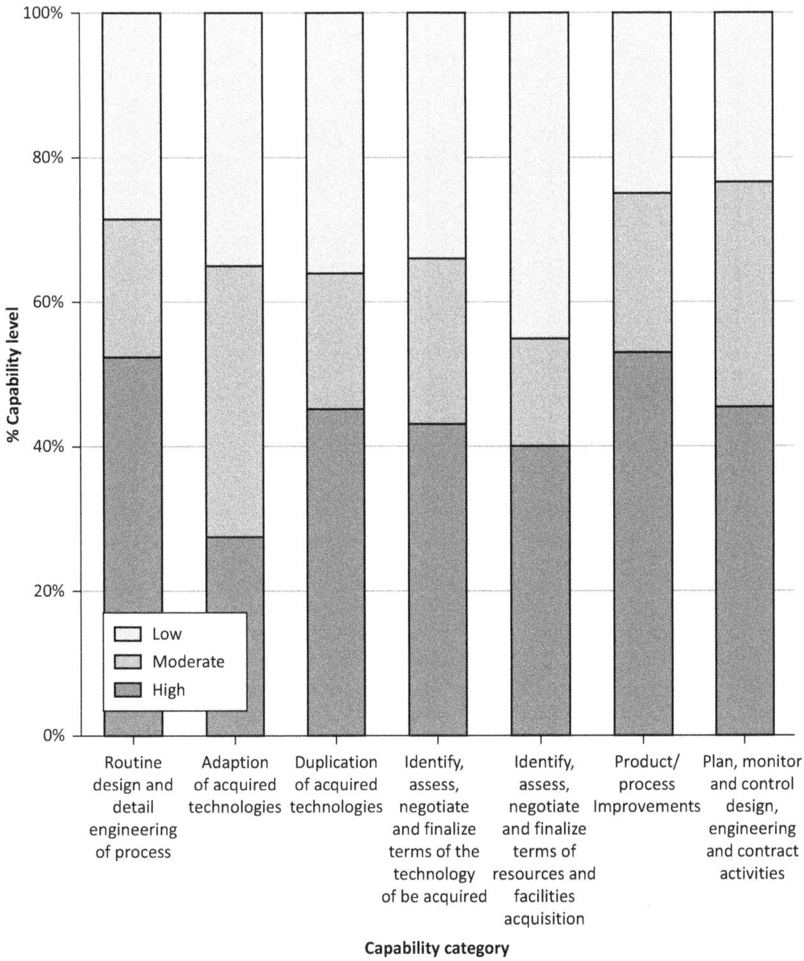

FIGURE 6.4 Analysis of design and engineering capabilities. *Source: authors.*

cited as the main cause of reduced capability to identify, assess, negotiate, and finalise the terms of technology acquisition.

Implementation capabilities in low carbon energy projects

The implementation capabilities in low carbon energy firms and projects in Kenya are presented in Figure 6.5. The capability to erect and commission equipment and the capability to plan, monitor, and control construction and commissioning activities were the most dominant. In contrast, the capability for inventory control of the major spare parts was the most conspicuously missing capability. Other missing capabilities included the capability to support project feasibility

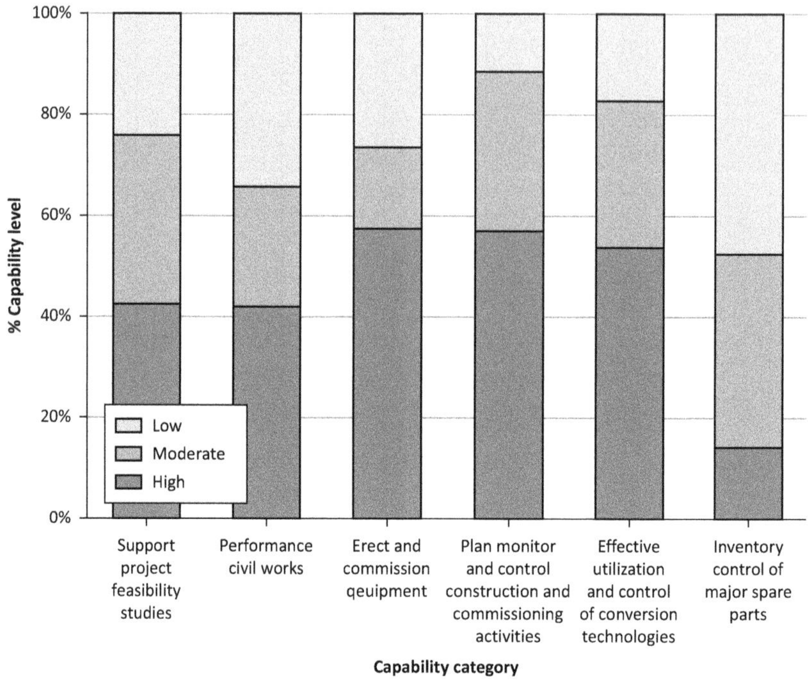

FIGURE 6.5 Analysis of implementation capabilities. *Source: authors.*

studies as well as the capability to perform civil works. Hence it is evident that, for all the missing capabilities, most respondents had noticeable dependence on external agents for the implementation of their energy projects. However, the capabilities for site selection, preparation of construction specifications, and cost estimation were observed to range from moderate to high.

Servicing and maintenance in low carbon energy projects

The servicing and maintenance capabilities in low carbon energy projects and firms in Kenya is presented in Figure 6.6. The results show that at a 60% high capability level, the most predominant servicing and maintenance capability was with respect to planning, monitoring, and coordination of service activities. An average time taken to attend to interruptions of less than four hours was also associated with the high capability level. In contrast, with over 75% of the interviewed firms reporting low capability levels; the most missing capability was with respect to undertaking maintenance. The predominance of low capability in undertaking maintenance was specifically manifested by low utilisation of maintenance hours whereby a vast majority of the firms reported that they utilised less than 70% of their total available maintenance hours. Lack of full

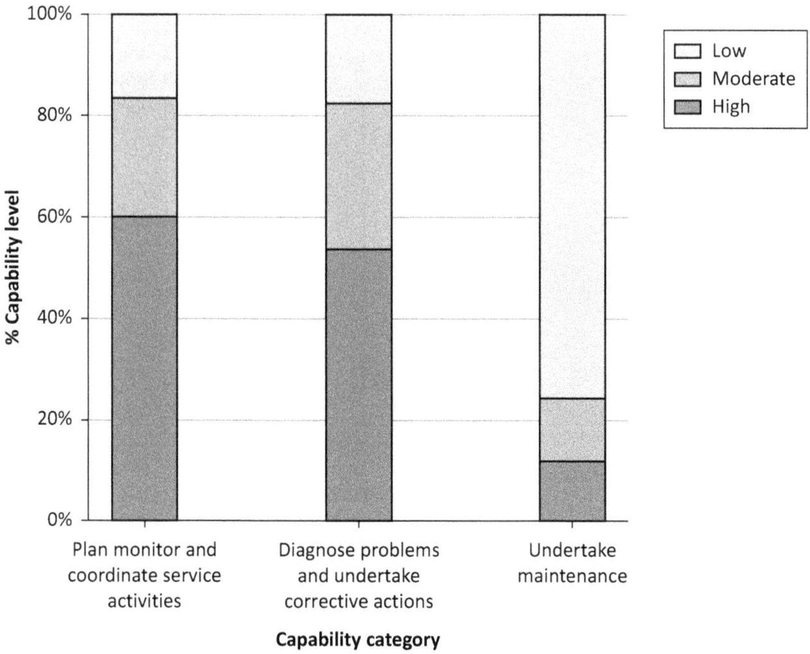

FIGURE 6.6 Analysis of servicing and maintenance capabilities. *Source: authors.*

utilisation of the allocated maintenance hours could possibly lead to fewer scheduled power interruptions in the short term but it portents an apparent risk of equipment failures in the long term. Indeed, optimum utilisation of maintenance hours is vital since it curtails potential breakdowns in RE projects and the resultant stresses on power distribution systems.

Supplementary capabilities in low carbon energy projects

Supplementary capabilities in low carbon energy projects and firms in Kenya are presented in Figure 6.7. It is shown that the acquisition of competent human resources was the most predominant supplementary capability. Conversely, the capability to acquire the necessary technology and raw materials (support facilities, spare parts, and consumables; identify, assess, negotiate, and finalise the terms of acquisition) was the most missing capability.

Conclusions and recommendations

This chapter sought to develop and implement a capability assessment framework to investigate the character of capabilities as well as whether the distribution of capabilities differs among the renewable energy projects in Kenya. Based on responses from

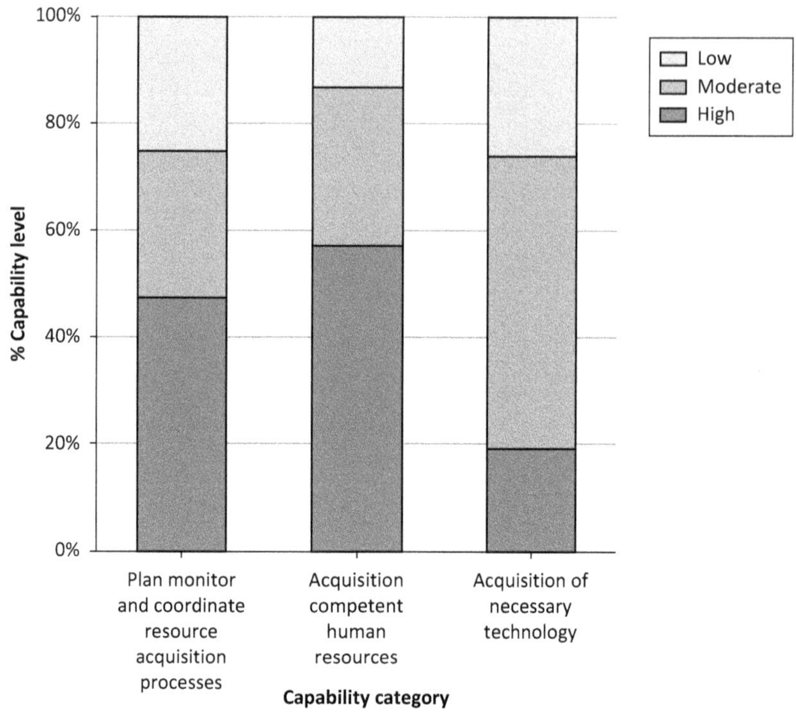

FIGURE 6.7 Analysis of supplementary capabilities. *Source: authors.*

71 RE firms and organisations in Kenya, the survey established the extent, location, and distribution of capabilities, which capabilities are in place, where the shortfalls are, and to what extent technology sectors (solar, wind, small hydro, geothermal, and hybrids) influence the development and deployment of capabilities.

Influence of business and project activities on the development of capabilities

The survey results revealed that the vast majority of the respondents had distinct involvement in all the main activities within the project lifecycle, that is project initiation, planning, implementation, monitoring, and evaluation, as well as follow-up and closure. This shows that most of the respondents have developed or at least have the requisite capabilities to participate in project execution. However, it was also revealed that there were certain activities within the project life cycle where the respondents had least involvement, especially in most activities related to project implementation, follow-up, and closure. It can therefore be concluded that there is slow development of capabilities with respect to project implementation, follow-up, and closure as well as in certain aspects of project initiation. The overall implication here is that most firms are not adequately specialised due to the apparent practice of spreading their existing capabilities thinly so as to have broad economies of scope. In addition, the question pertaining to the division of labour within the firms suffices.

Distribution of capabilities

Generally, an association between the category and distribution of capabilities has been observed and it can therefore be concluded that the RE capabilities related to deployment in Kenya are relatively high on average but with noticeable bottlenecks. This conclusion tallies with the findings in Hanlin and Okemwa (2022; this volume) as well as Kingiri and Okemwa (2022; this volume). The most predominant capabilities where the highest association between the category and distribution of capabilities was observed were the design and engineering capability and the financial and management capability. Within the two categories of capabilities, the highest association was observed in solar PV and small hydro. In contrast, servicing and maintenance capability was the most lacking capability (where the lowest association between the category and distribution of capabilities was observed). In this category, the lowest association was observed in biogas, geothermal, and wind energy, respectively. Consequently, it could be postulated that while the share of solar PV technology in Kenya's energy mix remains largely obscure, the reported higher development and distribution of capabilities in solar PV RE projects implies that the trajectory of solar PV may in future leapfrog other established RE technologies such as wind and small hydro. The potentially uncharted increased share of solar PV in the country's energy mix will present new opportunities as well as disruption to the country's energy mix.

Nature, extent, and role of the development of capabilities for sustained renewable electrification in Kenya

The survey results indicate that all the respondents had undertaken a variety of activities geared towards development of capabilities in their projects. However, the avenue for mitigating any shortfalls in the existing capabilities largely remains uncharted. The order of predominance of capabilities in the deployment chain of RE technologies was observed to be design and engineering, financial and management capabilities, planning and implementation, servicing and maintenance, and supplementary capabilities, respectively. The implications of these findings are that the studied firms possess divergent generic capability to individually mobilise different resources necessary for successful implementation of competitive strategy and value creation in the low carbon energy environment. The survey results also indicate that while the low carbon firms and projects in Kenya have high management capabilities, there are a number of missing capabilities especially with respect to the capability to plan, monitor, and control research and development activities. In addition, the moderate to complete dependency on external agents for undertaking various project activities expose the local firms to the influence by global actors devoid of any alternative recourse.

Pertaining to the deployment chain, the findings in this chapter further reaffirm that unlike the other low carbon energy categories, solar PV technology remains a significant player in the Kenyan energy market segment in spite of its near obscure contribution to the country's energy mix. Nevertheless, it is noteworthy to recognise that there is a distinct difference between large and small solar PV in terms of significant market shares whereby the former (as focused in this chapter) is mainly linked to national or mini-grids while the latter is predominantly linked to solar home systems. It can therefore be concluded that the reported high capability in the solar PV firms may enable the local firms to carry out activities in the entire project life cycle leading to a high degree of self-reliance in terms of capabilities. This enables the firms to gain a high level of value addition and provide a solid foundation for an increase in the share of solar PV in the country's energy mix. Whether this means that solar PV will leapfrog the other low carbon resources such as wind and small hydropower remains to be seen.

Acknowledgements

Support for research on which this chapter is based from the Danish Ministry of Foreign Affairs, Grant: DFC 14-09AAU is gratefully acknowledged.

References

Almeshqab, F. and Ustun, T.S. (2019) 'Lessons learned from rural electrification initiatives in developing countries: Insights for technical, social, financial and public policy aspects', *Renewable and Sustainable Energy Reviews*, 102, pp. 35–53. https://doi.org/10.1016/j.rser.2018.11.035

Archibugi, D. and Coco, A. (2005) 'Measuring technological capabilities at the country level: A survey and a menu for choice', *Research Policy*, 34(2), pp. 175–194. https://doi.org/10.1016/j.respol.2004.12.002

Augier, M. and Teece, D.J. (2007) 'Dynamic capabilities and multinational enterprise: Penrosean insights and omissions', *Management International Review*, 47, 175–192. https://doi.org/10.1007/s11575-007-0010-8.

Azzuni, A. and Breyer, C. (2018) 'Definitions and dimensions of energy security: A literature review', *Wiley Interdisciplinary Reviews: Energy and Environment*, 7(1), pp 1–34. https://doi.org/10.1002/wene.268

Bell M. (1984) '"Learning" and the accumulation of industrial technological capacity in developing countries', in: Fransman M. and King K. (eds) Technological Capability in the Third World. London: Palgrave Macmillan. https://doi.org/10.1007/978-1-349-17487-4_10

Bell, M. (1990) *Continuing Industrialisation, Climate Change and International Technology Transfer, A report prepared in collaboration with the resource policy group*, Oslo, Norway: Science Policy Research Unit, Brighton: University of Sussex.

Bell, M. and Figueiredo, P.N. (2012) 'Building innovative capabilities in latecomer emerging market firms: Some key issues', in: Amann, E. and Cantwell, J. (eds) *Innovative Firms in Emerging Market Countries*. New York and London: Oxford University Press, pp. 24–109. https://doi.org/10.1093/acprof:oso/9780199646005.003.0002

Bell, M. and Pavitt, K. (1993) 'Technological accumulation and industrial growth: Contrasts between developed and developing countries', *Industrial and Corporate Change*, 2(2), pp. 157–210. https://doi.org/10.1093/icc/2.2.157.

Bell, M. and Pavitt, K. (1995) 'The development of technological capabilities', *Trade, Technology and International Competitiveness*, 22(4831), pp. 69–101.

Byrne, R., Smith, A., Watson, J. and Ockwell, D. (2012) 'Energy pathways in low carbon development', In: *Low Carbon Technology. From Rhetorics to Reality*. London, p. 123. DOI: https://doi.org/10.4324/9780203121481

Byrne, R., Ockwell, D., Urama, K., Ozor, N., Kirumba, E., Ely, A. and Gollwitzer, L. (2014) *Sustainable Energy for Whom? Governing Pro-poor, Low-carbon Pathways to Development: Lessons from Solar PV in Kenya*, STEPS Working Paper 61, Brighton: STEPS Centre, University of Sussex.

Byrne, R., Mbeva, K. and Ockwell, D. (2018) 'A political economy of niche-building: Neoliberal-developmental encounters in photovoltaic electrification in Kenya', *Energy Research and Social Science*, 44, pp. 6–16. https://doi.org/10.1016/j.erss.2018.03.028.

Cole, P. (2018) 'Assessing the impact of a renewable energy programme in Bamyan, Afghanistan: The value of a capability approach', *Energy for Sustainable Development*, 45, pp. 198–205. https://doi.org/10.1016/j.esd.2018.06.014

Converse, P.D., Wolfe, E.W., Huang, X. and Oswald, F. L. (2008) 'Response rates for mixed-mode surveys using mail and e-mail/web', *American Journal of Evaluation*, 29(1), pp. 99–107. https://doi.org/10.1177/1098214007313228

De Bakker, F. and Nijhof, A. (2002) 'Responsible chain management: A capability assessment framework', *Business Strategy and the Environment*, 11(1), pp. 63–75. https://doi.org/10.1002/bse.319

DeSarbo, W.S., Anthony Di Benedetto, C., Song, M. and Sinha, I. (2005) 'Revisiting the miles and snow strategic framework: Uncovering interrelationships between strategic types, capabilities, environmental uncertainty, and firm performance', *Strategic Management Journal*, 26(1), pp. 47–74. https://doi.org/10.1002/smj.431

Draugalis, J.R., Coons, S.J. and Plaza, C.M. (2008) 'Best practices for survey research reports: A synopsis for authors and reviewers', *American Journal of Pharmaceutical Education*, 72(1), 1–6. https://doi.org/10.5688/aj720111

Dutrénit, G. (2004) 'Building technological capabilities in latecomer firms: A review essay', *Science, Technology and Society*, 9(2), pp. 209–241. https://doi.org/10.1177/097172180400900202

EAC (2016) *Renewable energy and efficiency: Regional Status Report*, p. 53. Available at: https://www.iea.org/policiesandmeasures/renewableenergy.

Eisenhardt, K.M. and Martin, J.A. (2000) 'Dynamic capabilities: What are they?', *Strategic Management Journal*, 21(10–11), pp. 1105–1121. https://doi.org/10.1002/1097-0266(200010/11)21:10/11<1105::AID-SMJ133>3.0.CO;2-E

Figueiredo, P.N. (2001) *Technological* Learning *and* Competitive Performance. Northampton, MA: Edward Elgar Publishing, number 2373. https://ideas.repec.org/b/elg/eebook/2373.html

Figueiredo, P.N. (2017) 'Micro-level technological capability accumulation in developing economies: Insights from the Brazilian sugarcane ethanol industry', *Journal of Cleaner Production*, 167, pp. 416–431. https://doi.org/10.1016/j.jclepro.2017.08.201.

Flüeler, T., Goldblatt, D.L., Minsch, J. and Spreng, D. (Eds) (2012) 'Energy-related challenges', in: *Tackling Long-Term Global Energy Problems*, Dordrecht: Springer, pp. 11–22. https://doi.org/10.1007/978-94-007-2333-7

Freeman, C. and Soete, L. (2009) 'Developing science, technology and innovation indicators: What we can learn from the past', *Research Policy*, 38(4), pp. 583–589. https://ideas.repec.org/a/eee/respol/v38y2009i4p583-589.html

Garcia-Muiña, F. and Navas-López, J. (2007) 'Explaining and measuring success in new business: The effect of technological capabilities on firm results', *Technovation*, 27(1–2), pp. 30–46. https://doi.org/10.1016/j.technovation.2006.04.004

Gill, J., Johnson, P. and Clark. M. (2010) *Research Methods for Managers*. 4th edition. London: Sage Publications, pp. 123–145.

Hanlin, R. and Okemwa, J. (2022) 'Interactive learning and capability-building in critical projects', in *Building Innovation Capabilities for Sustainable Industrialisation: Renewable Electrification in Developing Economies*. New York: Routledge. https://doi.org/10.4324/9781003054665-7

Karjalainen, J. and Byrne, R. (2022) 'Moving forward? Building foundational capabilities in Kenyan and Tanzanian off-grid solar PV firms', in *Building Innovation Capabilities for Sustainable Industrialisation: Renewable Electrification in Developing Economies*. New York: Routledge. https://doi.org/10.4324/9781003054665-9

Kasunic, M. (2005) *Designing an Effective Survey* (No. CMU/SEI-2005-HB-004). Pittsburgh, PA: Carnegie-Mellon University. Software Engineering Inst. https://doi.org/10.1184/R1/6573062.v1

Kingiri, A. and Okemwa, J. (2022) 'Local content and capabilities: Policy processes and stakeholders in Kenya', in *Building Innovation Capabilities for Sustainable Industrialisation: Renewable Electrification in Developing Economies*. New York: Routledge. https://doi.org/10.4324/9781003054665-11

Kirchherr, J. and Urban, F. (2018) 'Technology transfer and cooperation for low carbon energy technology: Analysing 30 years of scholarship and proposing a research agenda', *Energy Policy*, 119, pp. 600–609. https://doi.org/10.1016/j.enpol.2018.05.001

Lall, S. (1992) 'Technological capabilities and industrialization', *World Development*, 20(2), pp. 165–186. https://doi.org/10.1016/0305-750x(92)90097-f.

Lema, R., Hanlin, R., Hansen, U.E. and Nzila, C. (2018) 'Renewable electrification and local capability formation: Linkages and interactive learning', *Energy Policy*, 117, pp. 326–339. https://doi.org/10.1016/j.enpol.2018.02.011.

Lindbom, H., Tehler, H., Eriksson, K. and Aven, T. (2015) 'The capability concept: On how to define and describe capability in relation to risk, vulnerability and resilience', *Reliability Engineering and System Safety*, 135, pp. 45–54. https://doi.org/10.1016/j.ress.2014.11.007

Lundvall, B.Å., Joseph, K.J., Chaminade, C. and Vang, J. (Eds.) (2009) *Handbook of Innovation Systems and Developing Countries: Building Domestic Capabilities in a Global Setting*. Cheltenham, UK: Edward Elgar Publishing.

Momeni, M., Nielsen, S.B. and Kafash, M.H. (2015) 'Determination of innovation capability of organizations: Qualitative meta synthesis and delphi method', in *Proceedings of RESER2015: Innovative Services in the 21st Century*. 10–12 September 2015, Denmark. DTU.

Nzila, C. and Korir, M. (2020) *Capabilities in Renewable Energy Firms and Projects in Kenya*, IREK Working Paper. Available at: http://irekproject.net (Accessed on: 20/05/2020).

Nzila, C., Dewulf, J., Spanjers, H., Tuigong, D., Kiriamiti, H. and Van-Langenhove. H. (2012) 'Multi criteria sustainability assessment of biogas production in Kenya', *Applied Energy*, 93, pp. 496–506. https://doi.org/10.1016/j.apenergy.2011.12.020.

Ockwell, D. and Byrne, R. (2016) *Sustainable Energy for All: Technology, Innovation and Pro-poor Green Transformations*. Abingdon: Routledge. https://doi.org/10.4324/9781315621623

Oshiro, K., Kainuma, M. and Masui, T. (2016) 'Assessing decarbonization pathways and their implications for energy security policies in Japan', *Climate Policy*, 16(sup1), pp. 63–77. https://doi.org/10.1080/14693062.2016.1155042

Panda, H. and Ramanathan, K. (1996) 'Technological capability assessment of a firm in the electricity sector', *Technovation*, 16(10), pp. 561–568. https://doi.org/10.1016/S0166-4972(97)82896-9

Pedersen, M.B. (2016) 'Deconstructing the concept of renewable energy-based mini-grids for rural electrification in East Africa', *Wiley Interdisciplinary Reviews: Energy and Environment*, 5, pp. 570–587. https://doi.org/10.1002/wene.205.

Penrose, E. and Penrose, E.T. (1995) *The Theory of the Growth of the Firm*. New York: Oxford Scholarship Online, Oxford University Press. https://doi.org/10.1093/0198289774.001.0001

Reichert, F.M., Beltrame, R.S., Corso, K.B., Trevisan, M. and Zawislak, P.A. (2011) 'Technological capability's predictor variables', *Journal of Technology Management and Innovation*, 6(1), pp. 14–25. https://doi.org/10.4067/S0718-27242011000100002

Rolffs, P., Ockwell, D. and Byrne, R. (2015) 'Beyond technology and finance: Pay-as-you-go sustainable energy access and theories of social change', *Environment and Planning A*, 47(12), pp. 2609–2627. https://doi.org/10.1177/0308518X15615368.

Taherdoost, H. (2017) 'Determining sample size; how to calculate survey sample size', *International Journal of Economics and Management Systems*, 2, 2017. Available at SSRN: https://ssrn.com/abstract=3224205

Teece, D.J. (2006) 'Reflections on Profiting from Innovation', *Research Policy*, 35(8), pp. 1131–1146. https://doi.org/10.1016/j.respol.2006.09.009

Teece, D.J. (2007) 'Explicating dynamic capabilities: The nature and microfoundations of (sustainable) enterprise performance', *Strategic Management Journal*, 28(13), pp. 1319–1350. https://doi.org/10.1002/smj.640. www.un.org (accessed on 20.06.2020).

Teece, D.J., Pisano, G. and Shuen, A. (1997) 'Dynamic capabilities and strategic management', *Strategic Management Journal*, 18(7), pp. 509–533. https://doi.org/10.1002/(SICI)1097-0266(199708)18:7<509::AID-SMJ882>3.0.CO;2-Z

Ye, L., Zhang, C., Xue, H., Li, J., Lu, P. and Zhao, Y. (2019) 'Study of assessment on capability of wind power accommodation in regional power grids', *Renewable Energy*, 133, pp. 647–662. https://doi.org/10.1016/j.renene.2018.10.042

Zhang, F. and Gallagher, K.S. (2016) 'Innovation and technology transfer through global value chains: Evidence from China's PV industry', *Energy Policy*, 94, pp. 191–203. https://doi.org/10.1016/j.enpol.2016.04.014

7

INTERACTIVE LEARNING AND CAPABILITY-BUILDING IN CRITICAL PROJECTS

Rebecca Hanlin and Josephat Mongare Okemwa

Abstract

This chapter investigates how wind and solar energy parks (large and small) provide opportunities to build capabilities that have the potential to enable more inclusive economic and social development in Kenya. It analyses what types of capabilities are built through renewable energy projects and asks whether the size and shape of a project is important to the opportunities to build capabilities. It finds that size of the project is important but that valuable capabilities are built in both large- and small-scale renewable electrification projects. It finds that across both large and small projects, a specific set of capabilities are important: linkage capabilities or a series of dynamic capabilities at the strategic level relating to project management. The findings raise interesting questions on the way projects are designed and managed and the need for more research on the relative merits of different types of project management contracts. This has implications for the way policy is promoted not just in skills development but also local content rules and the wider issue of export-oriented sustainable industrialisation.

Introduction

The interest in wind and solar as alternative energy sources is predominately related to the need to increase energy access and particularly clean energy access. However, this chapter starts from a different hypothesis. We hypothesise that the promotion of wind and solar on- and off-grid energy projects, in addition to providing opportunities for improved clean energy access, has a valuable role in contributing to the economic growth and development of the country.

African countries are increasingly promoting large infrastructure projects with the hope of boosting economic development (Africa Renewable Energy

DOI: 10.4324/9781003054665-7

Initiative, 2016a). For example, the Kenyan Vision 2030 puts infrastructure among the key pillars that anchor economic progress. Therefore, the strategy puts infrastructure development as a key priority of action (Government of the Republic of Kenya, 2007). The more recent 'Big Four Agenda', which is now driving all government policy initiatives in Kenya, requires energy as an enabler of the four focus areas of growth (manufacturing, food security, universal health coverage, and affordable housing) (Hoka, Njogu, and Obiero, 2018).

There is a large body of evidence that shows a positive relationship between infrastructure development, economic growth, and development (for example, Calderon and Serven, 2010; Démurger, 2001; Khandker, Bakht, and Koolwal, 2009). Despite this, little research exists on the role solar and wind energy infrastructure, specifically, plays on economic growth. This chapter is therefore interested in understanding – through the investigation of various critical case studies – the ways in which renewable energy projects, both on- and off-grid, contribute to economic growth and development. We investigate how projects do this by building different types of capabilities within local firms through the opportunities afforded by collaborations with external and internal actors. These opportunities might be an explicit part of a collaborative agreement between two firms, a publicised commitment, or objective of one of the project partners or related stakeholders, or it could occur more implicitly as an 'added extra' as a result of routine project activities.

With the increased recognition of the importance of local content and local capacity building, such issues have now entered the policy discourse in Kenya. The revised Energy Act (2019) specifies the need for the development of local capabilities to manufacture, install, and maintain renewable technologies (clause 44 (1)(o)). In addition, the Act gives the Energy and Petroleum Regulatory Authority (EPRA) the authority to enforce local content requirements including the use of Kenyan contractors and Kenyan staff where qualified and skilled staff/companies are available. Firms involved in energy production are also expected to submit annual and long-term local content plans.

As such, this chapter asks the question; 'what types of capabilities are built at national level through the design and construction of renewable energy infrastructure projects in Kenya?' It further asks; 'how does the way a project is designed and managed impact on capability-building?' Through asking these questions, and discussing data collected from a review of four renewable energy solar and wind projects in Kenya (two large and two small scale), this chapter identifies areas where capabilities are built that could provide long-term opportunities for improved economic development in Kenya.

Why capability-building through projects?

Projectisation of the Kenyan energy sector

Kenya's attempts to use renewable energy supplies to 'reach the last mile' and improve energy access focuses on a range of different renewable energy strategies, some involving large-scale on-grid projects using solar and wind technology and other off-grid solutions. The development of these usually operates using a 'project' approach; the design, installation, and operation of the renewable technologies are managed by one lead partner and can involve a series of other partners to provide technological or other inputs as needed. This approach is very different to vertically integrated government infrastructure projects for energy electrification.

In fact, the move away from vertically integrated government infrastructure projects has taken place over the last 20 plus years in a range of sectors across the globe. The most often discussed examples are in health and education. Although very well researched in high income countries, the change in focus has been felt in Africa too with the rise of public-private partnerships and contracting out of service provision in various sectors (Liu, Hotchkiss, and Bose, 2008; Farlam, 2005).

The idea of 'project management' as a concept and practice has become very well established in the last 30 years (although the concept and approach started in the 1930s and 1940s with military and process engineering in the United States (Brady and Hobday, 2011)) and is seen by many as an important tool in ensuring projects – of all types – are managed effectively. Brady and Hobday (2011) refer to one project management style of 'project business' which has taken hold since the late 1990s and is focused on the creation of new markets through project activities.

Davies and Brady (2016) identify two types of projects in relation to dynamic capabilities: 'routine projects' and 'innovative projects'. Routine projects are projects that utilise existing and mature products and technologies to satisfy current customer demands. These types of projects require traditional and routine forms of project management capability. Their major strength is exploiting the 'economies of repetition'. Innovative projects, on the other hand, aim at identifying and experimenting with new ideas and approaches that create entirely new market segments, technologies, products, services, and approaches. These are highly risky and unpredictable endeavours being difficult to plan. Such projects require novel ways of organisational planning and a complete shift from existing and prior project routines and capabilities both at the strategic and operational levels (Davies and Brady, 2016).

Key to all is capabilities at project level

To study how project business successfully builds local capabilities places a focus not just on the ability to conduct routine firm related technology and business management but also focuses on a firm's ability to manage relationships with other firms involved in the project (either in their role as project manager or member of the project team). This requires not just firm level capabilities-building (firm and individual) but linkage level capabilities (Lall, 1994).

Bell (2009) sums up his understanding of the capabilities needed into two types of technological capabilities; first, that which is required for ongoing operations with existing *forms* of technology already in use (the routine technology and business management routines mentioned above). These capabilities are referred to in the literature as 'production capabilities'. The second type of capabilities referred to by Bell is that of 'innovation capabilities' or the ability to recognise the need for different forms of technology that are not currently in use in the firm.

In this context, the term 'technology' is not just referring to physical products but also to a range of different forms of knowledge. Archibugi and Coco (2005) group technological capabilities into three different sets of contrasts:

1. Either embodied physical technologies or disembodied technologies or knowledge.
2. Either codified or tacit in nature with regard to the degree to which the product or knowledge can be easily understood through written instructions, plans, or diagrams.
3. About generation and/or about diffusion of products and knowledge. Not all firms will generate new or modified technologies or knowledge but will gain new capabilities through the use of technologies and knowledge developed by others.

Others (see Hansen and Ockwell, 2014) focus on a continuum of technological capabilities from production through to innovation and focus on the way firms change from being able to conduct capabilities that are new to the firm, through to those that are new to the market and eventually to those that are new to the world i.e., ever increasing levels of innovativeness.

Technological capabilities and competitiveness

The development of technological capabilities provides firms with means to become increasingly competitive. There are two parts to this argument. First, that export-oriented firms are more likely to develop more relevant technological capabilities than those that are focused on a national market. Lall (1994) argues that inward-oriented firms learn to 'make do' or 'stretch' available resources while those that are outward looking try to reduce costs, raise quality, and introduce new knowledge.

Interestingly, a set of literature that has become prominent in the innovation and development field more recently argues that innovation in scarcity conditions is a positive attribute (regardless of a firm's outward or inward looking orientation), because it usually ensures that only those innovations and related capabilities are built that are needed by society (Srinivas and Sutz, 2006).

The second set of arguments here relate to the global value chains literature and are linked to the arguments of Lall above. The overarching argument of this literature is that developing country firms would benefit from insertion into global value chains and global production networks. This literature focuses on two main areas of thinking that are relevant here: (i) upgrading within the firm (process, product, or function) or (ii) linkage creation and capabilities-building/ upgrading along the chain (Haakonsson, 2009). We are interested in both but assume the first as a result of the second. Therefore, our primary interest is the second focus area. In addition, the focus of global value chain literature is predominately on upgrading in firms or linkages between firms because of exporting activity, but it is actually possible to learn from importing (Haakonsson, 2009). It is learning through importing that we are interested in.

Methodology

We focused specifically on several indicators for the assessment of capabilities and upgrading (see Table 7.1). We chose these based on a literature review undertaken (see Hanlin, Okemwa, and Gregersen, 2020).

TABLE 7.1 Indicators to be used to assess capabilities and upgrading

Inputs and outcomes	*Capabilities and upgrading*	*Indicators*
Micro-level inputs	Individual skills	1. Government minimum standards 2. Additional 'on the job' skills identified 3. Training opportunities
Meso-level inputs	Technological capabilities	4. New physical technologies (e.g., new piece of testing equipment) introduced into the firm that results in new business opportunities at any stage in the project cycle 5. New knowledge introduced into the firm that results in new business opportunity at any stage in the project cycle (e.g., recruitment of a staff member with EPC experience or training of existing staff in how to install a specific new invertor design)
	Core competences	6. Function as an EPC contractor 7. Evidence of ability to leverage new partnerships on the back of previous work
Outcomes	Upgrading	8. Process upgrading (e.g., increased efficiency of installation process – speed/ manpower requirement) 9. Product upgrading (e.g., from using Chinese to German inverters) 10. Functional upgrading (e.g., move from being a contractor to doing full EPC) 11. Chain upgrading (e.g., move from installing solar heaters to installing mini-grid systems)

Source: authors

We chose to study these indicators through a few case studies of renewable electrification projects in Kenya. These cases were chosen based on a review of the answers given to a question 'what have been the most important energy projects for Kenya in the last ten years?' during a survey conducted as part of the IREK project (Andersen et al., 2017). This survey found that Lake Turkana Wind Power (LTWP) Project was the most frequently cited large-scale project. For small projects, the survey identified the most frequently named project as Kitonyoni solar project (13.5 Kwp). In addition, we decided to choose a small number of mixed projects in terms of characteristics or what are sometimes referred to as 'maximum variation cases' (Flyvbjerg, 2006). As a result, we chose a selection of projects that differed in terms of the origin of the equipment, the project size, focus energy source, and grid connectivity. Further differences included: different sized lead firms, different organisational project set-ups, different investors, and different locations. Our final four case studies in Kenya were: LTWP, Kitonyoni solar project, Garissa solar park, and SOS Children's Village solar park in Mombasa.

While ideally we would have liked to have included a large-scale solar off-grid and a small-scale wind on-grid project, unfortunately we found it difficult to identify suitable project candidates in these two categories. The difficulties of finding small-scale wind projects in Kenya have been noted elsewhere (Wandera, this volume).

Research methods

We employed qualitative research methods to study these four projects. The methods used a combination of desk review of materials, observation, and interviews. While colleagues (Nzila and Korir, this volume) have utilised surveys to identify types of capabilities built in projects; qualitative methods provide us with the opportunity to explore the 'why' questions relating to capability-building. We recognise that, unlike with a survey approach, we are unable to argue that the findings are necessarily anything other than unique for the projects we have studied. We believe they provide important insights into how capabilities through project business are created for those conducting project activities in these areas and for those designing policies to encourage local content development. Full details of the research methods and data analysis techniques used are provided in Hanlin, Okemwa, and Gregersen (2020).

The projects introduced

The first large-scale project that we studied was LTWP project. The construction of the project was started in October 2014, and it was commissioned in November 2018. It is a 310-megawatt wind power park. The project is owned by a Kenyan registered company, the Lake Turkana Wind Power Ltd, with financing from a range of sources (local and international) and has a 30-year power purchasing agreement with the Kenyan electrical utility company. Engineering, procurement, and construction was managed through an Engineering Procurement and Construction Management (EPCM)[1] contract given to a South African branch of a global engineering project management firm, Worley Parsons. A range of different firms were then contracted for various project elements. Many of these firms were from Kenya or East African countries but during construction of the wind turbines two external firms (one Danish, Vestas, and one Greek, Anopsitiki) conducted most of the work. That said, Anopsitiki did work with a local Kenyan firm (SECO Engineering) using their equipment and personnel to assist them in erecting the turbine towers, and Vestas has trained up local engineers to conduct maintenance on the turbines during the operational phase of the project (despite having originally expected to need to use expatriate staff to operate and maintain the equipment). Other local or regional firms were contracted to conduct the groundwork and camp building (SECO), road construction (Civicon), and wind turbine plinth construction (Entreprise Générale Malta Forrest). The wind turbine technology was Danish with production done in the Vestas' manufacturing facilities in China. Other major equipment and technology was also sourced externally although locally produced consumables were utilised during the project construction period.

The second large-scale project that we studied was the Garissa solar park. It is a 55-megawatt park and was the result of an agreement with the government of Kenya and China signed in 2013. The construction of the project was started in January 2017 and the park was commissioned in October 2018. Rural Electrification and Renewable Energy Corporation (REREC), formerly known as Rural Electrification Agency (REA) have a 25-year power purchase agreement with Kenya Power at 5 KShs/KWh. REREC have contracted the Kenyan national electricity provider, KenGen, to operate the plant. Project financing was through China's EXIM bank. China Jiangxi Corporation for International Economic and Technical Cooperation Company (CJIC) were given an EPC (engineering, procurement, construction contract) to build the plant but did utilise a local firm (Maknes Consulting) as the local site agent and CJIC have trained up staff of REREC to operate the facility while still providing technical support for the first two years. Very few other Kenyan firms were involved other than a transporter (Landmark Port Conveyers) contracted to bring panels and stands to site. All equipment used was Chinese in origin other than some locally purchased consumables.

The first small-scale project studied was SOS Children's Village in Mombasa. The project idea was conceived in 2010, construction started in 2011, and the site was renovated in 2018. It consists of a 60-kilowatt solar park for a single organisation, but which was originally designed to be the first 'ready-for-net metering' project (although this has never materialised). The project was funded with German development partner assistance and involved a German

EPC contractor as the lead together with German equipment. However, the German EPC (Asantys) entered a joint venture with a local company (African Solar Designs or ASD) whose role at the SOS project was to audit and specify the system. More recently another Kenyan EPC contractor (Knight Energy and Apps Ltd) was awarded the renovation contract. Specifically, Knight Energy and Apps were involved in customer relations, planning, resource mobilisation, commissioning of the project, post installation, and maintenance of the project.

The last project studied was Kitonyoni solar plant. This project was conceived by Prof. Abubakr Bahaj, a lecturer at Southampton University, UK. The owner of the system is the Makueni County Solar Supply Co-operative Society. The project was conceived in 2012, construction started in 2012, and the park was commissioned in 2012. It is often used as a showcase project for small-scale off grid solar because of the length of time it has been running. It is a 13.5-kilowatt solar park that provides power for the local community. While it has strong local community ownership during the operation phase, the project was designed, engineered, and constructed predominately by UK engineers using a containerised 'ready-to-go' system that was put together in the UK, shipped out, and placed on site with minimal need for local construction support although a Kenyan battery supplier acted as a local partner and provided batteries for the system. The equipment used is from a range of places including the Netherlands (solar panels, charge controller, inverters); UK companies developed the mobile based payment system and provided the steel for the panel frames. As noted above, a Kenyan firm (Chloride Exide) provided locally manufactured batteries. Chloride Exide also sub-contracted another local firm (Gilgil Electrical) to work on the power distribution set up (cabling from solar park to houses).

It should be noted that while we took these projects as case studies through which to study the existence of capabilities, the analysis in the results section mostly focused on the firms involved in the projects. This is because the capabilities reside in the staff and resources of the firm and not in the projects themselves. The projects are the vehicles through which experience and capabilities are developed. As such, a firm can engage in a series of activities to build its capabilities as a result of sequential learning during the project cycle.

Research results

Overview of skills and capabilities built

An overview of the fieldwork results is presented in Table 7.2 which outlines benefits gained by each firm that was interviewed. It should be noted that the evidence of upgrading is not specifically related in all cases to the projects reviewed but as a result of general time spent in the industry.

TABLE 7.2 Overview of skills and technological capability-building in Kenyan firms through project activities

Kenyan firm	*Increased skills through informal in-house knowledge transfer*	*Increased skills through formal training (in-house or externally provided)*	*Increased firm capability through introduction of new physical technology*	*Increased firm capability through introduction of new staff members*	*Evidence of leveraging new partnerships to bring in skills/ knowledge/ physical technologies*	*Evidence of upgrading (process, product, functional, or chain)*
Firms involved in small-scale projects						
Knight Energy and Apps	X	X	X		X	X
Chloride Exide	X	X			X	
Gilgil Electricals	X		X		X	
ASD	X			X	X	X
Makueni Coop Energy Society	X	X				
Firms involved in large-scale projects						
KENGEN		X				
Maknes						
Bollore	X	X	X	X	X	
SECO	X	X	X	X	X	X
LTWP	X	X		X	X	

Source: authors. NB: Maknes was found not to have increased skills or capabilities during our analysis of interview data.

Skills

Due to Kenyan EPRA regulations the solar engineering firms all had qualified staff (always at least one per project) who had been certified to at least the minimum national standards. The same was true more generally of electrical and construction engineers who were involved in the case study projects. All of the firms interviewed and involved in the case study project sites had given training opportunities to their staff – some directly as a result of the need for new skills required for the project e.g., training to install a new inverter type or wind blade maintenance training. This was provided through in-house workshops (often where one member of staff would provide peer-to-peer feedback and training) or through formal off-site training. At times this was provided by equipment suppliers. A number of the projects provided opportunities for staff to visit other sites or the equipment manufacturers outside of Kenya. In this way they were provided with opportunities to see and learn about other contexts. That said, the majority of technicians interviewed during the fieldwork highlighted the importance of 'on the job' training which was often not formal in nature but the result of 'seeing others work' and 'sharing experiences'.

Capabilities

During fieldwork we found evidence of new physical technologies (new testing equipment for example) being introduced into firms which had resulted in new business opportunities at different stages of the project cycle. This included new IT based monitoring equipment for solar power systems and new IT data log management systems to specialist trucks and trailers. The use of imported equipment was widespread and often the choice of country of origin of the technology was due to the origin of the project finance or the links that the lead contractor(s) had already established with equipment suppliers. For the most part the use of Kenyan technology was confined to the use of consumables during project construction.

In addition to new physical technologies, firms involved in the projects studied needed to bring new knowledge into the firm. This new knowledge was bought in to exploit and develop new business opportunities at a relevant stage in the project cycle (e.g., recruitment of a staff member with EPC experience or training of existing staff in how to install a specific new invertor design). New knowledge has been a key element of the strategies of all the firms interviewed who were involved in the case studies we looked at. Joint ventures or partnerships were a key way in which new knowledge (especially from outside Kenya) was bought into a project. The new knowledge embodied in staff does not always stay at the same firm. In the case of the large wind turbine projects, we have seen a lot of staff movement across firms with project managers moving from one firm to another as projects finish and another one starts. As such, while their old firm might still work on a project, the knowledge that they have embodied within them is technically no longer available to their old firm.

Competences

Competence building is the ability to 'coordinate diverse production skills and integrate multiple steams of technologies'. It is essentially the skill of knowing what knowledge and technologies are needed and how to integrate them. As such, in this study we looked for evidence of firms functioning as an EPC (engineering, procurement, and construction) contractor, i.e., being responsible for all elements of the project cycle until handover for operation. This is because the role of an EPC is to manage multiple stages of the project cycle in house. We also looked for evidence of the ability of firms to leverage new partnerships through their activities in these projects in order to conduct more elements of the project cycle.

We find that there is evidence of firms moving into EPC contracting while we investigated our case studies; although not specifically as a result of our project case studies – although in some cases the projects did give them added experience which they will leverage on as they built their name as EPC contractors. In some cases we find that there is what might be termed 'nested EPC' whereby one company is given an overarching EPCM contract to oversee the whole project and others are brought in to conduct different elements of the project using an EPC approach. This was true on the large-scale wind project. One of the results is a huge and complex array of partnerships built up between firms across these projects. Partnerships (some informal and others more formalised, even to the point of becoming joint ventures) are entered into so that foreign companies can get access to local knowledge and equipment/human resource already on ground. In return, local Kenyan companies receive training, skills, and can leverage further projects from the exposure they get through these partnerships. These partnerships are more complex the larger the project. The result is multiple layers of contracts and sub-contracts with multiple interactions. For example, one company is hired to do earthworks, construction, or catering by one project partner and then is asked to do the same for another project partner.

The final area where we were looking for evidence of competence building was in the area of upgrading. Upgrading strategies not only require acquisition of competences but also changing relationships with buyers in the market. Upgrading offers many opportunities to firms including increased efficiency and output in accessing new market networks as well as industrial knowledge. We found little evidence of process upgrading (e.g., increased efficiency of the installation process for example in terms of the speed or manpower requirement of an install) by local firms involved in the four case study projects. That said, two firms mentioned the introduction of new project/time management mechanisms through working on the large-scale wind project. The story of product upgrading (e.g., from using Chinese to German inverters or invertors that are deemed to be of higher quality) is more complex as we find that equipment choice was highly personal to the firms and depended on experience with a product and/or general sector optimism for a new product. We found some evidence – as noted above – of functional upgrading (e.g., where a firm moves from being a contractor to doing full EPC) and a small amount of evidence of chain upgrading. Again, as noted above, these last two types of upgrading are not the result of the projects themselves but a longer-term shift in company trends. Examples included one company moving from installing solar heaters to installing mini-grid systems and another company from being a mainstream automotive battery manufacturer to an active participant in solar energy mini-grid projects. We find that this takes place mostly as a result of engagement in smaller scale projects, i.e., we find very little evidence of upgrading in the two large-scale projects where companies involved are essentially doing 'business as usual'; self-reporting as such during interviews.

The importance of project objectives

As can be seen from the case study overviews, despite Kenyan ownership of the power generation plants in all four case studies, we did not see a leadership role taken by Kenyan firms at the initial stages of project design and engineering. This is the situation in the case studies with the partial exception of ASD who were partners with the German firm who were the EPC for the SOS Children's Village initial design and install. There is more Kenyan firm participation at the installation/construction phase and even more still during the operation and maintenance phase. In large projects there are service contracts involving foreign firms in both case studies.

That said, the use of Kenyan technicians and labourers is higher than was originally expected when we started this study (and higher than expected by those involved in the case studies i.e., in the case of Vestas' expectations of needing to utilise foreign engineers). In fact, we feel we could go as far as argue that the issue is not so much the technology type. This is because the level of basic skills in the engineering field is sufficient in Kenya. Instead, the issue is knowledge of the finer points of specific pieces of technology.

Thus, most capabilities-building has taken place in the installation and operations phases of the projects. Kenyan engineers have received specific training – both in Kenya and abroad – on how to utilise specific 'pieces of kit' e.g., SMA inverter systems and/or the maintenance of such kit e.g., Global Wind Organization blade repair training (especially how to work at height). Those trained in these firms have fed this training back to colleagues through in-house firm level training where appropriate. In some cases, firms have introduced new pieces of kit which enhances their technological capabilities to offer enhanced services.

As outlined in Table 7.2, while the majority of capabilities built relate to skills training, a significant level of capabilities-building has taken place across the firms interviewed in the area of leveraging new partnerships to bring in skills, knowledge, and/or physical technologies. As such, a key competence that firms involved in all the case studies are gaining more experience in is around partnership relations and, in the case of EPC or EPCM and/or hands-on owners of the plants, how to manage multiple partners. We come back to this below.

Discussion

Does size and shape matter?

Hansen et al. (this volume) found that the degree to which a project is large scale or small scale in size matters more than the shape of the project in terms of the type of technology involved in the renewable electrification project (e.g., wind vs. solar technologies). They argue that large-scale projects have a focus on EPC and turnkey solutions, predominately, by foreign firms. They did not find the same in firms working on small-scale projects. Our findings echo this but also dispute this finding somewhat. Specifically, we see that small-scale projects do provide opportunity for EPC experience for local firms and, over time, dominance in this area by local firms. However, at the same time, we find that many local firms in the LTWP manage EPC contracts but do so in a nested format, i.e., no local firm is responsible for overall project management (EPCM).

Project types, specifically the focus on EPC vs. EPCM

One of the interesting observations from this analysis of the four case studies has been the increasing opportunity for EPC roles by Kenyan firms in small solar PV plant projects. We have noted in an allied paper (Hanlin, Okemwa, and Gregersen, 2020) that this change has occurred in a relatively short time (one respondent during our study argued it had occurred in the last three years only, i.e., since 2016). It highlights a significant change from 2001 when Murphy, in a well-cited paper (Murphy, 2001), stated that the low technological capability of project managers led to the abandonment of many solar PV systems.

At the same time, we note that in the large-scale projects different models were evident. The Garissa project involved a Chinese-led EPC style of project management while the LTWP involved an adapted EPCM style of project management. In all four cases reviewed in this chapter, foreign EPC firms have played a central role either across the whole project cycle or for a significant part of the project cycle. This mirrors the finding of Hansen et al. (this volume).

Vidican (2012) noted in the case of Egyptian solar PV plant projects that having a local EPC facilitated more local company involvement in the solar PV project value chain because there was more chance that local sub-contractors would be utilised. Morris, Kaplinsky, and Kaplan (2012) also noted the same with regard to resource extraction projects in Africa, finding that where there was a local lead firm, the projects are more embedded into the local economy and are more committed to local development.

While we do not see any cases during the initial design/engineer and install/construction phases where there is a significant role played by local firms in any of the four cases, it must be mentioned that the LTWP did have (and maintains for operational purposes) a site and project owner who were hands-on throughout the project cycle. This owner identifies as being Kenyan and the LTWP consortium is incorporated as a Kenyan company. At the same time, this project – as opposed to our other large-scale renewable electrification project example of Garissa solar park – also involved a significant number of sub-contractors; many of which were Kenyan companies or sub-contracted Kenyan companies. The LTWP, as noted above, had a sizeable local content element to the project in terms of firm involvement.

We would argue that a key enabler for this was the type of contract that was given to the EPC firm, even though they were a foreign firm. Specifically, Worley Parsons were given an EPCM (engineer, procurement, and construction *management*) contract and not the usual EPC contract. The rise of the EPCM has come about as a result of the difficulty of finding a single supplier who can provide an EPC role (including bank-rolling this activity) for large-scale projects – which have a high degree of risk and uncertainty (Loots and Henchie, 2007).

Therefore we would argue that a key enabler was not just having a local hands-on owner of the project, but also that the projects specifically focused on local content clauses which included a desire to see – whether implicitly or explicitly – the use of local firms through a decision to hire a construction management firm and not one turnkey firm to conduct the whole project cycle in-house. We have not investigated the reasons behind this in any depth during this study. It has been found elsewhere (although not in a large study) that utilising an EPCM contract has advantages over EPC forms of contract strategy in terms of enhancing the potential involvement of local suppliers in the project process (Awuzie and McDermott, 2016). It is therefore an area for further research.

This chapter took as its starting point a set of literature on project management and the role of project management as a means of enhancing work organisation and innovative activity. However, our case studies have highlighted that the question to ask isn't just over the type of project management (outlined by Brady and Hobday (2011) as matrix, functional, adhocracies, Scandinavian, or project business) but also the contractual relations involved within these project management approaches.

In investigating this issue, there appears to be a set of work around project contract types and the creation of social value through infrastructure projects (c.f. Awuzie and McDermott, 2016). Further research would be useful to interrogate the relationship between this literature and that of innovation, project management, and capabilities-building.

In addition, Loots and Henchie (2007) in their definition of the difference between an EPC and an EPCM contract note that an EPCM contract is a 'professional service contract' because the provider is not the principal. As such it is not responsible for the actual construction of the solar PV or wind generation plant.

Vidican (2012) for example notes that in Egypt the solar PV plants she investigated required project execution capabilities as well as production capabilities and innovation capabilities. Similarly, Figueiredo and Piana (2018) specifically focus on 'knowledge intensive service' enterprises in the mining sector (e.g., geotechnical engineering firms or the environmental services firms that have a strong research and development unit). Davies and Brady (2016) therefore talk of project capabilities at the operational level and dynamic capabilities at the strategic level. Thus, recognising the importance of different types of capabilities for different elements of project management (phases and types of project management contract) appears to be more important than first expected.

There are two major implications here. Firstly, more consideration is perhaps needed as to whether certain types of project management contract should be promoted to enhance local content/involvement of more local firms. Secondly, the importance of project management and experience of EPCM project management is needed if Kenya is to stop having to rely on foreign firms to manage this activity, especially for large-scale projects.

Another dimension of shape of technology: origin of technology

One of our starting premises for the IREK project as a whole (i.e., not just this chapter) was that both large-scale projects that we studied used foreign technology; one predominately Chinese technology and the other Danish technology. For the most part this turned out to be correct. Similarly, the small-scale projects we studied we assumed would be dominated by European technology. However, we found a more complex picture in the smaller projects. Table 7.3 outlines the differences.

Despite the finding that a mix of technologies was in place, the overwhelming response received during interviews (and borne out by the fact that the projects are being implemented predominately by local Kenyan technicians and not those from the country of origin of the equipment) is that, in the case of renewable energy technologies in Kenya, the origin of the equipment isn't an issue. The

TABLE 7.3 Technology types

	Starting premise	*What we found*
Garissa	Chinese technology	Only Chinese technology
LTWP	Danish technology	Danish technology but also Chinese technology and local consumables
SOS Children's Village	German technology	German technology but increasing use of Chinese technology following renovation
Kitonyoni	EU technology	A combination of UK, Dutch, and Kenyan technology

Source: authors

reason given was the level of engineering training available in Kenya. It was considered the case that, once you learnt the underlying principles of how the technology worked, where the technology came from was not a problem. The only time this was described as an issue (and one at both the large and small scale) was when the instructions were not available in English. However, in all cases, training by the vendors of the technology was available and was given (at some point in the project or as part of general in-house capacity building by firms).

Firms in the mini-grid sector appear to prefer certain types of equipment and align with manufacturers of that equipment to receive training. For example, in the case of inverters we found a preference for European as opposed to Chinese inverters. This is little different to other sectors such as car mechanics or farming where a mechanic or farmer tends to buy a particular make of tyre or fertiliser to use either due to the cost or the customer service received from the supplier. In the case of these renewable energy firms, the reason given was the quality of the equipment.

Project size and potential for capability-building

The capability-building story has turned out to be much more complicated than expected. Our original proposition was that 'small is beautiful', i.e., that small-scale projects are more likely to be developed by indigenous owned and run firms. We also expected firms involved in the design and construction of the projects would benefit substantially in terms of skills gained and then utilise those skills in other jobs. This has not run true in all cases. The Kitonyoni project is similar to a number of other what might be called 'containerised projects' whereby stakeholders design the project, and the kit is installed with minimal local staff content. An internet search and literature review found: two more projects like this also developed by the same UK university; four containerised solutions produced and sold through an energy supplies firm in Nairobi; and at least two other containerised solutions introduced to two communities in Turkana region by a Kenyan company. It is unclear how much local[2] input goes into these containerised solutions. This differs, however, from SOS Children's Village which fits our proposition and where local firms have benefited (including through upgrading) as a result of their involvement in small-scale mini-grid renewable electrification projects. However, nuances arrive when you start looking at employment figures and longer-term employment opportunities.

Based on employment figures, large-scale projects we have investigated provide – on paper – the highest opportunities for capabilities-building. For example, we were informed by interviewees from LTWP and its contractors that there was a workforce ranging from 180 to 400 people involved in the construction of the project (depending on how many sub-contractor staff are included in the calculations). The Garissa project was also found to employ around 300 people during the main construction period. As one of the EPC contractors on the LTWP project noted, many of the day labourers left the project with skills they didn't have before (such as carpentry or masonry skills) and which they will take with them when they leave. That said, these jobs are time bound; once the project is completed, many of these workers do not move with the contractors to the next job; especially the day labourers. Unfortunately, due to the timing of this study, whether or not they have been able to use the skills acquired in different projects or contexts has not been investigated here.

When we consider only engineers and related job positions in contractor companies working on the large sites, we find that the numbers required were much less (around 20 – mostly from Kenya; especially in the case of the Lake Turkana Wind Project). In small-scale mini-grid projects the numbers of engineers involved was less than ten in each project. In both large- and small-scale projects, engineers (as opposed to less skilled workers) were also more likely to be retained by the company after the project. We also find that engineering capability is highly mobile – with staff moving from project to project and company to company on a regular basis.

The larger projects have a high level of staff movement and a focus of project management on ensuring that sub-contractors with the knowledge required are brought into complete specific tasks within limited timeframes. This restricts the interaction with other actors; interaction is rather narrowly focused on issues of timing and avoiding budget overruns and delays. Different organisations may focus internally on building capabilities within their teams to complete their tasks rather than on more interactive types of knowledge transfer or capability-building. Future research may explore the implications of individual vs. wider organisational capability-building. At the broader macro level, questions may also be raised as to whether and how these capabilities may be diffused to future wind and solar projects. Will it be carried on through individuals, firms, or other types of organisations? Could a wider intra-organisational interaction be encouraged to ensure a wider transfer of knowledge within the innovation system for renewable energy technologies in Kenya?[3]

What does this mean for capability-building and subsequently economic development?

The four case studies highlight strong evidence of learning from partnerships and interactions with other firms, especially those from outside the country. As noted earlier, these case studies highlight evidence of 'learning from importing' (Haakonsson, 2009; Blalock and Veloso, 2007). These studies focus predominately on the importation of technologies (defined as physical products). These are deemed to create opportunity for new knowledge creation, result in improved productivity or reduced cost of domestic production, or change the way goods are made/what goods are made in country (Foster-McGregor and Stehrer, 2013). New knowledge creation can be within the firm receiving the technology, about how the technology works, is maintained, and how it can be utilised to improve productivity, reduce costs, or produce a new or improved good (Foster-McGregor and Stehrer, 2013). Alternatively, it can be arms-length knowledge spill-overs to peers in business, who see how their competitor is utilising a new piece of equipment (Bisztray, Koren, and Szeidl, 2018).

This links to a broader set of literature on how (a) linkage building is a tool for economic growth (Hirschman, 1981) and (b) technology transfer is beneficial to firm performance and therefore economic performance of countries – what is known as 'technology gap theory' (Fernandez and Gavilanes, 2017; Coe and Helpman, 1995). An increasing focus on developing countries achieving economic growth through innovation is based on the idea that they will innovate new technologies (Zanello et al., 2016). However, fundamental innovation of radical new technologies is difficult for developing countries because of a lack of absorptive capacity (skills and knowledge) plus often social, economic, and political conditions that are not conducive, which means institutional arrangements are insufficient. The result is that 'external sources of technology account for a large component of productivity growth in most developing countries' (Zanello et al., 2016, p. 2). Therefore, what is put forward as being needed is to tap into 'existing knowledge and know-how from foreign countries' or to 'facilitate the exchange of both external and local knowledge within a country' (Zanello et al., 2016).

The learning from importing that we are seeing in these case studies is not simply related to what is available through a 'piece of kit' – what Haakonsson and Slepniov (2018) have discussed as 'learning by suppliers' through 'technology transmission'. It is more congruent with a broader definition of technology as outlined above – as embodied in a physical entity but also defined as skills, knowledge, and know-how. It also highlights the importance – in the context of the discussions of the type of technology and knowledge that is important for economic growth – of these linkages and interactions that are playing out on the ground in these case studies.

In fact, it might be that these case studies are not so much about learning from importing as much as about the importance of learning through interacting. It reiterates the importance from a policy perspective of encouraging linkages between diverse groups of actors in a supply chain and providing appropriate incentives. Haakonsson (2009) highlighted the importance of joint ventures and foreign investors' role in promoting upgrading through upstream linkages with suppliers in the Ugandan pharmaceutical industry. Haakonsson and Slepniov (2018) note the importance of local content and import substitution policies in China on technology upgrading in the Chinese wind turbine industry. Hanlin and Hanlin (2012) highlight the importance of 'facilitatory policy' such as local content rules relating to procurement of goods (not just employment of local staff) in the mining industry in Tanzania and the Democratic Republic of Congo. Our four case studies are in Kenya, and Kenya has different contextual situations in which businesses work compared to Uganda, Tanzania, Democratic Republic of Congo, and China, however, the importance of linkages promotion by government policy cannot be downplayed.

Conclusion

The case studies we have examined in this chapter highlight the importance of a set of 'base skills' and educational standards which allows more opportunity for Kenyan firms to benefit and build capabilities and competencies. It also highlights the important role of linkages – whether as formal contractual partnerships or more informal exchanges or linkages. Providing the right policy on capabilities-building is crucial for the promotion of a continued set of base skills and educational standards as well as promotion of linkage opportunities.

The case studies reveal that a firm can engage in a series of activities to build its capabilities; it can engage in sequential learning through a project cycle. The case studies also highlight that what is important, from a policy perspective, is not so much the learning that can occur from importing technology but the importance of learning through interacting with project partners, including those from outside the country. This has important implications for policy on the export-oriented sustainable industrialisation path which a country may want to take.

Specifically, the findings of this study of four different sized and technology focused projects highlight that a range of capabilities can – and should – be built. Perhaps the most important are those relating to project management – the dynamic capabilities at the strategic level as noted by Davies and Brady (2016). In particular, the study highlights the importance that needs to be given to researching and understanding the difference between EPC and EPCM contract types and the skills and capabilities needed to manage these. This dovetails the results of our first research question on the type of capabilities that are important with our second research question on how the design and management of a project matters. Design and management of a project matter, and therefore firms must have the capabilities at this level and not just ordinary capabilities at the project operation level.

Acknowledgements

Support for research on which this chapter is based from the Danish Ministry of Foreign Affairs, Grant: DFC 14-09AAU is gratefully acknowledged. We also acknowledge all the interviewees and reviewers of this chapter during IREK project meetings.

Notes

1 An EPCM contract follows the EPC contract style but differs in that the same firm is also in charge of project management.

2 We note – as Gregersen (2020) does elsewhere – that there are different definitions of 'local' i.e., local to the surrounding area and local as in within national boundaries plus a series of options in between. The lack of information about who is providing what and from where (a general finding throughout the IREK project under which

this study was undertaken) is a key hindrance in developing a clear understanding of what capabilities are being built and by whom.

3 We are grateful to discussions with a co-author, Cecilia Gregersen, of a working paper (Hanlin et al., 2020) on which this chapter is based, for helping us think through the development of these questions for future research.

References

Africa Renewable Energy Initiative (2016) *AREI Action Plan.* [Online]. Available at: http://www.arei.org/wp-content/uploads/2018/03/AREI-Action-Plan-Nov-2016.pdf. (Accessed: 30/08/2020).

Andersen, M.H., Lema, R., Hanlin, R., Tigabu, A. and Kingiri, A. (2017) *Collaboration and Capabilities in Kenya's Wind and Solar Industries: Perception of Stakeholders.* IREK Report number 3 [Online]. Available at: https://www.irekproject.net/portfolio/report3/ (Accessed: 11/06/ 2020).

Archibugi, D. and Coco, A. (2005) 'Measuring technological capabilities at the country level: A survey and a menu for choice' *Research Policy*, 34, pp. 175–194. https://doi.org/10.1016/j.respol.2004.12.002

Awuzie, B.O. and McDermott, P. (2016) 'The role of contracting strategies in social value implementation', *Proceedings of the ICE-Management, Procurement and Law*, 169(3), pp. 106–114.

Bell, M. (2009) *Innovation Capabilities and Directions of Development*, STEPS working paper 33. Brighton: STEPS Centre.

Bisztray, M., Koren, M. and Szeidl, A. (2018) 'Learning to import from your peers', *Journal of International Economics*, 115, pp. 242–258. https://doi.org/10.1016/j.jinteco.2018.09.010

Blalock, G. and Veloso, F.M. (2007) 'Imports, productivity growth, and supply chain learning', *World Development*, 35(7), pp. 1134–1151. https://doi.org/10.1016/j.worlddev.2006.10.009

Brady, T. and Hobday, M. (2011) 'Projects and innovation: Innovation and projects' in: Morris, P.W., Pinto, J.K. and Söderlund, J. (eds.) (2011) *The Oxford Handbook of Project Management*. Oxford: Oxford University Press.

Calderon, C. and Serven, L. (2010) 'Infrastructure and economic development in Sub-Saharan Africa', *Journal of African Economies*, 19, pp. i13–i87. https://doi.org/10.1093/jae/ejp022

Coe, D.T. and Helpman, E. (1995) 'International R&D spillovers', *European Economic Review*, 39(5), pp. 829–859. https://doi.org/10.1016/0014-2921(94)00100-E

Davies, A. and Brady, T. (2016) 'Explicating the dynamics of project capabilities', *International Journal of Project Management*, 34(2), pp. 314–327. https://doi.org/10.1016/j.ijproman.2015.04.006

Démurger, S. (2001) 'Infrastructure development and economic growth: An explanation for regional disparities in China?', *Journal of Comparative Economics* 29, pp. 95–117. https://doi.org/10.1006/jcec.2000.1693

Farlam, P. (2005) 'Working together: Assessing public–private partnerships in Africa. The South African Institute of International Affairs'. [Online]. Available at: https://www.oecd.org/investment/investmentfordevelopment/34867724.pdf (Accessed: 05/06/2020).

Fernández, J. and Gavilanes, J.C. (2017) 'Learning-by-importing in emerging innovation systems: evidence from Ecuador', *The Journal of International Trade &*

Economic Development, 26(1), pp. 45–64. https://doi.org/10.1080/09638199.2016.1205121
Figueiredo, P.N. and Piana, J. (2018) 'Innovative capability building and learning linkages in knowledge-intensive service SMEs in Brazil's mining industry', *Resources Policy*, 58, pp. 21–33. https://doi.org/10.1016/j.resourpol.2017.10.012
Flyvbjerg, B. (2006) 'Five misunderstandings about case-study research', *Qualitative inquiry*, 12(2), pp. 219–245. https://doi.org/10.1177/1077800405284363
Foster-McGregor, N., and Stehrer, R. (2013) 'Value added content of trade: A comprehensive approach', *Economics Letters*, 120(2), pp. 354–357. https://doi.org/10.1016/j.econlet.2013.05.003
Government of the Republic of Kenya (2007) *Kenya Vision 2030*. Nairobi: Government of Kenya.
Gregersen, C. (2020) 'Dynamic sublimes as drivers of a Kenyan renewable energy megaproject' in: Paper presented at the DRUID Academy Conference 2019 at Aalborg University, Denmark, January 16-18, 2019. [Online]. Available at: https://conference.druid.dk/acc_papers/17gmx4987wjkyqke4epbpvs421fk0h.pdf (Accessed: 06/06/2020).
Haakonsson, S.J. (2009) '"Learning by importing" in global value chains: upgrading and South–South strategies in the Ugandan pharmaceutical industry', *Development Southern Africa*, 26(3), pp. 499–516. https://doi.org/10.1080/03768350903086861
Haakonsson, S.J. and Slepniov, D. (2018) 'Technology transmission across national innovation systems: The role of Danish suppliers in upgrading the wind energy industry in China', *European Journal of Development Research*, 30(3), pp. 462–480. https://doi.org/10.1057/s41287-018-0128-5
Hanlin, R. and Hanlin, C. (2012) 'The view from below: 'Lock-in' and local procurement in the African gold mining sector', *Resources Policy*, 37(4), pp. 468–474. https://doi.org/10.1016/j.resourpol.2012.06.005
Hanlin, R., Okemwa, J. and Gregersen, C. (2020) *Building Competences and Capabilities through Projects: Examples from Kenya's Renewable Energy Sector*, IREK Working Paper 8. [Online]. Available at: http://irekproject.net (Accessed: 06/06/2020).
Hansen, U.E. and Ockwell, D. (2014) 'Learning and technological capability building in emerging economies: The case of the biomass power equipment industry in Malaysia', *Technovation*, 34(10), pp. 617–630. https://doi.org/10.1016/j.technovation.2014.07.003
Hirschman, A.O. (1981) *Essays in Trespassing: Economics to Politics and Beyond*. New York: Cambridge University Press.
Hoka, H., Njogu, H. and Obiero, B. (2018) *Realising the Big Four Agenda*. Nairobi: KIPPRA Working Paper.
Khandker, S.R., Bakht, Z. and Koolwal, G.B. (2009) 'The poverty impact of rural roads: Evidence from Bangladesh', *Economic Development and Cultural Change*, 57, pp. 685–722. https://doi.org/10.1086/598765
Lall, S. (1994) 'Technological capabilities' in: Solomon, J.J. et al. (eds.) *The Uncertain Quest: Science, Technology and Development*. Tokyo: UN University Press.
Liu, X., Hotchkiss, D.R. and Bose, S. (2008) 'The effectiveness of contracting-out primary health care services in developing countries: a review of the evidence', *Health policy and planning*, 23(1), pp. 1–13. https://doi.org/10.1093/heapol/czm042
Loots, P. and Henchie, N. (2007) 'Worlds apart: EPC and EPCM contracts: Risk issues and allocation', *International Construction Law Review*, 24 (1/4), p. 252.
Morris, M., Kaplinsky, R. and Kaplan, D. (2012) 'One thing leads to another: Commodities, linkages and industrial development', *Resource Policy* 37(4), pp. 408–416. https://doi.org/10.1016/j.resourpol.2012.06.008

Murphy, J.T. (2001) 'Making the energy transition in rural East Africa: Is leapfrogging an alternative?', *Technological Forecasting and Social Change*, 68(2), pp. 173–193. https://doi.org/10.1016/S0040-1625(99)00091-8

Srinivas, S. and Sutz, J. (2006) *Economic Development and Innovation Problem-solving in Scarcity Conditions*. CID Graduate Student and Postdoctoral Fellow Working Paper number 13, Sustainability Science program. Cambridge, MA: Center for International Development, Harvard University.

Vidican, G. (2012) 'Building domestic capabilities in renewable energy: A case study of Egypt', *German Development Institute Studies Series*, 66. Bonn: Deutsches Institut für Entwicklungspolitik (DIE). ISBN 978-3-88985-503-9.

Zanello, G., Fu, X., Mohnen, P. and Ventresca, M. (2016) 'The creation and diffusion of innovation in developing countries: A systematic literature review', *Journal of Economic Surveys*, 30(5), pp. 884–912. https://doi.org/10.1111/joes.12126.

8

INTERACTIVE LEARNING SPACES

Insights from two wind power megaprojects

Cecilia Gregersen and Birgitte Gregersen

Abstract

Kenya and Ethiopia are frontrunners in the region when it comes to adding wind power to their power generation capacity and there is high interest from project developers. The chapter uses the lens of 'interactive learning spaces' to understand how interactions between different stakeholders in a megaproject can lead to the accumulation of technological and managerial capabilities. The two projects offer interesting and different examples of the types of learning spaces in which the transfer of both formalised and tacit knowledge can occur. The chapter argues that it is important to understand and deliberately create and nurture such interactive learning spaces in order to spur and sustain local skills upgrading and capability-building in connection to large infrastructure projects based on imported key technologies.

Introduction

The Lake Turkana Wind Power (LTWP) project in Kenya and Adama II in Ethiopia are two of Africa's largest wind power plants in terms of megawatt (MW) installed. When fully deployed and running, they will contribute substantially to secure better access to reliable energy to households and businesses in Kenya and Ethiopia using sustainable sources of energy such as wind. However, will the two turnkey projects based on imported key technologies also generate local skills upgrading and local capability-building? This question has its roots in a long tradition of technology transfer and development literature emphasising the potential of a variety of flows of knowledge and technologies following large turnkey projects (Bell 2007, 2012). Often such large infrastructure projects generate several local low-skilled jobs related to the construction phase but very

DOI: 10.4324/9781003054665-8

few local high-skilled jobs. Management and engineering jobs are often supplied from abroad together with the key technologies. When a turnkey project is delivered and the foreign experts have left the country, the sustaining local capability-building is often very limited as Rennkamp and Boyd (2015) confirmed in their study of technology transfer in relation to wind and solar projects in South Africa. Nevertheless, in this chapter we show that a deliberate creation of interactive learning spaces can be one way to establish, maintain, and further develop local high-skilled jobs in relation to large turnkey infrastructure project even with key technologies imported.[1]

In short, 'interactive learning spaces' are defined as 'situations in which different actors are able to strengthen their capacities to learn while interacting in the search for the solution to a given problem' (Arocena and Sutz, 2000, p. 1). Interactive learning spaces integrate the coexistence of learning capabilities and learning opportunities in a specific context. An interactive learning space is therefore a social space created as an opportunity for knowledge producers and users to build innovation capacity, and to devise solutions to specific social and economic problems through interaction.

> Relevant learning processes related with problem solving include the capacity to recognise the useful existing knowledge, to detect the missing knowledge needed, to organise the search process to acquire it, to integrate new knowledge into the previous base and the whole into current practices.
>
> *(Arocena and Sutz, 2000, p. 7)*

There are clear overlaps to Cohen and Levinthal's absorptive capacity concept defined as 'the ability of a firm to recognize the value of new, external information, assimilate it, and apply it to commercial ends' (Cohen and Levinthal, 1990, p. 128). Learning is cumulative and path-dependent or in other words, absorptive capacity depends on the level of prior related knowledge. Introducing the interactive learning space concept in the current chapter underlines the focus on when and under what institutional settings absorptive capacity may develop and how it can be supported by a deliberate process. The institutional settings within and around the projects, and the ability to shape these to foster capability accumulation, are key in shaping the path from technology adoption to learning and innovation (Lema, Iizuka, and Walz, 2015). Furthermore, the idea of creating deliberate learning spaces within projects relates to the literature which looks at the criticality of inter-project learning and cross-project learning (Davies and Hobday, 2005) and how projects may stimulate learning and function as arenas for learning (Lundin and Midler, 2012).

Creation of interactive learning spaces can emerge and develop as a process where actors identify and solve relevant problems – as a reactive process. Interactive learning spaces can also be created as a deliberate and proactive strategy to build capacities and create learning opportunities (Johnson and Lundvall, 1994; Johnson and Andersen, 2012; Petersen et al., 2018). In practice, the two forms can interact and mutate into new mixed forms, for instance if a concrete university-industry collaboration project involving a couple of staff develops into a broader collaboration framework between the university and the external partner for starting more student projects and scholar engagement in the future. In both the LTWP and the Adama II case, we focus on two examples of interactive learning spaces that fall into the category of being created as a proactive strategy to capability-building.

Assuming that learning spaces are embryonic points in the development of innovation systems (Arocena and Sutz, 2000), it is relevant to identify and study them empirically – how they emerge, grow, and disappear. In situations where technologies are imported as turnkey projects including agreements on operation and maintenance, the learning opportunities and capability-building for local companies and organisations may be very limited, if a proactive approach to creating learning opportunities is not applied. Even when a proactive approach is in play it still takes continuous investments in learning and capability-building to maintain and accumulate new knowledge.

Deployment of large wind parks is a complex process involving a very broad range of skills and knowledge types, technologies, people, procedures, and organisational arrangements within the different phases from the planning and project development phase, to the production and construction phase, to the final electricity production and maintenance phase. The two wind power projects (LTWP and Adama II) show variations in their set-up, the partners engaged, and the energy systems in which they are embedded, but using the lens of interactive learning spaces on the two case study projects helps us understand how interactions between different stakeholders can lead to accumulation of technological and managerial capabilities. The distinction between multiple learning spaces in these projects bears a resemblance to the ideas of Davies and Brady (2000) that an organisational learning cycle must be put in place to learn from the multiple sets of capabilities required in complex projects.

The analysis of the two wind power projects draws on data collected during site visits to the Adama II project in November 2017 and the LTWP project in December 2017.[2] In addition, secondary data such as policy documents, press releases, journal papers, and project webpages support the analysis.[3]

The analysis is structured according to two types of interactive learning spaces. One is a project management interactive learning space related to the project development and construction stages of the wind parks. The other is an interactive learning space related to the operations and maintenance phases of the projects. In each case we:

1. Introduce the specific context and institutional settings of the megaprojects, including identifying the key actors – who is interacting with whom.
2. Analyse how a proactive strategy of creating an interactive learning space can spur capability-building in project management as an example of local high-skilled capability-building.

3. Analyse how a proactive strategy of creating an interactive learning space can spur capability-building within operating and maintenance of the wind turbines as an example of local medium to high-skilled capability-building.

In the following, first the Adama case and then the Lake Turkana Wind Power case is presented. The main learning from the two cases is discussed followed by a conclusion.

The Adama II wind power case

Key actors in the Adama wind power project

The Adama wind power project, Adama I and II, in Ethiopia is owned by the state-owned electricity producer Ethiopian Electric Power (EEP). It is a joint venture between the Chinese turnkey contractor HydroChina and the CGCOC group, a Chinese construction company. Phase I was finalised in 2012 and added 51 MW to the electricity grid. Phase II was commissioned in 2015 adding an additional 153 MW to the grid.

The total investments of US$ 462 million (US$ 117 million in phase I and US$ 345 million in phase II) of the projects were financed by preferential export buyer's credit from the China Exim bank (85%)[4] and own capital of EEP and the Government of Ethiopia (15%).[5] The financing agreements specified that Chinese wind turbine generator (WTG) technology was to be used. For Adama I, a Goldwind direct drive model (GW77/1500) was used while Adama II was completed with a gear box model from Sany (model SE7715). The following presentation of findings will focus specifically on interactive learning spaces occurring in Adama II's overall management of the construction phase and the succeeding maintenance phase.

As a turnkey contractor, HydroChina was responsible for the entire industry chain, from design and financing right through to engineering construction, equipment, and project contracting. They have multiple design and construction teams in China, and HydroChina's project manager for Adama II explained how they were able to work with the teams with the most experience required for this type of project (e.g., turbine model and construction requirements). The investment model, design, and blueprints from the project were proposed by HydroChina and CGCOC and negotiated with EEP. The final Engineering, Procurement, and Construction (EPC) contract included the design, manufacturing, supply, installation, testing, and commissioning of the project, including all ancillary works and civil works.

Following on the practice of the Ashegoda wind power project and Adama phase I, the Government of Ethiopia requested that Ethiopian universities submit proposals to act as owners' consultants on the project. For phase II, EEP hired a team of consultants from two Ethiopian universities (from Adama Science and Technology University (ASTU) and Mekelle University (MU)) as construction supervisors and contract administrators. According to the terms of reference, the aim of bringing in the university consultants was to:

- Build the capacity to implement construction contracts based on foreign technologies and suppliers,
- Build the capacity to manufacture main components such as towers and blades, and
- Eventually to build the capacity to manufacture and develop own technology.

At the peak construction period, HydroChina is estimated to have had over 200 employees working on site. The managing team was around 20–30 employees from HydroChina's head office, including subsidiaries. The construction teams were specialised in for example transmission lines, sub-stations, and turbine erection. Sub-contractors included Beijing Engineering Corporation Limited, 'Bureau no. 5', and SinoHydro – all under the HydroChina mother company. While all sub-contractors were Chinese, a large number of Ethiopians were employed during the project construction. The large number of Chinese employees during this phase reflects that the job types varied and that the project management (also based on the CVs of HydroChina's key personnel) was mainly carried out by Chinese employees. The key project management personnel counted approx. 13 Chinese staff for phase II, ten of which had already worked on phase I. Figure 8.1 illustrates the key actors involved in the Adama II project.

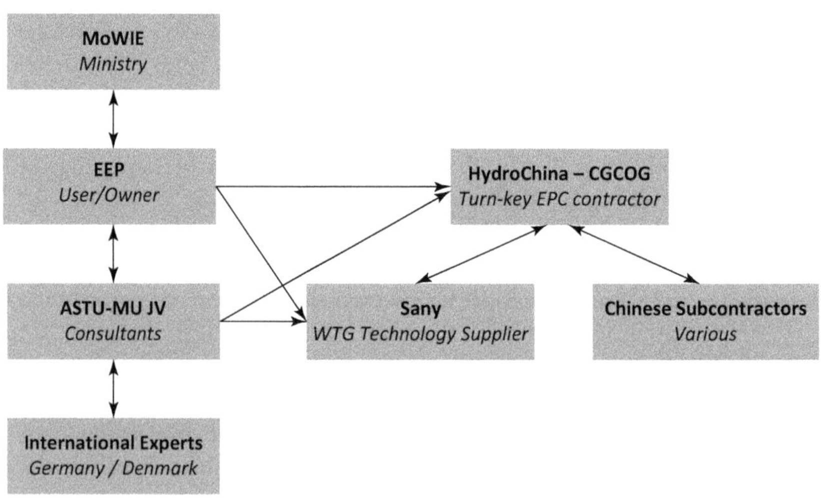

FIGURE 8.1 Key actors involved in Adama II project. *Source: authors*

Interactive learning spaces and capability-building in project management

In Adama II the team of consultants from Adama Science and Technology University (ASTU) and Mekelle University (MU) signed a joint venture to engage in the consultancy contract. The consulting team was made up of 17 academics from the two universities, working as project managers, a resident engineer, and three teams of engineers: civil engineers (structural, geo-technics, and a surveyor); power/electrical engineers (SCADA, communication, control, machine); mechanical engineers (structure and aerodynamics); and one environmental expert. These three teams mirrored the set up on EEP's team, while HydroChina's teams included the design team, construction team, and the managing team.

The Chinese teams were brought in to complete their respective tasks during implementation for short periods of time, to save time. For some civil works, for example ditch construction, only a Chinese foreman was involved to instruct workers based on the overall planning and design. In terms of choice of employees, locals who were affected by the land use were offered employment first, e.g., in civil works or as guards. According to a project manager, HydroChina's salary was two to three times higher than an average salary would have been for these workers.

The university consultancy team's main tasks were to manage the overall supervision of the implementation of the project in contract administration and design verification, including:

- Optimised energy prognosis
- Approval of WTG selection
- Substantiation of micro-siting for turbine layouts
- Construction and erection supervision
- Acceptance testing start up, commissioning, and initial operation of the plant
- Handover of the project and preparation of project manuals, reports, etc.

As specified in the contract, the university consultants hired international experts, from companies such as the Danish wind turbine technology company, Norwin, and German rotor blade specialists, CP Maxx, who possessed the required knowledge in wind energy and wind turbine technologies. These international experts conducted training sessions with the university consultant team in their areas of expertise, including on issues regarding international standards, quality control, and inspection and reported on issues such as control of blades after transportation.

There were weekly meetings between the EEP manager, the consultants, and various teams from HydroChina. They would discuss progress made and plans for the following week. Sometimes deadlines were given for evaluations, negotiations about extensions on certain parts of the work, as well as negotiations about technical issues. There was a reporting mechanism to the Ministry of Water, Irrigation, and Electricity (MoWIE) and meetings with government officials, where every team head had to report their experiences. 'It's a kind of not only consulting, it was also an experience sharing, searching for us. Because it is a new project and the government is planning to expand it. So, a pool of experts was needed' (interview with university consultant, 11 November 2017).

The majority of university consultants came from technical backgrounds in thermal, industrial, and mechanical engineering, but they had not worked on wind energy projects before. The university consultants as well as EEP staff on the projects had received a number of training courses, including at the manufacturers' location in China, as well as on site. The desired skills transfer to the university consultants was specified as consultancy and project management skills. For EEP, the major skill to learn was how to control the contractor, e.g., what kind of reporting is most important, and what clauses should be included in the contracts in the future.

Bringing in university researchers as part of the knowledge transfer is specific for wind energy projects in Ethiopia and has not been done for example in the big hydro power projects. As mentioned, EEP has a duty to report to the Ministry (MoWIE) on the progress of the project and they pay particular attention to the issue of knowledge transfer:

> We will focus on knowledge transfer and how that is happening. And we will ask the employees there, EEP employee, whether they acquired desired knowledge or not. In that case there was for example documentation issues. The documentation issue and I think they say they don't reveal some design document or something like that. So, we try to solve that kind of problem and also, we will see also with their quality of material is up to the standard or not. We will ask our EEP partners about the quality of their Chinese work.
>
> *(Interview with a ministry official, 13 November 2017)*

However, challenges were outlined in the institutionalisation of such knowledge transfer, due to employee turnover from project to project:

> I think the problem with knowledge transfer is that there is turnover of employee, that is the main problem. Like after they acquired some basic knowledge, there is a turnover of employees.
>
> *(Interview with a ministry official, 13 November 2017)*

Interactive learning spaces and capability-building related to maintenance

Part of the Engineering, Procurement, and Construction (EPC) contract specified that EEP staff were to receive training from HydroChina and Sany in order to hand over the maintenance and plant management tasks swiftly once operations started. There was a relatively short handover period from Sany (as technology suppliers in phase II) and HydroChina with only an Operation and Management (O&M) support agreement rather than the standard practice in the industry with a service agreement of five years or more. The required training in operations and maintenance will however have increased the skills transfer.

HydroChina had a team on site for three years for training purposes, particularly for EEP's engineers to train them on sub-station management, for example adjustment of power. Furthermore, a team from Sany was on site during the warranty period of the nacelles and to hand over and conduct continuous training in maintenance.

The training began already in the construction phase where EEP engineers and university consultants were invited to China for one month of training. According to interviewees, between 20–30 persons (engineers and supervisors) attended this training. The planned activities included factory visits, power plant visits, and classroom teaching. Once operations began there was a four-month training on site at Adama II. Two dedicated trainers from HydroChina remained on site after installation to conduct these trainings, one focused on WTG training and one focused on sub-station management. This training included classroom teaching as well as on-the-job training. The overall handover from HydroChina to EEP staff entailed the sharing of manuals and technical drawings of the WTGs and sub-station design, basic knowledge of how to run the WTG and the plant, standard processes for troubleshooting and reparations, as well as how to manage a maintenance team. As an interviewee recounted, the troubleshooting process aims to tell engineers to 'follow this ticket' next time so the engineers have 'no need to think by themselves' (interview with a project manager, 9 November 2017). A challenge highlighted by HydroChina was how to create training programmes when levels of education varied to a much greater extent than expected or when it was unclear whether the counterparts were certified engineers or interns not yet finished with their education. In fact, HydroChina's project manager recounted how company training in HydroChina China is a long-term and continuous process including job rotation schemes, monthly examinations, and mandatory courses before promotions and operation codes exist for every employee on a power plant. Transferring such a plant management scheme from one organisation to another may be very challenging and the interviewees raised some challenges in the transfer of skills listing; e.g., differences in work culture between Chinese and Ethiopian engineers as a major hurdle, the level of acceptance of the Chinese '24/7' work culture, as well as inevitable lost in translation issues (interview with a project manager, 9 November 2017). It was reported, however, that one of HydroChina's long-term plans is to open a training centre in Addis Ababa.

Summary

Overall, the case of Adama II illustrates how the Government of Ethiopia specifically created and institutionalised an interactive learning space by bringing in the university consultants. The aims of technology transfer were clearly outlined, and distinctive types of interaction arose between multiple actors in the project management of construction. As indicated in Figure 8.1, interactions were manifold between all key actors. During the operations and maintenance phase, a different type of interactive learning space occurred as defined by the support agreement between HydroChina, Sany, and EEP. This learning space was defined by standardised learning opportunities related to handover of WTG operations and sub-station management including classroom teaching and on-the-job training for EEP engineers.

Despite the efforts to be proactive and design these interactive learning spaces, several challenges arose in the interactions and the subsequent transfer and use of the knowledge generated by the consultants involved in the projects. New teams were formed for each wind project without handover from the previous project other than EEP's own project reports. In addition, HydroChina and Sany, the project developers, were responsible for the design, installation, and construction from beginning to end, with different units from headquarters fulfilling each task. Local staff was hired for some construction jobs but otherwise the staff was largely Chinese. Some of the challenges mentioned for the actual knowledge transfer include:

- Communication difficulties, including the use of translation during the training courses.
- Problems in relation to sharing documentation from the manufacturer and labelling in Chinese rather than English.
- High turnover of EEP staff – one of the reasons for continued training courses for new employees.

Further, a number of sources of conflict strained the relations between suppliers and users and the consultants as intermediaries, including; disputes over the verification of parts of turbines delivered being new or used, e.g., the installation of old generators on the project painted to look new, and general suspicion of the quality of Chinese products and unplanned changes for cost reduction. The university consultants recounted that while Chinese project managers maintained that things were done to plan, local staff shared different information regarding how the project was progressing. Similar challenges occurred when discussing whether manufactured goods and design of the sub-station followed international or Chinese standards; Sany's production in China follows the Chinese national standards for the industry which was according to the equipment contract.

The Lake Turkana Wind Power case

Key actors in the LTWP wind power project

The LTWP project in Kenya was the largest wind power project in Africa with its 310-MW installed capacity upon commissioning in 2019. The total costs of the project reached EUR 678 million and covered the installation of 365 wind turbines in a remote area in the Marsabit region in Northern Kenya. The project furthermore entailed the construction of more than 428 km new transmission line as well as the upgrading of over 200 km of roads and various bridges. The project is operated as an independent power producer (IPP) owned by the LTWP consortium.

The project owners hired an international engineering consulting company, Worley Parsons, for the engineering, design, and construction management (EPCM) contract. In essence, this contract was an overall project management contract to ensure 'interfacing', i.e., managing budgets and avoiding delays between work sub-contracted out to different suppliers. This is a typical kind of organisational arrangement in megaprojects where turnkey contractors are difficult to find (Steen, Ford, and Verreyne, 2017). As indicated in Figure 8.2, in the LTWP case, the project was divided into five main contractors: Vestas (wind turbine generators), Siemens (grid and sub-station), RXPE (Statcom – Static synchronous compensator), SECO (camp construction), and CIVICON (road construction). Worley Parsons acted as LTWP's 'eyes and ears' on site, ensuring the smooth collaboration between the five major contractors engaged for the construction phase. Each contractor hired sub-contractors to complete parts of their work, and local Kenyan firms were engaged by e.g., Siemens for part of the electrical cabling works. Other sub-contractors for e.g., Vestas included regional firm EGMF for work on the foundations and Bollore Logistics for the specialised

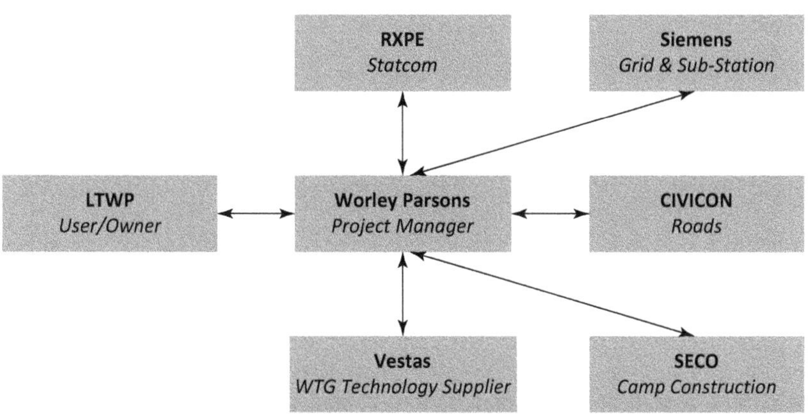

FIGURE 8.2 Key actors involved in LTWP project. *Source: authors*

transportation from Mombasa Port to the site. However, while Worley Parsons acted as project managers during the construction phase, in the many years leading up to the financial close of the project, the LTWP consortium and 'founding fathers'[6] of the project planned and designed the project in great detail. Thus, the choice of technology supplier, simulations of grid stability for a 310-MW wind power plant as well as road construction dilemmas and a number of other problem-solving activities were carried out by the owners.

Interactive learning spaces and capability-building in project management

Speaking to the project managers of the LTWP the project appears as very mission driven; they strived to complete an unparalleled project in a very challenging geographical location in order to prove to the world that such a project is possible. The project was developed as an Independent Power Project (IPP), and as the first major wind power project in Kenya it required new knowledge and skills for both project developers and regulators. The project developers faced and overcame a range of challenges from working with local communities (obtaining and maintaining their social licence to operate), negotiating the first Power Purchase Agreement (PPA) for wind in Kenya, to facing delays in externally managed critical parts for project commissioning (the transmission line) and subsequent critiques. The LTWP special purpose vehicle was established in 2006 to develop the feasibility studies, planning, and negotiations for the project which ran until financial agreement was reached in 2014, a period of eight years. Over this time and in response to multiple critiques the project was designed to specifically reflect commitment to involving the local communities both through employment plans and corporate social responsibility (CSR) activities. For example, contractors, such as Vestas with a long-term involvement, were asked to include training programmes.

> It was a requirement of LTWP to have some form of training, we couldn't tell them what to train them in, but had to say, you are here for 15 years doing O&M, so you are one of the companies that is here long term, we need to see you do some training, so they selected what they wanted. We didn't say you will do it on turbine maintenance but it's the obvious.
>
> *(Interview with LTWP manager, 5 December 2017)*

Despite the focus on CSR and community engagement, the project management approach described by interviewees was focused on interface management: identifying the critical paths of all the contracts, how they interlink and where risk of delays would be critical for the completion of the project. As indicated in Figure 8.2, this created a hierarchical design of interactions. However, within this structure there was a focus on intra-organisational learning through community recruitment. Sub-contractors were mandated to follow LTWP's aims of engaging local communities and prioritising job opportunities for the communities in the concession area before engaging Kenyan nationals from other regions of the country. While this was not guided by governmental requirements on for example local content, it was an approach that was constructed by LTWP in collaboration with key local actors as a strategy for 'earning and retaining' a social licence to operate. At peak construction, approximately 1,700 people were employed, the majority of whom were local (LTWP, 2017). Beyond the construction project management by Worley Parsons, LTWP can be identified as the key actor and repository of knowledge for capabilities on wind power project management. The 'founding fathers' accumulated the necessary knowledge through different activities of problem solving and searching as a result of their interaction with many different actors in the value chain of the project. According to one of the LTWP managers, none of the original team had previous experience in the wind industry. They simply had to learn on the job and bring in expertise:

> we just hit the ground running and said that this is what we are going to do and who can do what. We all decided and then we all went off and did our own bits and then we met once a month. We'd come back here from the field, sit, talk, this is what we have got to do and then we disappear again and come back again and meet the next month and just that's how we just got the ball running to start with.
>
> *(Interview with LTWP manager, 5 December 2017)*

LTWP's interactions with upstream and downstream actors proved an excellent channel for interactive learning for LTWP as an organisation. LTWP could be seen as an intermediator attaining the ability to translate codified analytical and engineering knowledge of suppliers in the wind industry to their downstream partners (Kenya Power and Lighting Company, the Ministry of Energy and Petroleum, the Energy Regulatory Commission, the Kenya Transmission Company).

> Everybody involved had a huge learning curve because we employed or hired the cream of the crop across the globe on grid stability. KEMA for example, which is a Dutch company – we actually got them to do a study on the national grid system to see if it could cope with the power and they gave that report to the government so they had to base plate to grow on and work on. And now KEMA is actually continuing to consult for them to make sure the grid works for all the other projects that are coming online. It's been great for Kenya. It's a fantastic project and so many people have learned so much.
>
> *(Interview with LTWP manager, 5 December 2017)*

During the planning and development phase (2006–2014), LTWP as an organisation accumulated experience by interacting with a very heterogeneous group of stakeholders, from the local authorities who had little experience in the type of negotiations for such a large scale wind energy project, to contacting suppliers upstream in the value chain and convincing them of the business case. Furthermore, LTWP hired a large number of local unskilled labourers around the site for manual labour and site preparations. International expertise was brought in by hiring consultants and specialists to advise on the planning of the project, e.g., the experts from KEMA who made simulations of the grid integration. Both LTWP as well as local authorities were able to use this as a learning experience and implemented their experiences when bringing the project forward (Gregersen, 2020).

Interactive learning spaces and capability-building related to maintenance

As mentioned above, the Danish wind turbine producer Vestas was contracted to supply and install 365 WTGs during the construction phase as well as to manage the operations and service of the WTG for a 15-year period once the project was commissioned. As an industry leader, the knowledge required to perform these tasks already exists within the organisation. However, formation of an interactive learning space can be identified in the process of recruiting and training a team of engineers that will work on the service contract for the first 15 years of this.

Vestas' philosophy is to have an interim phase between installation and operations, with an overlap between the two teams taking care of each of these phases. Part of the service team was therefore recruited before the full operations started, in order to ensure association with the construction and to assist during the construction. This strategy aims to ensure a smooth transition from the construction to the operations phase.

LTWP followed a recruitment policy that favoured the recruitment of as many workers as possible from the communities in the immediate geographical constituency of the project. This was translated into contractual agreements for all contractors and sub-contractors including a target of 20% of the total employment being from the communities in the region. Vestas set additional targets, to recruit up to 95% of their employees from the Northern Kenyan region (interview, 4 February 2019).

The service team was recruited in teams of six. The first two teams recruited Kenyan technicians and diploma engineers with backgrounds in mechanical engineering, electronic communication skills, and higher level of experience (eight to ten years) within heavy engineering industries (e.g., with generators or in oil fields) (interview with service team manager, 4 February 2019). For the third and fourth teams there was a focus on hiring as many new university graduates from the immediate region as possible. In both teams, four or five of the selected technicians were from the communities living within and surrounding the project concession while the final recruitments for each team were made at the national level.

> So, these are the guys right now, the guys that you see walking around here in blue and black with Vestas on their back. They are all local, be it local local or from up country, who are currently maintaining the turbines so they have all had training already. [...] they were taken off to Denmark and [received] training on how to maintain this specific type of turbine. So, it is basically gearbox maintenance, checking oil, dust leaks, oil leaks, bearings of the nose cone of the turbines, the electrical to a degree.
>
> *(Interview with an LTWP manager, 5 December 2017)*

Vestas technicians worldwide are required to undergo a standard global wind (GW) organisation training. Furthermore, Vestas has developed programmes for vocational and theoretical coaching and has a simulator on site at LTWP to train the service team in troubleshooting and maintenance of the turbines. Thereafter, ex-post training takes place on the job, both through a buddy programme pairing junior technicians with senior colleagues and by bringing in experienced service technicians from other Vestas departments. In the case of LTWP, technicians from Greece and South Africa were brought in to support the service team at the upstart of operations.

An on-site GW training kit on safety practices and 'train the trainer' programmes enable service team supervisors to undertake training for new recruits. Additional training needs based on skills and certification levels are available at Vestas' global training facilities. The key actors in the learning space that was created to train the Vestas service team are thus all within the global organisation Vestas, including the service team itself, the training facilities in Germany and Denmark, as well as the experienced service technicians who were brought in from other departments.

Summary

The project management interactive learning space in the LTWP project is characterised by its mission driven and problem-solving approach. While Figure 8.2 illustrates a more hierarchical type of interaction this is limited to the construction management phase of the project. In fact, the interactive learning space for project management originated with the 'founding fathers' of the project who took on the role as the key actor and repository of knowledge. For construction management, project management was then outsourced to Worley Parsons and interactions among sub-contractors was limited to issues of interfacing and time management. The project management interactive learning space is therefore more broadly viewed as spanning from the project's conception and managing its development on a more holistic level, while project management of the construction period in itself was a different space where more limited learning may have been shared between actors. The LTWP interactive learning space on project management is not characterised by a proactive strategy on behalf of the government of Kenya.[7] Rather it is embedded in the existing energy system where IPPs are encouraged and therefore LTWP themselves had to create a space in which they could learn how to manage an IPP. Within this space they acquired the necessary knowledge about issues ranging from conducting feasibility studies for the site, road surveys, and grid simulations to what clauses to include in a power purchase agreement for wind power plants.

The maintenance learning space in the LTWP project is bounded by the organisational borders of Vestas (Gregersen, 2020). Because of the 15-year service contract the learning space is highly intra-organisational as Vestas needs to recruit and train a team of engineers to fulfil this contractual task. Although the team is project based, it is a long-term investment to train the employees which is backed up by the highly standardised educational programmes of the company, including the GW trainings, simulations, and on-the-job training.

Interesting questions arise as to whether the experience-based learning in the LTWP case results in 'local' knowledge, especially as the learning space in the maintenance phase is defined as exclusive to Vestas employees. Furthermore, the learnings accumulated by the LTWP developers is bounded by the project-based nature of the power plant and the uniqueness of the project. The prospective wind power plans in Kenya have been limited to projects that are much smaller in size and there are no concrete plans for LTWP to develop and own more wind projects at the time of writing.

Learning from the two cases

Looking across the two cases there are interesting similarities and differences concerning where and under what institutional setting the two large wind power projects have created local interactive learning spaces with opportunities for skills upgrading and local capability-building. In large complex infrastructure projects like Adama II and LTWP, multiple organisations and complex interactions are involved, and in principle all actors may gain experience and obtain new or adjusted knowledge that may be accumulated and used within the project as it develops and/or is transferred to another context. While such learning by doing, using, and interacting is key as it emerges and takes place everywhere all the time during a concrete project, it also raises an important question, as to whether learning spaces can be deliberately designed to support skills upgrading and local capability-building in the long run. Based on the analysis earlier in this chapter, two parallel examples in each of the two wind farm projects were selected to serve as illustrations of such deliberately designed learning spaces. One learning space is connected to managing the process and the other to maintenance. The different phases have different involvement of actors, activities, key technologies, and requirements of knowledge domains. While other studies have introduced the concept of project capabilities, referring to important activities of bid preparation and project execution (Davies and Brady, 2000), this chapter shows that it is useful to make even further distinctions in the organisational learning cycle of wind power projects.

Table 8.1 summarises the main characteristics of the four selected interactive learning spaces.

The Adama II and LTWP cases have raised interesting questions regarding the promotion of learning within and across organisations. Jensen et al. (2007) argue that firms can promote the doing, using, and interacting mode of learning

TABLE 8.1 Main characteristics of selected learning spaces

Management learning space	**Adama**	**LTWP**
Context and institutional setting	• EEP-owned power project designed, constructed, and handed over by Hydro-China-CGCOC	• Independent power project developed, designed, and operated by LTWP
Key actors	• EEP • HydroChina-CGCOC • ASTU and MU	• LTWP • Kenyan authorities
Capabilities in focus (direct/indirect skills)	• To manage and implement construction contracts	• To manage and implement construction contracts
Reactive or proactive by design	• Designed by GoE to involve universities in the contract management and supervision	• Emerging with elements designed by financial stakeholders to involve training of local workforce
Inclusive or exclusive	• Inclusive	• Inclusive
Maintenance learning space	**Adama**	**LTWP**
Context and institutional setting	• Short term handover contract	• 15-year service contract
Key actors	• Sany • EEP	• Vestas
Capabilities in focus (direct/indirect skills)	• Operations and maintenance of the WTG and plant management	• Operations and maintenance of the WTG
Reactive or proactive by design	• Designed by HydroChina/Sany	• Designed by Vestas
Inclusive or exclusive	• Exclusive	• Exclusive

Source: authors

by building structures and relationships which enhance and utilise learning by doing, using, and interacting (e.g., project teams, problem-solving groups, and job and task rotation, all of which promote learning and knowledge exchange). Project-based construction is thus necessarily interactive and problem solving. However, the two wind power project cases show important differences in the way interactive learning spaces can be designed and shaped to proactively contribute to a desired future. The case of Adama II has an interesting institutional setting supporting high skilled knowledge transfer. From the very beginning, Ethiopian universities became involved on a contractual basis with the explicit aims to secure knowledge transfer and local capability-building on wind technologies. The LTWP project did not have a similar involvement of universities or other national public knowledge institutions. Instead, skills upgrading and capability-building were regulated by contractual agreements between LTWP and a number of different sub-contractors. To secure that knowledge transfer and experience-based learning become locally rooted may be more difficult under this institutional construction.

In both the Adama II and LTWP cases the learning spaces for maintenance are characterised by efforts to codify knowledge through manuals and tailored training programmes. However, the need for other modes of learning is shown in the complementarity of on-the-job training programmes and 'buddy' systems, that foster informal communication and sharing. This mobilises the tacit knowledge of senior technicians as well. Empirical work has shown that both tacit and codified modes of learning and innovation play a role and in fact the combination may promote more innovation than either or (Jensen et al., 2007).

Johnson and Andersen (2012) point to the importance of inclusivity of learning. On a general level, inclusion refers to broad and active participation in a process of change. The project management learning spaces may possibly be viewed as more open and diverse in terms of the actors involved. Other empirical studies have proposed that the type of relational activities of project management include capability-building exercises, as the process itself becomes a learning experience as the team gradually develops its resource base (Söderlund, Vaagaasar, and Andersen, 2008; Hanlin and Okemwa, 2022; this volume). The case of Adama was explicitly designed to include universities (staff and students) while LTWP engaged many different stakeholders in a problem-solving process driven by the developer's interest. The maintenance learning spaces were more exclusively operated between trainers and engineers with a hierarchical structure. In the case of Adama this involved an inter-organisational transfer of knowledge while in LTWP this consisted of the accumulation of capabilities by a team within the organisation. The cases of learning spaces in maintenance highlight that despite their exclusivity, they are in fact spaces in which experience and knowledge can be applied in a formalised and tested learning culture.

Inclusion of universities as a proactive strategy is a way to ensure that knowledge and experience is shared in a key renewable electrification effort. However, while the inclusion was formalised in terms of a contract and specific tasks being outlined, it is also important to pay attention to the quality of the interactions and linkages. Particularly, problems of trust between actors can arise when the vision or mission has not been created together. For example, the university consultants in Adama expressed feelings of not being able to change anything that was already agreed or designed between HydroChina and EEP. Their mandate limited their role beyond objecting and waiting for rectifications during the construction phase. Johnson and Andersen (2012) do note that interactive learning spaces give rise to learning linkages mostly within the boundaries of the interactive space itself. As a consequence, reflections about inclusivity/exclusivity is important through all phases of such projects. Circling back to the mission setting of an interactive learning space one could question what opportunities exist for learning in exclusive learning spaces to be used beyond the learning spaces boundaries. For example, what opportunities do the service engineers of Vestas have to use their new knowledge beyond maintenance of the WTGs in Lake Turkana? Does any discussion of their experience feed back to Vestas' headquarters and training facilities? What opportunities do the university consultants have to use their acquired skills in project and contract management? How realistic are the efforts taken to ensure technology transfer for the longer term aims of component manufacturing in Ethiopia? Should one learning space be followed by another once it has been 'shut' (for example after the end of the contractual obligations binding HydroChina, EEP, and the university consultants' interactions)? These questions relate to discussion on the importance of avoiding 'de-learning', i.e., when interactive learning spaces are shut down or disappear (Arocena and Sutz, 2000).

Conclusion

In this chapter we showed that by a deliberate creation of interactive learning spaces it is possible to establish and further develop local high-skilled jobs in relation to large turnkey infrastructure projects when key technologies are imported.

The two large wind power projects (Adama II and LTWP) formed the point of departure to examine and engage with the concept of interactive learning spaces in global collaborative efforts towards renewable electrification. Interactive learning spaces have provided a way to understand micro-level interactions between different group of actors in specific contexts. In particular, the way in which future infrastructure projects are conceived in policy, as well as designed, developed, and implemented in practice. Issues of directionality, distribution, and diversity of learning spaces need to be raised and considered – is a learning space designed to be inclusive or exclusive? What efforts can be made to identify, foster, and protect interactive learning spaces? The ways to do this are manifold, depending on the problems and the actors around which the learning places are constituted within renewable electrification efforts at large. In particular, thinking of such wind power projects as opportunities to search for and apply knowledge is part of creating systemic learning from project to project. This has implications for policy making for a learning-based industrialisation, where focus is rather on collective capabilities and job creation, rather than catering only to those engaged in the individual projects.

The overall argument here is that opportunities to learn must be open and kept open and not only rely on temporary and fleeting learning spaces bounded to investment projects where key technologies and expertise are 'flown in' from abroad. The long-term role and linkages of these projects with local actors in the systems must be put in focus (Lema et al., 2018). However, this requires deliberate policy decisions and actions to make sure that skills upgrading and capability-building are institutionalised and grounded in local organisations. As the Adama II and LTWP cases show, this can be done in different ways.

Acknowledgements

This research was supported by the Danish Ministry of Foreign Affairs grant DFC 14-09AAU. The authors are very grateful for the constructive comments and ideas of the editors.

Notes

1 See Andersen and Lema (this volume) for a broader discussion of three key elements in the renewable electrification process: learning, development of capabilities, and the resulting outcomes.
2 A total of 37 semi-structured interviews with key actors were conducted between February 2017 and February 2019 focusing on the employees' roles, relationships with project members, and practices of collaboration, coordination, and interaction.
3 A more detailed analysis of the relations and interactions of the LTWP case study can be found in Gregersen (2020), while the Adama case study is also featured in Lema et al. (2021). The findings presented in this chapter draw upon the analysis of these studies but views and discusses them through the lens of interactive learning spaces.
4 At an interest rate of 2%.
5 The investment estimation did not include permanent and temporary land compensation expenses.
6 The project was developed by a group of Dutch, Kenyan and Norwegian entrepreneurs who have been labelled as the 'founding fathers' of the project. They worked together with Dutch-registered KP&P, a company with history of developing and operating wind power projects.
7 At the time of LTWP's development there were no local content regulations beyond the oil and gas sector in Kenya, however, the 2019 Energy Act has emphasised the need to develop local capabilities to manufacture, install and maintain renewable energy and stipulates that firms are expected to submit local content plans, including the use of Kenyan contractors and staff were qualified and available (Hanlin, Okemwa and Gregersen, 2019).

References

Arocena, R. and Sutz, J. (2000) 'Interactive learning spaces and development policies in Latin America', *DRUID Working Paper* No. 00-13. https://wp.druid.dk/wp/20000013.pdf (Accessed: 06/02/2020).

Bell, M. (2007) 'Technological learning and the development of production and innovative capacities in the industry and infrastructure sectors of the least developed countries: What roles for ODA?', *Background Paper for the UNCTAD Least Developed Countries Report 2007.* https://unctad.org/Sections/ldc_dir/docs/ldcr2007_Bell_en.pdf (Accessed: 06/02/2020).
Bell, M. (2012) 'International technology transfer, innovation capabilities and sustainable directions of development' in Ockwell, D. and Mallett, A. (eds.) *Low Carbon Innovation and Technology Transfer. Low Carbon Development: Key Issues*, pp. 109–128. London and New York: Routledge. DOI:10.4324/9780203121481
Cohen, W.M. and Levinthal, D.A. (1990) 'Absorptive capacity: A new perspective on learning and innovation', *Administrative Science Quarterly*, 35, pp. 128–152. DOI:10.2307/2393553
Davies, A. and Brady, T. (2000) 'Organisational capabilities and learning in complex product systems: Towards repeatable solutions', *Research Policy*, 29(7–8), pp. 931–953. DOI: 10.1016/S0048-7333(00)00113-X
Davies, A. and Hobday, M. (2005) *The Business of Projects: Managing Innovation in Complex Products and Systems.* Cambridge: Cambridge University Press. DOI:10.1017/CBO9780511493294
Gregersen, C. (2020) 'Local learning and capability building through technology transfer: Experiences from the Lake Turkana Wind Power Project in Kenya', *Innovation and Development*, pp. 1–22. DOI: 10.1080/2157930X.2020.1858612.
Hanlin, R., Okemwa, J. and Gregersen, C. (2019) 'Building competences and capabilities through projects: Examples from Kenya's renewable energy sector', IREK Working Paper. Available at: http://irekproject.net (Accessed: 06/02/2020).
Hanlin, R. and Okemwa, J. (2022) 'Interactive learning and capability-building in critical projects', in *Building Innovation Capabilities for Sustainable Industrialisation: Renewable Electrification in Developing Economies.* New York: Routledge. https://doi.org/10.4324/9781003054665-7
Jensen, M.B., Johnson, B., Lorenz, E. and Lundvall, B.Å. (2007) 'Forms of knowledge and modes of innovation', *Research Policy*, 36(5), pp. 680–693. DOI:10.1016/j.respol.2007.01.006
Johnson, B. and Andersen, A.D. (2012) *Learning, Innovation and Inclusive Development: New Perspectives on Economic Development Strategy and Development Aid.* Globelics Thematic Report, 2011/2012. Aalborg: Aalborg Universitetsforlag. https://www.globelics.org/publications/globelics-thematic-review/ (Accessed: 06/02/2020).
Johnson, B. and Lundvall, B.Å. (1994) 'The learning economy', *Journal of Industry Studies*, 1(2), pp. 23–42. DOI: 10.1080/13662719400000002
Lema, R., Bhamidipati, P. L., Gregersen, C., Hansen, U. E., and Kirchherr, J. (2021) 'China's investments in renewable energy in Africa: Creating co-benefits or just cashing-in?', *World Development*, 141(105365), pp. 1–18, https://doi.org/10.1016/j.worlddev.2020.105365
Lema, R., Iizuka, M. and Walz, R. (2015) 'Introduction to low-carbon innovation and development: Insights and future challenges for research', *Innovation and Development*, 5(2), pp. 173–187. DOI:10.1080/2157930X.2015.1065096
Lema, R., Hanlin, R., Hansen, U.E., and Nzila, C. (2018) 'Renewable electrification and local capability formation: Linkages and interactive learning', *Energy Policy*, 117, pp. 326–339. DOI:10.1016/j.enpol.2018.02.011
LTWP (2017) 'LTWP's construction labour summary report. August 2017'. Available at: https://ltwp.co.ke/newsite/wp-content/uploads/20170912-LTWP-Construction-Labour-Summary-Report-FINAL.pdf (Accessed: 06/02/2020).

Lundin, R.A. and Midler, C. (eds.) (2012) *Projects as Arenas for Renewal and Learning Processes*. New York: Springer Science & Business Media. DOI:10.1007/978-1-4615-5691-6

Petersen, I.H., Kruss, G., Gastrow, M. and Nalitava, P.A. (2018) 'Innovation Capacity-building and inclusive development in informal settings: A comparative analysis of two interactive learning spaces in South Africa and Malawi', *Journal of International Development*, 30(5), pp. 865–885. DOI:10.1002/jid.3232

Rennkamp, B. and Boyd, A. (2015) 'Technological capability and transfer for achieving South Africa's development goals', *Climate Policy*, 15(1), pp. 12–29. DOI: 10.1080/14693062.2013.831299

Söderlund, J., Vaagaasar, A.L. and Andersen, E.S. (2008) 'Relating, reflecting and routinizing: Developing project competence in cooperation with others', *International Journal of Project Management*, 26(5), pp. 517–526. DOI:10.1016/j.ijproman.2008.06.002

Steen, J., Ford, J.A. and Verreynne, M.L. (2017) 'Symbols, sublimes, solutions, and problems: A garbage can model of megaprojects', *Project Management Journal*, 48(6), pp. 117–131. DOI:10.1177/875697281704800609

9

MOVING FORWARD?

Building foundational capabilities in Kenyan and Tanzanian off-grid solar PV firms

Joni Karjalainen and Rob Byrne

Abstract

In this chapter, we ask how firms in the off-grid solar photovoltaics (PV) sectors in Kenya and Tanzania have accumulated their innovation capabilities. This enables us to provide a novel categorisation of solar PV companies in the two countries according to their levels of innovativeness. Further, and with reference to latecomer theory, we develop an argument on the importance of nurturing a 'pre-latecomer phase' in which foundational capabilities are built that could support sustainable industrialisation. In the off-grid PV sector in Kenya and Tanzania, this has taken about 30 years, consisting of various types of cumulative learning processes. With this foundation in place, in the 2010s, a number of start-up firms began to innovate in the countries' off-grid PV sectors to the extent that some of them are now world-leading. While most highly innovative companies are of foreign origin, this creates opportunities for local firms to strengthen their learning and potentially enter an 'early latecomer' phase where they could build increasingly complex capabilities, including for manufacturing. We end the chapter by discussing the policy issues and uncertainties relevant to nurturing these more complex capabilities.

Introduction

Sustainable industrialisation is based on a progressive increase of environmentally friendly and enduring industrial activities firmly rooted in the local economy (see Hanlin et al., in this volume). And, as argued in this book, building renewable electrification capabilities can support sustainable industrialisation. In this chapter, we analyse how innovation capabilities have been accumulated in the off-grid solar photovoltaics (PV) sectors in Kenya and Tanzania. Most of

DOI: 10.4324/9781003054665-9

the solar firms' activities are in servicing off-grid electrification demand, and their innovativeness is primarily in new business models facilitated by technologies designed and manufactured outside Africa (Rolffs et al., 2015; Byrne et al., 2018). However, there are signs of some features of original equipment manufacturing (OEM) present in Kenya, suggesting we may be close to seeing so-called latecomer firms emerge there (Hobday, 1995, 2001). If so, and something similar occurs in Tanzania, the experience of how innovation capabilities have been accumulated in off-grid solar PV may form a 'pre-latecomer' story that offers insights on how to foster industrialisation in countries with currently limited industrial bases. As such, our analysis helps us identify a range of issues policy makers must address if they wish to promote sustainable industrialisation in the poorer countries of the Global South.

The latecomer literature assumes the presence of local manufacturing, especially the literature drawing on Hobday's (1995) analysis of electronics firms in East Asia and his characterisation of capability-accumulation through the OEM-ODM-OBM[1] sequence. Analysis in this literature focuses on how such firms use international technology transfer mechanisms – e.g., licensing, joint ventures, and others – to establish an international market presence and gradually upgrade their capabilities until they can manufacture their own internationally competitive products. In countries with few manufacturing firms who can attract investments for local OEM activities, as is the case in much of Africa, latecomer theory is of limited direct value. For countries in this situation, we need analyses that can illuminate how to foster what we are calling a pre-latecomer phase.

Entrepreneurial firms are witnessing increasing turnover and sales in the off-grid PV sector (Lighting Global, 2020). Global investment for off-grid electricity access start-ups intensified in the 2010s – growing from USD 20 million (2013) to nearly USD 400 million (2018) – especially targeting solar home systems, and more recently mini-grids (REN21, 2020, p. 156). We guide our analysis by asking 'how have solar PV firms in Kenya and Tanzania accumulated their innovation capabilities?' From this basis, we argue that we can see significant effort over about 30 years in both countries to build foundational capabilities relevant to the evolution of their off-grid solar PV markets. This period of foundational capability-formation is what we call the pre-latecomer phase, and the presence of entrepreneurial firms could signal an early latecomer phase. The growth in entrepreneurial firms who are developing variously innovative business models and technologies is exemplified by the use of mobile finance – an important component of pay-as-you-go (PAYG) business models (Rolffs et al., 2015) – and part of the rise of start-up culture. Our analysis helps us reflect on the policy implications for countries looking to develop new sectors to promote sustainable industrialisation.

After briefly reviewing the literatures on entrepreneurship and latecomer theory, we explain our analytical framework, which focusfes on firms' innovation capabilities, and describe our methodology. We then provide a historical overview of initial efforts to promote the adoption of off-grid solar PV in Kenya and

Tanzania. In our analysis, we elaborate on the processes of learning revealed by the activities of off-grid solar PV firms, after which we reflect on the economic value of such endeavours. We then discuss the relevance of these Kenyan and Tanzanian developments, their position with reference to a potential latecomer phase, and the policy issues and related uncertainties they raise. Summarising remarks conclude our chapter.

Analytical framework and methodology

To explain our analytical framework, we first describe the features of entrepreneurial firms, who we position as key actors in the pursuit of sustainable industrialisation, and reflect briefly on what latecomer theory says about how developing-country firms can build their capabilities. We then present our framework, which characterises firms in terms of levels of innovation capabilities, and we finish the section with a description of our methodology.

Entrepreneurial firms

Entrepreneurs stem from specific contexts and so hold diverse profiles (Audretsch, 2012). An entrepreneurial firm searches for new business opportunities (Schumpeter, 1934) while a start-up, more specifically, is an entrepreneurial firm who uses innovation to thrive and grow (Ries, 2011). Social entrepreneurs also use this logic (Miller, 2009) to generate ideas for products or services (Picken, 2017). Firms benefit from experience (Shane, 2000), as opportunities may arise from technological and value shifts (Kim and Mauborgne, 1997; Rohrbeck, 2010). Enabling conditions and a supportive environment can help them (Grilli et al., 2018) as they are easily affected by discontinuities and changes in the policy environment (Georgallis and Durand, 2017). A challenge for entrepreneurship-oriented policies in African countries (Poole, 2018) is that capability gaps in these contexts may affect entrepreneurs more than their international peers (Gabriel et al., 2016).

Firms in Tanzania suffer from a low technological base and weak opportunity-recognition (Goedhuys, 2007), while some small and medium-sized enterprises in Kenya are in the early stages of internationalisation (Osano, 2019). Thus, entrepreneurial off-grid solar PV firms in Kenya and Tanzania face a range of challenges when seeking to benefit from the opportunities presented to them by, for example, the rising investments earlier noted. Amongst these challenges are enhancing the firms' innovation capabilities. If cultivated successfully, these capabilities will benefit an individual firm and can also benefit the wider national economy, if many local firms similarly develop their innovation capabilities. We next briefly discuss evidence and analyses in the latecomer literature that describes how firms can become internationally competitive.

Latecomer theory

Latecomer theory is interested in how developing-country firms aim to compete in international markets. In this vein, much academic theorising has been done on how to build innovation capabilities, and mostly on the basis of Asian and Latin American experiences (e.g., see Hobday, 1995; Bell, 1997, 2009; Bell and Figueiredo, 2012).

Hobday (1995) describes the experience of electronics firms in East Asia and how they succeeded in developing their capabilities. Initially, in East Asia's pre-latecomer phase, foreign firms possessed a technological advantage, and few local firms performed any manufacturing activity. The foreign firms began to act as examples for them after the local firms had acquired necessary basic-level capabilities. Some local firms were able to start performing assembly services for the foreign ones and thereby capture more of the value-added for themselves. Others went further, learning basic manufacturing skills, supported by national efforts and strategies (Hobday et al., 2004).

Acquiring such capabilities was achieved through different kinds of relationships with foreign firms: e.g., subcontracting, joint ventures, and licensing. Under subcontracting relationships, leading companies in East Asia trained local managers, engineers, and technicians to build important knowledge and skills for the future. For instance, in South Korea, firms benefited from visits by foreign engineers and visits by Koreans to overseas factories. Joint ventures, in turn, are strategic partnerships where the partners have a relatively equal footing. In licensing, a local firm pays for the right to manufacture, and the foreign firm transfers the required technology to that local firm. Licensing can be deep or shallow (Lall, 1992) but may require more complex capabilities than a joint venture. Other examples of these kinds of learning and technology transfer relationships can be seen across Asia: e.g., in India and China (Lema and Lema, 2013), in China (Watson et al., 2015), and in Thailand (Reinauer, 2019), amongst others.

The latecomer approach has not, to the best of our knowledge, been applied in the Kenyan and Tanzanian contexts, where there is relatively little manufacturing, and perhaps not in the African context more generally. To overcome the challenge of engaging in latecomer analysis, we begin by considering how firms in these environments can acquire foundational capabilities. As elaborated later, there have been considerable efforts in Kenya and Tanzania to build such capabilities in the off-grid solar PV sector. According to latecomer theory, this could represent a pre-latecomer phase. Once foundational capabilities are built, a specific question concerns how local firms can further raise their competitiveness and build more complex capabilities. Although the OEM-ODM-OBM learning sequence characterises the accumulation of production capabilities (Bell, 2009), the latecomer analysis also implies that firms in general benefit from becoming more innovative. In fact, given the aim of sustainable industrialisation, entirely novel configurations and constellations of environmentally friendly innovations are expected from a wide range of firms (Bell, 2012, p. 25).

With this in mind, we now consider what the literature states on different levels of innovation capabilities and how they are built, as the basis of our analytical framework.

Innovation capabilities of firms

Capabilities can be those that a firm already has or competences it needs to cultivate and/or acquire from other actors. Knowledge, as a firm's most significant capability, is embedded in human resources, procedures, and routines, making knowledge contextual, firm-specific, and tacit (von Hippel, 1994). To create novel ideas, firms rely on technological capabilities that enable them to master a specific technology, but they also need non-technological capabilities such as those in management, design, or foresight. And a feature of a globalised economy is that innovation is also global (Liu, 2017), implying that a firm's innovation capabilities are also affected by related dependencies such as the position of the firm in global networks.

Bell and Figueiredo (2012) reviewed 25 years of research on learning and innovation capability-building in firms in developing economies, providing an illustrative framework of capability levels through which firms might move over time. As can be seen in Table 9.1, Bell and Figueiredo identify four distinct levels of innovation capabilities: 'basic', 'intermediate', 'advanced', and 'world leading'. In their assessment, 'basic' innovative activity includes capabilities to make minor product, process, or organisational adaptations, often in informal conditions. Moving down the table, we see characterisations of innovative activity that refer to increasingly complex and formal innovation capabilities. These chime with the Oslo Manual (2018) definitions of change: 'intermediate' innovative activity (Bell and Figueiredo) is similar to activity that is 'new to the firm' (Oslo Manual); 'advanced' is similar to 'new to the economy/market'; and 'world leading' is similar to 'new to the world'. We make use of these heuristic categorisations but, considering that the framework is based primarily on research in Asia and Latin America, the illustrative elements of the capability levels may not be entirely applicable for our study context. For this reason, we adapt the characterisations given in the centre column into more general statements, given in the right-hand column.

Looking at the Bell and Figueiredo characterisations, they each describe the kinds of knowledge, skills, and actor-networks associated with the different levels of innovation capabilities. We use these categories across each of the levels and provide descriptions that indicate the differences we might expect to see in each category and level. For example, a firm who possesses world leading innovation capabilities may have highly specialised, formal, and frontier knowledge; and skills to create such knowledge imaginatively and with originality. Leading firms tend to act in substantial, professional, internationally recognised, and collaborative networks. It is these four sets of general descriptors we use as our analytical framework to categorise off-grid solar PV entrepreneurial start-ups in Kenya and

TABLE 9.1 Levels of innovative activity

Level of innovative activity (novelty)	Illustrative capability elements *(Human capital, knowledge bases, etc.)*	*General descriptors*
Basic ***(Limited innovation)***	Groups of engineers and qualified technicians working informally on experiments and incipient or informal R&D activities. Dedicated groups of engineers and qualified technicians and well-trained operators working on the implementation of minor adaptations in products, production processes, and organisational and/or automated systems.	**Knowledge** Disciplinary, 'applied', informal **Skills** Ability to apply and incrementally adapt existing knowledge **Actor-networks** Small, narrow, professional, and operational, single-unit intra-firm
Intermediate ***(New to the firm)***	Increased number of specialised engineers and technicians allocated in different and dedicated organisational units involved in product development, product re-design, process engineering, and automation systems. These professionals work on activities such as duplicative and/or creative imitation to advanced modifications to products, large-scale production systems, software. Firms tend to give preference for professionals with good technical skills and some cognitive skills (problem solving and framing) for creative imitation.	**Knowledge** Specialised, 'applied' **Skills** Ability to duplicate, creatively imitate, or substantially modify existing knowledge and problem-solve **Actor-networks** Narrow, professional, single- or few-unit intra-firm
Advanced ***(New to the economy/ market)***	Various types of design and development engineers, researchers and other specialised professionals in different functional areas within and outside the firm. Among these are those with additional skills for new knowledge-sharing and external knowledge screening/searching and leveraging, knowledge-bridging people, 'multilingual managers', technological gatekeepers. These professionals implement applied research, design, and development of complex products/services and production systems that are *close to the international innovation frontier.*	**Knowledge** Specialised, near-frontier, formal **Skills** Ability to screen, share and apply new knowledge (absorptive capability), bridge knowledges, choose technologies **Actor-networks** Broad, professional, multi-unit intra-firm, inter-firm collaborative

(Continued)

TABLE 9.1 *(Continued)*

Level of innovative activity (novelty)	Illustrative capability elements *(Human capital, knowledge bases, etc.)*	*General descriptors*
World-leading *(New to the world)*	A substantial body of internationally recognised R&D personnel with a number of teams of highly specialised engineers and related professionals working on cutting-edge research. Some teams may be engaged in precompetitive forefront research. Large incidence of people with sophisticated cognitive skills for generating imaginative and original innovations. These are distributed across different organisational units in the firm and also work on a collaborative basis with professionals from other organisations.	**Knowledge** Highly specialised, frontier, formal **Skills** Ability to create and apply frontier knowledge imaginatively and with originality **Actor-networks** Substantial, professional, internationally recognised, multi-unit intra-firm, inter-firm collaborative

Source: authors, adapted from Bell and Figueiredo (2012) and Oslo Manual (2018).

Tanzania. However, the 'levels' should be understood as indicative steps along a continuum rather than rigidly distinct categories.

Methodology

The materials on the historical evolution of the off-grid PV sector were gathered over the period 2007 to 2014, with fieldwork in two visits (see Byrne, 2011; Ockwell and Byrne, 2017): in both Kenya and Tanzania between 2007 and 2008; and in Kenya during 2013. In addition to desk-based work, the material included over 100 hours of interviews and information gathered in two stakeholder workshops. The analysis of firms' innovation capabilities draws on the observation of the Kenyan and Tanzanian off-grid solar PV sectors from 2013 onwards, a database of off-grid solar PV firms, field work in three visits between 2015 and 2019, three stakeholder workshops, and selected interviews. In line with the aim of understanding how innovation capabilities affect firms (Oslo Manual, 2018, p. 103–126), we analysed firms' revealed capabilities: i.e., what entrepreneurs and firms actually do.

The firms were identified from interviews, electronic platforms such as the Crunchbase and Owler for venture capital, the Energy and Environment Partnership project site, and industry reports. Selected case studies were built from the interviews and secondary data. The 2019 update provided a sample of off-grid solar PV firms, with 63 active (and three non-active) firms in Kenya, and 52 active (and six non-active) in Tanzania, yielding 94 active firms altogether after duplicates of those present in both markets were removed. The sample reveals that the sector is relatively young: with the exception of retail firms, 78% of the firms in the Kenya sample and 75% of those in Tanzania were established in 2007 or later.

In Kenya, 39% of the firms operate in solar lanterns and/or solar home systems (SHSs), sometimes having started in the lanterns market segment; 20% are in consultancy or engineering, procurement, and construction (EPC); and 8% are in the mini-grid sector. Firms in Tanzania are typically in lanterns and/or SHSs (32%), consultancy/EPC (31%), retail (15%), or mini-grids (13%). Some firms apply solar for productive uses (in agriculture, refrigeration, water pumping) or for other activities. In the database, 36% of the firms in Kenya and 29% of those in Tanzania are of local origin.[2] Local firms generally operate in one country, whereas foreign firms typically have a presence in numerous markets. The sample includes manufacturing and non-manufacturing firms, and many firms interact with one another.

Descriptions were written to elaborate the firms' skills (business model, characteristics, innovation, value proposition), knowledge (age, evolution, ownership, team, technology), and actor-networks (background, finance, geography, partnerships). Using this detail, we organised the firms according to the descriptors introduced in Table 9.1, enabling us to develop characterisations of their capabilities and how they have acquired them. These findings are summarised in Table 9.2. We then were able to reflect on the economic significance of their innovation activities by reviewing industry reports from 2015 to 2019. And, building on this, we could extend our discussion to sketch some of the features of what may be a pre-latecomer phase in off-grid solar PV in Kenya and Tanzania, and to identify a range of policy issues that may need addressing if the sector is to be further developed.

Accumulation of foundational capabilities

To understand the historical evolution of the off-grid solar PV sector in Kenya and Tanzania, we briefly explain how foundational capabilities in the sector were built. More detailed descriptions can be found in Byrne (2011) and Ockwell and Byrne (2017).

Although solar PV equipment first arrived in East Africa around the late 1970s, it was not until the mid-1980s that the SHS concept emerged (Ockwell and Byrne, 2017). A market for SHSs developed almost immediately in Kenya, but it took practically another two decades before it started growing in Tanzania. Over three decades, many interventions – usually donor-funded – helped build a range of foundational capabilities in the off-grid PV sector in both countries (Byrne, 2011). These interventions addressed issues such as a lack of awareness, finance, maintenance, and service delivery (Ahlborg and Hammar, 2014; Hansen et al., 2015) and intensified after several development actors became interested from the late 1990s. Over time, their interventions consisted of various capacity-building and learning experiments with the technology and

TABLE 9.2 Levels of innovation capabilities in Kenya and Tanzania's off-grid PV sector based on categorisation in Table 9.1

Level of innovative activity	*Illustrative capability elements*	*Examples of knowledge, skills, actor-networks*	*Examples of firms*
Firms with limited innovation capabilities (*Limited innovation*)	• Limited abilities to invest in or implement innovation activities • Adapt to new requirements, make few changes in the business model • Attract little to no foreign investment	• Focus in the business and management of the supply chain • Weak or no linkages to global innovation networks, possible linkages to product networks (e.g., foreign exporters, retailers) and technical experts • Tanzanian solar retailer mainly focuses on sales (Zara Solar)	Multiple firms (distribution, retail, supply, installations)
Experimentation, build-up, and international partnerships (*New to the firm*)	• Attempt innovation and perform search activities, but accumulation and scaling up may take time • Many local actors, start-ups, or SMEs, some nurtured in local incubators • Emerging access to global innovation networks, some may attract seed financing for pilots and demonstrations	• A pioneer who has evolved with the sector, moving from lanterns to solar home systems and into mini-grids (Ensol) • Tanzanian founder started as a teenager by selling Chinese-imported mobile phones, studied in China, took a loan to import solar products into Tanzania (Helvetic Solar) • Early pioneers in Kenya who developed their skills through international partnerships (SolarWorks, SunTransfer) • Social enterprise established hubs, conducted surveys, and trained employees to expand work with rural farmers on clean energy solutions with a grant from an international donor (Boma Safi)	Baraka Solar, Boma Safi, Ensol, Helvetic Solar, Juabar, Sikubora, SolAfrique, SolarWorks, SunTransfer

(*Continued*)

TABLE 9.2 *(Continued)*

Level of innovative activity	*Illustrative capability elements*	*Examples of knowledge, skills, actor-networks*	*Examples of firms*
Advanced capabilities and technological sophistication *(New to the economy/ market)*	• Drive sector evolution with incremental innovation and technological sophistication • Imitated and adopted the pay-as-you-go model, may operate in several markets • May train local entrepreneurs and create employment to build skills • Attract substantial investment for demonstrations and scaling up	• Solar-powered roof tiles designed in Kenya, hosted in local accelerators and incubators (Strauss Energy) • Partnerships and financing from donors and development banks (JUMEME) • Founders followed an initial idea, left their expert positions, performed years of search activities, moved to Tanzania (Devergy) • Construction of mini-grids in Kenya, Tanzania, and Zambia (PowerGen)	Angaza, Devergy, JUMEME, Powergen, Powerhive East Africa, Steama.co, Strauss Energy
Leading frontier capabilities *(New to the world)*	• Find key solutions for fundamental barriers with complex problem-solving activities despite uncertainty • Push the innovation frontier, set benchmarks • Begin to attract large-scale venture capital and private investment in millions, over multiple funding rounds	• Extensive piloting of solutions, a partnership with Safaricom, Kenya's telecommunications giant (M-KOPA Solar) • Machine-learning algorithm adjusts solar power output according to average system usage, weather patterns, and battery power (Azuri) • Grow to attract investment worth millions of USD from impact funds, private equity funds, and banks (BBOXX, ZOLA) • Establishing a partnership with e-waste and battery recyclers for recycling off-grid solar components (Mobisol)	Azuri, d.light, BBOXX, M-KOPA Solar, Mobisol, ZOLA Electric

Source: authors

microfinance, all of which also expanded and strengthened the off-grid solar PV actor-networks. Publications of reports, market information, and market surveys in part contributed to advocacy for favourable energy policy, which included the removal of tax on PV equipment, the introduction of PV standards and regulations, and more. These activities also built up necessary capabilities. Lessons learned were publicly shared and awareness of SHSs grew amongst many Kenyans and Tanzanians.

Market activity was nevertheless limited for a long time. For example, in Kenya in 1997, the International Finance Corporation (IFC) tried unsuccessfully to implement a USD 5 million market transformation initiative (IFC, 2007). But the markets were immature and local technicians, although trained, lacked funding, so few local businesses emerged. Results were more successful ten years later when the IFC began implementing the Lighting Africa project that promoted solar PV lanterns (Lighting Africa, 2008). In addition to capacity-building, Lighting Africa conducted consumer education campaigns to drive demand in Kenya, and supported the private sector seeking to enter the market. In Tanzania, intensified capacity-building activities resulted in the eventual growth of the PV market within just a few years from almost nothing to a value of USD 2 million or more in 2008. These successes can be partially attributed to the many donor-funded activities that aimed to build knowledge, skills, and actor-networks in the sector.

In 2011 in Naivasha, Kenya, a joint venture between a Dutch investor and a Kenyan holding company began operating a module assembly plant[3] (Ockwell and Byrne, 2017). Growth of the solar lantern market has spurred further growth of the SHS markets, and the private sector has begun to provide pico-scale solar systems on a commercial basis in Kenya and Tanzania (Nygaard et al., 2016; Davies, 2018). The use of mobile money, emerging technologies, and the PAYG model have become benchmarks in the nascent industry (Rolffs et al., 2015). Under the social entrepreneurship ethos, new activities continue to emerge according to Odarno et al. (2017), Tanzania leads in East Africa for mini-grid development, and applying solar PV for productive uses is a novel frontier. Owing to these diverse activities, a range of products from lighting to appliances has emerged in multiple categories (Lighting Global and GOGLA, 2018).

Overall, this pre-latecomer phase saw the building of foundational capabilities, the growth of several market segments, and the introduction of a range of new technologies. It is into this context that many new firms, as entrepreneurial start-ups, have entered.

Innovation capabilities of off-grid solar PV firms

Our analysis of the innovation capability levels of firms in off-grid solar PV in Kenya and Tanzania begins with those who have limited innovation capabilities. Firms at the second level seem to accumulate capabilities through experimentation and through learning from international partnerships. Advanced firms, placed at the third level, can use emerging technologies. At the highest level are leading firms who, with sophisticated capabilities, perform highly complex problem solving. See Table 9.2 for a summary of our analysis.

Firms with limited innovation capabilities

Firms at this first level offer products and services such as retail, wholesale, distribution, dealership, and installations, and they constitute the majority of firms in the sector. Any firm must learn about the technology, whether a small or medium-sized shop or retail firm selling products and equipment for cash (KCIC, 2017). One example is Zara Solar, an early pioneer, launched in Tanzania in 2005. Initial learning efforts won it the Ashden Award in 2007. Operating a small core staff, it employed a network of freelance technicians who were first trained through a United Nations Development Programme (UNDP) project. The firm continues to perform installations and sales in Mwanza and Dar es Salaam.

The firms in this category use technical, management, and marketing capabilities, but the limited nature of their capabilities means few of them attract investment or can connect with frontier knowledge. Operating in a nascent sector is in itself a kind of innovative behaviour but, without an ability to innovate further, they will likely remain at this level. Nevertheless, even firms who do not perform more innovative activities provide an essential function in the sector through the economic activity they generate. For example, after an increasing diversity of off-grid solar products and services came onto the market, between 2014 and 2016 alone, around five million devices were sold in Kenya and Tanzania (Lighting Global and GOGLA, 2018). Many of these devices would have been sold through firms with limited innovation capabilities.

Experimentation, build-up, and international partnerships

Firms placed at the second level are those who aim to adapt their business models, suggesting they are more entrepreneurial and that they possess more innovative capabilities than those at the first level. For example, Sollatek Electronics, a wholesale and distribution firm established in 1985, has experimented with crowdfunding. In Tanzania, Ensol was one of the early PV companies, established in 2001, and has since been involved with solar lanterns, SHSs, and mini-grids. The Tanzanian founder of Helvetic Solar, founded in 2007, started his first business as a teenager in Arusha selling Chinese-imported mobile phones. Having studied in China before a return to Tanzania, he got a loan to import solar products. After initial success, the firm established a philanthropic foundation who partnered with international agencies. International exposure seems to have benefited the capability-acquisition of firms in Kenya such as SunTransfer, a distributor of solar products, and SolarWorks, a project company, both established in 2009.

An especially interesting case is Chloride Exide, a customary battery manufacturer and retailer in Kenya. According to the website of the owning company,

ABM Group, the manufacturer began operations in 1963 to produce batteries for the Chloride Group, its UK-based founder, and for others.[4] It has innovated in several ways since beginning operations, including introducing a modified car battery for use in SHSs, achieved through a donor-funded project in the late 1990s. Now, it produces a variety of own-brand batteries, has manufacturing facilities in Tanzania and Uganda, and claims to be the largest battery and renewable energy distributor in the East Africa region. It invested in the Naivasha solar PV module assembly plant (mentioned earlier) when it was being set up in 2011. Majority-owned in Kenya by Solinc, the plant does not fabricate solar cells but has made incremental process innovations to double its initial production capacity of solar modules (Ockwell and Byrne, 2017). In 2018, Chloride Exide partnered with the industry leader M-KOPA Solar.

Certain start-ups belong at this level. A Kenyan entrepreneur learned about 'green' entrepreneurial models at Strathmore Business School when carbon finance was introduced to Kenya in the late-2000s. After initial search activities, the entrepreneur founded a solar micro-grid firm, serving clients through a power purchase agreement.[5] In Tanzania, a Stanford-educated director and a small local team worked in a start-up called Juabar. In its franchise model, the firm has a network of solar kiosks in off-grid areas, which are leased to local entrepreneurs who then sell electricity for phone charging. In Boma Safi's model, founded by a Kenyan, rural women sell products through an order and delivery system, enabled by energy and distribution hubs and village credit organisations. Firms placed at this level may attract seed funding for pilot and demonstration efforts but are distinguished from more advanced firms by their smaller size, slower growth trajectory, and less complex and technology-intensive activities.

Advanced capabilities and technological sophistication

In the early 2010s, the industry pioneers introduced the PAYG model, and many foreign early-stage firms with advanced innovation capabilities entered the market in their tail. The typical founder of an advanced firm has a degree from a world-leading university combined with past consultancy, finance, and IT experience. These companies' advanced marketing capabilities help in brand differentiation for customer-acquisition, market-expansion, and scalability. A noticeable feature is their ability to integrate emerging technologies, which makes them attractive for financing. Some test new products with dedicated R&D facilities outside of the countries.

In this sense, mini-grid companies who incorporate the PAYG model and latest technologies can be placed at this level. Devergy, a mini-grid operator in Tanzania, piloted its model also in Ghana. JUMEME builds solar-hybrid micro-grids in Lake Victoria. PowerGen, established in 2011, has constructed mini-grids in Kenya, Tanzania, and Zambia. Funding from CrossBoundary Energy Access, a mini-grid project finance facility, contracted the firm to build 60 mini-grids in Tanzania. Power Corner Tanzania and Rafiki Power have raised the interest of large energy corporations. And, given the undeveloped nature of the productive-uses market segment, firms aiming to use PV for agriculture, refrigeration, and water pumping can also be placed in this category.

Almost no firms of local origin operate at this advanced level. An exception is Strauss Energy in Kenya, who have piloted solar PV roof tile manufacturing and experimented with compressed-air energy storage. The experienced founder learned about solar energy already in 2002 during an MSc-level project (Ciambotti et al., 2019), designing a PV roof tile from low-cost materials. Strauss Energy has benefited from the accelerator and incubation programmes at Kenya Climate Innovation Center (KCIC) in Strathmore University. Although the solar tiles are now produced and assembled by Shenzen Solar in China, Strauss Energy's technological capabilities are 'new to the market'.

Leading frontier capabilities

Firms at the 'highest' level are able to conduct complex problem-solving efforts. M-KOPA Solar – the PAYG pioneer – was established around 2010. The firm's founders joined forces with a microfinance expert and learned about mobile money when it was in its infancy, innovated with potential customers, and enrolled in technology competitions. Before commercial sales, the firm attracted international finance and established a partnership with Safaricom, Kenya's leading mobile network operator. M-KOPA Solar designed a user-friendly 'plug-and-play' solar kit that included a two-year warranty and a durable battery. The data chip in the kit was connected to a technology platform that handled customer payments, inventory, accounting, and customer relations.

Achievement of industry milestones may place a firm in this category. D.light and Greenlight Planet were early pioneers in solar lanterns, and the UK-based Azuri Technologies entered the PAYG space directly in 2012. Azuri's system includes a machine-learning algorithm, which adjusts house lighting according to usage, weather patterns, and battery power. The firm trains local employees in finance and marketing but has a product design factory in the UK. Its research teams study potential customers and pilot different products locally. It partners with an airtime distribution network and an experienced local agent network that has distributed jewellery beads.

Mobisol, a German firm who had social impact investors and international development agencies as initial owners and lenders, established an academy in Tanzania that trains local entrepreneurs, contractors, and staff. It also established partnerships for battery recycling and electronic waste. ZOLA Electric, established in 2011 and originating from San Francisco in the United States, has a proprietary software platform that provides a personalised service for customers and real-time data to track product sales, service and installation teams. It claims it learned from the electric vehicle and large-scale solar industries, and its investors include Tesla, GE, EDF, and Helios.

Discussion: innovation capabilities in the pre-latecomer phase

Having differentiated the innovation capabilities of the off-grid solar PV firms by their knowledge, skills, and actor-networks, we can see that interactive learning is an important activity for any firm and that firms serve complementary market roles. The local Kenyan firms are within the first two 'levels' of capabilities, except for Strauss Energy. In Tanzania, most firms operate at the basic (or the intermediate) level, and no Tanzanian firm possesses advanced or world-leading capabilities. PV module assembly and battery manufacturing, both in Kenya, are the only production capabilities.

In Kenya, those firms who are more involved in innovative activities have learned through joint ventures and partnerships with foreign firms. Indeed, any local innovative firm seems to have international linkages. Still, in the sampled firms, 64% in Kenya and 71% in Tanzania had a non-local CEO and/or founder, often from the United States or UK, but also of German, Dutch, French, or Italian origin, including expatriates who have long lived in these countries. This suggests that the advanced and world-leading capabilities largely stem from outside of Kenya and Tanzania. Advanced and leading firms are also employers who can contribute to local capabilities in support, sales, services, and management.

Once a firm succeeds in establishing a customer base, a revenue model and can repeatedly solve challenges, they are able to attract (sometimes substantial) funding or finance. The current industry leaders evolved in the 2010s by first attracting seed capital from development partners, then social impact funds and, eventually, venture capital. According to Wood Mackenzie (2019), M-KOPA Solar went through six finance rounds and had raised USD 190 million by 2018. Globally, only four firms attracted two thirds of all the financing that went into building the off-grid PV sector: ZOLA Electric (USD 261m), D.Light (USD 188.5m), Lumos (USD 108m), and Greenlight Planet (USD 82m) (Lighting Global and GOGLA, 2018).

The value of the off-grid solar lighting products sold in Kenya and in Tanzania suggests that PAYG firms are attractive businesses. Altogether, almost six million products were sold in Kenya and 1.5 million units in Tanzania in the late-2010s. The lines rising in 2017 show that the value of a PAYG unit sold is considerably more than that of traditional PV products, shown by the lower lines (Figure 9.1). Industry figures for late 2018 estimated the value of cash sales in Kenya at USD 13 million but the value of PAYG services amounted to USD 59 million. The difference is starker in Tanzania, where the value of cash sales was USD 1.5 million and PAYG sales were USD 21.7 million. In East Africa, the value of cash products was USD 25 million compared to USD 110 million for PAYG units. Globally, PAYG firms accounted for 24% of the sales volume but 62% of the revenue (GOGLA, 2019).

Cash sales may generate much less revenue than the more lucrative PAYG business models, but they continue to play an important role in at least three ways. First, the number of PV units sold for cash remains greater than under PAYG terms. Therefore, many small firms – who may be widely distributed across both

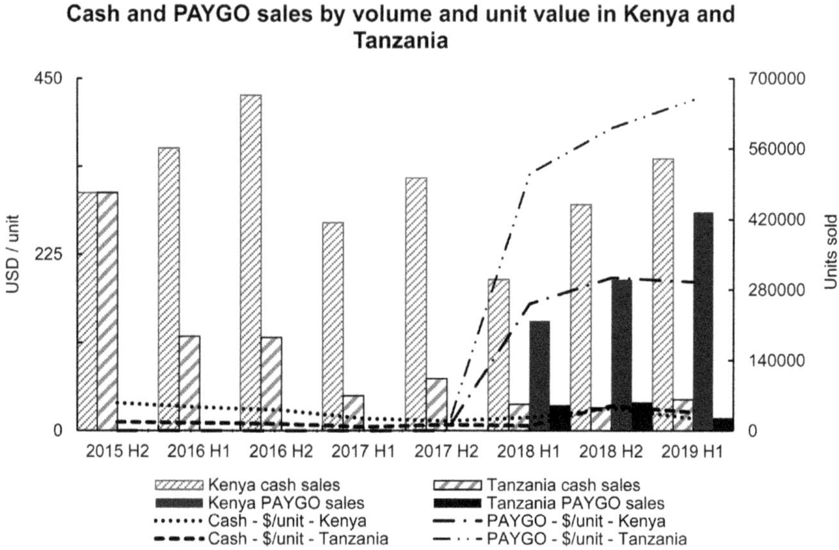

FIGURE 9.1 Off-grid solar lighting products sold in cash and PAYG in Kenya and Tanzania by volume and unit value. *Source: GOGLA bi-annual market reports (2015–2019).*

countries – are benefiting from at least some business activity compared to the more concentrated PAYG segment. And it may also mean that more people are getting electricity access through cash sales than through PAYG. Second, the firms selling PV for cash are owned locally whereas PAYG firms tend to be foreign owned. As such, more of the profits are likely to remain in the local economy compared with those from PAYG businesses. Third, the price per PV unit would appear to be much lower for cash sales than for PAYG terms, although we should note this is a crude comparison based on aggregate sales revenue per units sold and so the observation is open to further analysis. Nevertheless, if this price differential is real, it means those who cannot afford upfront purchase of PV systems pay a higher cost for electricity access, raising social justice concerns. There may also be other useful metrics to develop a fuller picture of the relative importance of cash versus PAYG market segments and their associated advantages and disadvantages, such as the number and quality of jobs created and sustained in each segment. These metrics are important for achieving a more refined understanding of current capabilities in Kenya and Tanzania, and of those that could be enhanced or created.

Policy issues for off-grid solar PV and sustainable industrialisation

Building on the analysis of the capabilities of off-grid solar PV firms operating in Kenya and Tanzania and related economic dynamics, we can reflect on the policy issues that arise for how to enhance the gains of the countries' pre-latecomer phase with a view to entering a latecomer phase and encouraging sustainable industrialisation. Our discussion begins with the implications of cash versus PAYG market segments. We then consider the policy issues relevant to building innovative capabilities centred around technologies and manufacturing, and argue that other perspectives besides latecomer theory may need to be marshalled if the various complexities involved in capability-building efforts are to be fully understood and addressed.

Cash sales, PAYG, and foundational capabilities

Based on the importance of cash sales, we could argue it is essential to maintain and strengthen this segment, which raises several policy issues. The capabilities of actors in the supply chains need to be built or enhanced to maximise both the quality of PV technologies supplied and the quality of customer services provided (sales, technical support, etc.). And customers themselves need help to build their own capabilities for understanding the technology and service options offered to them, as well as for using PV systems. These issues have been long-standing in Kenya and Tanzania, and various efforts have been made to address them. In Kenya, for example, the IFC market transformation initiative, mentioned in the historical section, funded capacity-building efforts that included training of PV technicians and vendors alongside publication of information manuals tailored to each of these groups and to customers. Kenya also established PV standards – later strengthened to regulations – and has more recently introduced a national-level PV training curriculum and certification of PV market actors (Ockwell and Byrne, 2017). It is unclear how these measures have affected capabilities in Kenya as, to the best of our knowledge, no research has been done on their impacts. Nevertheless, they provide examples of the kinds of action that policy can take to help build or strengthen at least some of the foundational capabilities in off-grid solar PV. Thinking more broadly about supply chains, there may be potential to build innovative capabilities for increasing efficiencies, lowering costs, and improving sustainability throughout supply chain operations. Included in these capabilities are likely to be finance and management competences, among others, that could be transferrable to sectors beyond off-grid PV and so contribute more generally to sustainable industrialisation.

Technological innovation capabilities

Policy interventions and the efforts of a variety of actors aimed at building technologically centred innovative capabilities will tend to be more complex, expensive, time-consuming, and uncertain. Firms and other actors with varying capabilities need to learn to interact with each other in collaborative as well as competitive ways across varied innovation contexts. Building innovative capabilities requires a great deal of formal training throughout education systems as well as on-the-job training and other lengthy processes of experiential learning. This is all expensive and time-consuming. Such features pose a great deal of uncertainty about how developments will unfold. Indeed, as we write, the world is trying to respond effectively to the coronavirus pandemic, showing us an especially dramatic example of how uncertain the future can be. One of the most immediate effects of the global response to the pandemic has been the drastic reduction in economic activity. According to the Managing Director of the International Monetary Fund,[6] as a result, the world seemed to be facing its worst recession since the Great Depression. No-one can know the longer-term social, political, and economic consequences of such a recession, nor how long any consequences will take to be fully realised. But, taking an optimistic view, we may return to some kind of normality over the next few years, enabling once again the kinds of international efforts required to tackle the challenges of building innovative capabilities.

In thinking about what these efforts would look like, we can turn to what the latecomer and other literatures tell us about past experiences. For Kenya and Tanzania directly, the groundwork to establish the pre-latecomer phase was achieved through a multitude of donor-funded projects intervening on a range of challenges – e.g., technical skills, technological products, consumer finance, market research, standards, policies – and included some attention to local manufacture. According to Ockwell and Byrne (2017), for example, in the late 1990s and early 2000s in Kenya there were donor-funded attempts to develop the local manufacture of battery charge regulators for SHSs. Despite early promise such manufacture would succeed, it failed because of better quality and cheaper products imported from China. We do not have a deep analysis of these local manufacturing efforts, but we could argue they may have failed because they were ad hoc, involved just one firm, and were implemented without any national-level strategic cooperation. If they had taken a more systemic nurturing approach, such as working with policy makers to establish a degree of protectionism while investing in the capability-building of several local firms, they might have been more successful. Kenya does, however, have at least two examples of manufacturing success related to off-grid PV, already noted earlier. One is the Solinc module assembly plant in Naivasha and the other is the battery manufacturer Chloride Exide (whose Kenyan owner part owns Solinc[7]).

We can also learn from the East Asian experience, where local electronics firms upgraded their capabilities through various kinds of relationships with foreign firms. The general policy issue arising from an understanding of this experience is how to translate lessons to contexts such as those in Kenya and Tanzania in the present time. Answering this question will require answering a number of other questions. We would need to know, for example, the extent to which there are similarities and differences between the pre-latecomer phases of the East Asian countries and those of Kenya and Tanzania, and other poor countries in the Global South. The kinds of relationship investments in East Asia that helped nurture innovation capability-building are unlikely to be easily established in the poorer countries of the Global South, especially those in Africa. Leading PV system technology companies such as those manufacturing PAYG equipment may not see any benefit in moving their manufacturing operations to countries such as Kenya and Tanzania; indeed, they may only see huge risks and a range of disbenefits. Enticing these companies to invest in what may be drawn-out processes of capability-building would likely need complementary and strategic public sector investments. These investments may be beyond the capacity of poorer countries to achieve, suggesting there may have to be a significant role played by donors.

This raises further and perhaps more complex questions probably better answered by drawing upon politically attuned analyses rather than relying solely on the primarily technical streams of innovation studies. For example, under the still dominant neoliberal orthodoxy informing much development cooperation, the preference is to let market forces determine where manufacturing takes place. The problem with this approach is it favours the status quo and may even entrench it. That is, those countries now benefiting from manufacturing comparative advantages are more likely to continue doing so while those countries with no manufacturing comparative advantage are unlikely to ever develop any (e.g., see Reinert, 2007). As Reinert and many others argue, upsetting the status quo will require interventionist public sector action. The challenge for policy makers in the Global South is how to fund innovation capability-building interventions when relying on money from development partners who remain wedded to neoliberal orthodoxy and whose willingness to fund long-term interventions may be weak in the aftermath of the coronavirus pandemic. And a further challenge for Southern policy makers may stem from the potentially disruptive consequences of the so-called fourth industrial revolution, in which the use of robotics for manufacturing is predicted to become pervasive. Despite certain gains, if such predictions prove realistic, the need for human-embodied capabilities will be severely reduced and the types of skill sets will be altered, with drastic consequences for the number and type of jobs available in an economy.

Are we there yet? Building innovation capabilities and sustainable industrialisation

So, we are not there yet. Our analysis paints a mixed picture of the conditions across Kenya and Tanzania – and relevant to other Global South contexts – in relation to the current pre-latecomer phase, and the prospects for fostering latecomer entry that could nurture sustainable industrialisation. The gloomier part of the picture shows a range of complex uncertainties facing Southern policy makers and development partners. But there is also a bright side to the picture from which to draw hopeful inspiration.

The foundational capabilities in off-grid solar PV already present in Kenya and Tanzania can help in accumulating more sophisticated capabilities for mastering the technologies aligned with the aim of sustainable industrialisation. Individual start-up endeavours may collapse, but human-embodied capabilities remain, can be developed, and can be transferred into new ventures. The experience of off-grid solar PV in Kenya and Tanzania is testament to this view. However, world-leading innovativeness so far primarily results from new business models facilitated by technologies designed and manufactured elsewhere. In line with the thrust of latecomer theory, our chapter calls for much more concerted efforts to build the necessary capabilities for local assembly and manufacturing. Firms, policy makers, development partners, and others will have to strategically build these capabilities over the long-term while navigating a range of uncertainties. If they do not, firms and others involved in the sector will be left to innovate on their own without the necessary capabilities. Under such conditions, they will 'merely keep up rather than catch up' (Hobday, 1995, p. 1186). If relevant actors do succeed in building the necessary capabilities, there are likely to be more opportunities on the part of local firms to assume diverse roles in innovating in the sector, its emerging segments, and realistic hopes of creating sustainable industrialisation development pathways. And lessons from successful demonstration of such pathway creation will be valuable to other sectors and countries in the Global South.

Conclusions

Given the interest in innovation capabilities for renewable electrification that can contribute to sustainable industrialisation, in this chapter we analysed the accumulation of capabilities in the off-grid solar PV sector in Kenya and Tanzania. As an attempt to apply latecomer theory in these contexts when innovating with environmentally friendly technologies, we explained how foundational capabilities in the off-grid solar PV sector have been built. This underscores how important it is to begin by building a sustainable base of capabilities. We disaggregated the innovation capabilities of off-grid solar PV firms active in Kenya and Tanzania and showed the value of these capabilities under a rapidly changing market structure, increasingly dominated by PAYG business models. Some local firms – more so in Kenya than in Tanzania – are manifesting increasingly innovative behaviour, but only simple assembly activities exist in Kenya.

Our analysis generally recommends Kenya and Tanzania to build on the gains of the foundational capability-building period, which we have called a pre-latecomer phase. This means enhancing the ways for local entrepreneurial firms to gain from the innovative developments in the off-grid PV sector. More specifically, these efforts would need to build local production capabilities, enabling firms to move 'up' the value chain. Over time, the achievement of more local assembly and manufacturing would signify entering an early latecomer phase, aligning with the aim of sustainable industrialisation. As a challenge, any capability-building efforts in the sector should be reflected against future uncertainties, such as the potential impacts of the fourth industrial revolution. Finally, experiences in the sector may provide valuable lessons on a range of issues for policy makers who wish to promote sustainable industrialisation in the Global South.

Acknowledgement

This research was supported by Tekes – the Finnish Funding Agency for Innovation, Grant: 40101/14 and the Danish Ministry of Foreign Affairs, Grant: DFC 14-09AAU. The authors acknowledge this support.

Notes

1 OEM is original equipment manufacture; ODM is own-design and manufacture; OBM is own-brand manufacture.
2 On what is a 'local' or 'indigenous' firm, see Jackson et al. (2008).
3 In the mid-2010s, this formerly Dutch-owned joint venture Ubbink East Africa became a Kenyan-owned company called Solinc East Africa.
4 Chloride Exide's history is briefly described on its website at www.chlorideexide.com/about/ (accessed 24 April 2020) and on the ABM Group (Chloride Exide's owner) website at www.abmeastafrica.com/about-us (accessed 24 April 2020).
5 Interview, SME director, 16 May 2016.
6 Kristalina Georgieva's statement of 15 April 2020, 'Exceptional Times, Exceptional Action', Opening Remarks for Spring Meetings Press Conference, available at www.imf.org/en/News/Articles/2020/04/15/pr20162-exceptional-times-exceptional-action-opening-remarks-for-spring-meetings-press-conference (accessed 28 April 2020).
7 See the ABM website describing majority ownership of Solinc at www.abmeastafrica.com/about-us (accessed 24 April 2020).

References

Ahlborg, H. and Hammar, L. (2014) 'Drivers and barriers to rural electrification in Tanzania and Mozambique: Grid-extension, off-grid, and renewable energy technologies', *Renewable Energy* 61, pp. 117–124. https://doi.org/10.1016/j.renene.2012.09.057

Audretsch, D. (2012) 'Entrepreneurship research', *Management Decision* 50(5), pp. 755–764. https://doi.org/10.1108/00251741211227384

Bell, M. (1997) 'Technology transfer to transition countries: Are there lessons from the experience of the post-war industrializing countries?' in Dyker, D. (ed.) *The Technology of Transition: Science and Technology Policies for Transition Countries*. Budapest: Central European University Press, pp. 63–94. Available at: http://sro.sussex.ac.uk/id/eprint/23027

Bell, M. (2009) 'Innovation capabilities and directions of development', *STEPS Working Paper* 33. Brighton: STEPS Centre. Available at: https://steps-centre.org/wp-content/uploads/bell-paper-33.pdf (Accessed: 16/09/2020).

Bell, M. (2012) 'International technology transfer, innovation capabilities and sustainable directions of development' in Ockwell, D. and Mallett, A. (eds.) *Low-Carbon Technology Transfer: From Rhetoric to Reality*. London: Routledge, Earthscan, pp. 20–47. https://doi.org/10.4324/9780203121481

Bell, M. and Figueiredo, P.N. (2012) 'Building innovative capabilities in latecomer emerging market firms: Some key issues' in Amann, E. and Cantwell, C. (eds.)

Innovative Firms in Emerging Market Countries. New York: Oxford University Press, pp. 24–110. https://doi.org/10.1093/acprof:oso/9780199646005.003.0002

Byrne, R. (2011) 'Learning drivers: Rural electrification regime building in Kenya and Tanzania', DPhil, SPRU: University of Sussex. Available at: http://sro.sussex.ac.uk/6963/ (Accessed: 16/09/2020).

Byrne, R., Mbeva, K. and Ockwell, D. (2018) 'A political economy of niche-building: Neoliberal-developmental encounters in photovoltaic electrification in Kenya', *Energy Research & Social Science* 44, pp. 6–16. https://doi.org/10.1016/j.erss.2018.03.028

Ciambotti, G., Sydow, A. and Sottini, A. (2019) *The Strauss Energy Business Model: Affordable Technology Innovation to Empower Kenya and Light Up the World*. London: SAGE Publications. SAGE Business Cases Originals. https://doi.org/10.4135/9781526488091

Davies, G. (2018) 'Clean energy product markets in sub-Saharan Africa: Complex market devices and power asymmetries', *Energy Research & Social Science* 42, pp. 80–89. https://doi.org/10.1016/j.erss.2018.03.009

Gabriel, C.A., Kirkwood, J., Walton, S. and Rose, E.L. (2016) 'How do developing country constraints affect renewable energy entrepreneurs?', *Energy for Sustainable Development* 35, pp. 52–66. https://doi.org/10.1016/j.esd.2016.09.006

Georgallis, P.P. and Durand. R. (2017) 'Achieving high growth in policy-dependent industries: Differences between startups and corporate-backed ventures', *Long-Range Planning* 50(4), pp. 487–500. https://doi.org/10.1016/j.lrp.2016.06.005

Goedhuys, M. (2007) 'Learning, product innovation, and firm heterogeneity in developing countries: Evidence from Tanzania', *Industrial and Corporate Change* 16(2), pp. 269–292, https://doi.org/10.1093/icc/dtm003

GOGLA (2019) 'Global off-grid solar market semi-annual sales and impact data', Global Off-Grid and Lighting Alliance and Lighting Global. Available at: https://www.gogla.org/global-off-grid-solar-market-report (Accessed: 01/12/2019).

Grilli, L., Mazzucato, M., Meoli, M. and Scellato, G. (2018) 'Sowing the seeds of the future: Policies for financing tomorrow's innovations', *Technological Forecasting and Social Change* 127, pp. 1–7. https://doi.org/10.1016/j.techfore.2017.10.021

Hansen, U.E., Pedersen, M.P. and Nygaard, I. (2015) 'Review of solar PV policies, interventions and diffusion in East Africa', *Renewable and Sustainable Energy Reviews* 46, pp. 236–48. https://doi.org/10.1016/j.rser.2015.02.046

Hobday, M. (1995) 'East Asian latecomer firms: Learning the technology of electronics', *World Development* 23(7), pp. 1171–93. https://doi.org/10.1016/0305-750X(95)00035-B

Hobday, M. (2001) 'The electronics industries of the Asia-pacific: exploiting international production networks for economic development', *Asian-Pacific Economic Literature* 15(1), pp. 3–30. https://doi.org/10.1111/1467-8411.00092

Hobday, M., Rush, H. and Bessant, J. (2004) 'Approaching the innovation frontier in Korea: the transition phase to leadership', *Research Policy* 33, pp. 1433–1457. https://doi.org/10.1016/j.respol.2004.05.005

International Finance Corporation (IFC) (2007) 'Selling solar', International Finance Corporation and Global Environmental Facility. Available at: https://www.ifc.org/wps/wcm/connect/topics_ext_content/ifc_external_corporate_site/sustainability-at-ifc/publications/publications_loe_sellingsolar__wci__1319577385747 (Accessed: 16/09/2020).

Jackson, T., Amaeshi, K. and Yavuz, S. (2008) 'Untangling African indigenous management: Multiple influences on the success of SMEs in Kenya', *Journal of World Business* 43, pp. 400–416. https://doi.org/10.1016/j.jwb.2008.03.002

Kenya Climate Innovation Center (KCIC) (2017) *Kenya Solar PV Market Assessment*. Nairobi: Kenya Climate Innovation Center. Available at: https://www.kenyacic.org/sites/default/files/publications/KCIC%20Solar%20Survey-3.pdf (Accessed: 16/09/2020).

Kim, C.W. and Mauborgne, R. (1997) 'Value innovation: The strategic logic of high growth', *Harvard Business Review* 75(1), pp. 102–112. Available at: https://hbr.org/2004/07/value-innovation-the-strategic-logic-of-high-growth (Accessed: 17/09/2020).

Lall, S. (1992) 'Technological capabilities and industrialization', *World Development* 20(2), pp. 165–186. https://doi.org/10.1016/0305-750X(92)90097-F

Lema, A. and Lema, R. (2013) 'Technology transfer in the clean development mechanism: Insights from wind power', *Global Environmental Change* 23(1), pp. 301–13. https://doi.org/10.1016/j.gloenvcha.2012.10.010

Lighting Africa (2008) *Off-Grid Energy Has Key Role in Kenya's New Electrification Strategy* [21 December]. Lighting Africa. Available at: https://www.lightingafrica.org/off-grid-energy-has-key-role-in-kenyas-new-electrification-strategy/ (Accessed: 21/06/2019).

Lighting Global (2020) *Off-Grid Solar Market Trends Report 2020 Report Summary*. Washington, DC: Lighting Global, Global Off-Grid Lighting Alliance and ESMAP. Available at: https://www.worldbank.org/en/topic/energy/publication/off-grid-solar-market-trends-report-2020 (Accessed: 22/06/2020).

Lighting Global and GOGLA (2018) *Off-Grid Solar Market Trends Report 2018 Full Report*. Lighting Global and Global Off-Grid Lighting Alliance. Available at: https://www.lightingafrica.org/wp-content/uploads/2018/02/2018_Off_Grid_Solar_Market_Trends_Report_Full.pdf (Accessed: 22/06/2019).

Liu, J. (2017) 'The global innovation networks and global production networks of firms: Conceptualization and implication', *African Journal of Science, Technology, Innovation and Development* 9(3), pp. 229–40. https://doi.org/10.1080/20421338.2017.1309809

Miller, D. (2009) *Selling Solar: The Diffusion of Renewable Energy in Emerging Markets*. London: Earthscan. https://doi.org/10.1108/meq.2009.08320fae.001

Nygaard, I., Hansen, U.E. and Larsen, T.H. (2016) 'The emerging market for pico-scale solar PV systems in Sub-Saharan Africa: From donor-supported niches toward market-based rural electrification', Copenhagen: Technical University of Denmark, UNEP DTU Partnership. Available at: https://www.gogla.org/sites/default/files/recource_docs/market-pico-solar_web.pdf (Accessed: 17/09/2020).

Ockwell, D. and Byrne, R. (2017) *Sustainable Energy for All: Innovation, Technology and Pro-Poor Green Transformations*. Abingdon: Routledge. https://doi.org/10.4324/9781315621623

Odarno, L., Sawe, E., Katyega, M.J.J. and Lee, A. (2017) 'Accelerating mini grid deployment in Sub-Saharan Africa: Lessons from Tanzania' Washington, DC: World Resources Institute and TATEDO. Available at: https://www.esmap.org/node/140593 (Accessed: 16/09/2020).

Osano, H.M. (2019) 'Global expansion of SMEs: Role of global market strategy for Kenyan SMEs', *Journal of Innovation and Entrepreneurship* 8(1), pp. 1–31. https://doi.org/10.1186/s13731-019-0109-8

Oslo Manual (2018). *Oslo Manual 2018: Guidelines for Collecting, Reporting and Using Data on Innovation*. 4th Edition. Paris: Eurostat, Luxembourg: OECD Publishing. https://doi.org/10.1787/9789264304604-en

Picken, J.C. (2017) 'From startup to scalable enterprise: Laying the foundation', *Business Horizons* 60(5), pp. 587–595. https://doi.org/10.1016/j.bushor.2017.05.002

Poole, D.L. (2018) 'Entrepreneurs, entrepreneurship and SMEs in developing economies: How subverting terminology sustains flawed policy', *World Development Perspectives* 9, pp. 35–42. https://doi.org/10.1016/j.wdp.2018.04.003

Reinauer, T. (2019) 'Learning among latecomer firms in low-carbon energy technology: the case of the Thai biogas industry', PhD thesis, University College London. Available at: https://discovery.ucl.ac.uk/id/eprint/10078261/

Reinert, E. (2007) *How Rich Countries Got Rich…and Why Poor Countries Stay Poor.* London: Constable & Robinson.

Renewable Energy Policy Network for the 21st Century (REN21) (2020) *Renewables 2020, Global Status Report.* Paris: Renewable Energy Policy Network for the 21st Century. Available at: https://www.ren21.net/wp-content/uploads/2019/05/gsr_2020_full_report_en.pdf (Accessed: 24/06/2020).

Ries, E. (2011) *The Lean Startup: How Today's Entrepreneurs Use Continuous Innovation to Create Radically Successful Businesses.* London & New York: Penguin Books.

Rohrbeck, R. (2010) *Corporate Foresight: Towards a Maturity Model for the Future Orientation of a Firm.* Heidelberg: Physica-Verlag, Springer. https://doi.org/10.1007/978-3-7908-2626-5

Rolffs, P., Ockwell, D. and Byrne, R. (2015) 'Beyond technology and finance: Pay-as-you-go sustainable energy access and theories of social change', *Environment and Planning A* 47 (12), pp. 2609–27. https://doi.org/10.1177/0308518X15615368

Schumpeter, J.A. (1934) *The Theory of Economic Development: An Inquiry Into Profits, Capital, Credit, Interest, and the Business Cycle.* Cambridge, MA: Harvard Economic Studies, 46.

Shane, S. (2000) 'Prior knowledge and the discovery of entrepreneurial opportunities', *Organization Science* 11(4), pp. 448–469. https://doi.org/10.1287/orsc.11.4.448.14602

von Hippel, E. (1994) '"Sticky Information" and the locus of problem solving: Implications for innovation', *Management Science* 40 (4), pp. 429–439. https://doi.org/10.1287/mnsc.40.4.429

Watson, J., Byrne, R., Ockwell, D. and Stua, M. (2015) 'Lessons from China: building technological capabilities for low carbon technology transfer and development', *Climatic Change* 131(3), pp. 387–99. https://doi.org/10.1007/s10584-014-1124-1

Wood Mackenzie (2019) 'Strategic investments in off-grid energy access: Scaling the utility of the future for the last mile', *Wood Mackenzie and Energy* 4 Impact [28 February] Available at: https://www.energy4impact.org/sites/default/files/strategic_investments_in_off-grid_energy_access_final.pdf (Accessed: 02/12/2019).

10

CHINESE GREEN ENERGY PROJECTS IN SUB-SAHARAN AFRICA

Are there co-benefits?

Padmasai Lakshmi Bhamidipati, Cecilia Gregersen, Ulrich Elmer Hansen, Julian Kirchherr, and Rasmus Lema

Abstract

Investments in renewable energy are increasing rapidly in sub-Saharan Africa. An interesting trend to note is the rapid increase and likely future growth of Chinese involvement in large-scale renewable-energy infrastructure projects. Our focus in this chapter is to determine the extent of co-benefits created when renewable-energy projects are developed by Chinese investors. For this, we undertake an in-depth micro-level analysis of three Chinese renewable-energy investment projects in hydro (Ghana), wind (Ethiopia), and solar photovoltaic (PV) (Kenya), based on primary data. Overall, we find evidence of 'bounded benefits'. On the one hand, we can identify some newly created jobs, linkages generated with actors in local systems of production, and training activities involving local staff. On the other hand, the extent of these benefits is very limited. The results suggest that policymakers should be wary of overly optimistic expectations when it comes to assessing the co-benefits of renewable energy projects in the context of scarce pre-existing capabilities. However, the adoption of pro-active strategies and the implementation of carefully designed policies can increase the local economic co-benefits.[1]

Introduction

The electricity generating capacity in sub-Saharan Africa (SSA) will double over the next 20 years, with renewables accounting for three-quarters of new generation, the majority of that coming from solar, hydro, and wind (IEA, 2020). The purpose of this book chapter is to explore to what extent and under what conditions these massive investments in renewable energy (RE) have economic co-benefits. Additional benefits going beyond countering climate change in

DOI: 10.4324/9781003054665-10

sub-Saharan Africa include job creation, improvement of local skills and creation of income-generating activities. The RE sector can become an integral part of local economies, integrated both through upstream supply chain, such as production of equipment components, and downstream energy related services, such as maintenance (IRENA, 2013, p. 15; see also Sperling, Granoff and Vyas, 2012).

In this chapter we focus on investments made by China as it accounts for the single largest investment portfolio in SSAs power sector.[2] According to the International Energy Agency (IEA, 2016, p. 7), projects in which a Chinese firm is the main contractor alone account for 30% of new capacity additions in SSA; of these projects, 56% are in renewable energy, with the vast majority being in hydropower, but increasingly also in wind and solar energy. Insights from other infrastructure, utility, and resource-extraction sectors in SSA suggest that China is pursuing a specific Chinese model of investments consisting of enclave characteristics, including finance, turnkey project development, and imports of labour and equipment from China (Kaplinsky and Morris, 2009; Sanfilippo, 2010; Wegenast et al., 2019). Hence our focus in this chapter is to what extent economic co-benefits arise in sub-Saharan Africa when renewable-energy projects are developed by Chinese investors: *what is the potential for benefiting from Chinese renewable-energy investments in terms of employment, localisation of the value chain and technological learning?* In order to seek insights into this question, we focus on investments in hydro, wind, and solar energy for electricity generation.

Despite the increasing attention paid to Chinese renewable-energy investments in SSA and the economic opportunities associated with them, there are few studies, let alone systematic analyses, in the existing literature (Shen and Power, 2016). In addition, there is very little evidence of the real economic opportunities associated with green investments and policies in low and lower-middle income countries. In this chapter, we undertake an empirical examination of three specific Chinese projects in hydro, wind, and solar energy. We also address the research gap by gathering insights about economic opportunities and developmental effects from three case studies of frontrunner green energy projects in SSA.[3] By providing in-depth analysis of co-benefits in terms job creation, value-chain localisation, and capability-building, we hope to stimulate an informed discussion of the conditions and policy measures which may maximise the local benefits of these investments.

We develop a conceptual framework and employ co-benefits approach to unravel the three case studies and the explanatory determinants for the respective outcomes. While the analytical framework is elaborated in detail in another article (Lema, Fu, and Rabellotti, 2020), we focus mainly on the empirical findings in this book chapter. In the next section, we highlight China's involvement in the renewable energy deployment in SSA, followed by in-depth findings from the case studies in detail. This is followed by a section which analyses the explanatory factors and summarises the co-benefits in a comparative way to gauge the similarities and differences among the three cases. The chapter ends with a conclusion and policy implications.

China's involvement in renewable energy deployment in SSA

This section provides an overview of China's involvement in renewable energy deployment in sub-Saharan Africa in relation to the three technologies discussed in this article. Discussing the patterns of capital and technology flows from China allows us to examine the macro-evidence for the existence of a 'Chinese model' of green energy investments. The purpose is to provide a backdrop for the project-level analyses in subsequent sections.

China's overall role in the SSA energy sector

Shen (2020) estimates that Chinese finance for the energy sector in Africa, including North Africa, amounted to a total of more than USD 30 billion over the sixteen-year period from 2000 to 2016, but this includes all energy sources, both black and green. However, according to the IEA (2016), in an analysis of Chinese greenfield energy investment projects which had been completed, were under construction, or were planned for completion over the 2010–2020 period, 56% of Chinese energy-generation projects were found to use sources of renewable energy. The total investments involved amounted to USD 13 billion across 37 countries.

We analysed the available data on the installed capacity in SSA across the three energy sources. (IRENA, 2013, 2019). In the hydropower sector, Chinese investors accounted for 60% of investments in sub-Saharan projects. The Chinese are also significantly involved in both the solar PV investments (108 MW in 2009 to 6.1 GW in 2018) – which surpassed investments in hydropower for the first time in 2019 – and the wind-energy sector (739 MW in 2009 to 5.5 GW in 2018), which is forecast to grow rapidly in SSA, in particular in countries with high altitudes or locations at some distance from the equator (IEA, 2016, 2020). However, there are no data sources which can give a complete picture of the relative degrees of Chinese involvement across the three technologies (Shen, 2020).

Roles of Chinese actors as financiers, EPC contractors, and technology suppliers

This section analyses the role of various Chinese actors in the development of hydropower, solar PV, and wind-power projects, focusing specifically on:

i) *Financial institutions* – the Export-Import Bank of China is by far the main investor in projects constructed by Chinese contractors, providing finance to more than 60% of the projects analysed in IEA (2016). The main investment model is based on preferential loans and export credits provided to project developers. In addition, direct equity-based investments, commercial loans, and grants are also provided, in particular from the financial institutions mentioned in Table

10.1. Financial institutions are powerful actors in the transnational investment-production complexes in which green energy infrastructure projects are embedded, and they may specify 'foreign content requirements' involving Chinese Engineering, Procurement, and Construction (EPC) contractors and technology providers as part of financing deals.

ii) *EPC contractors* – The main Chinese contractors involved in renewable-energy projects in sub-Saharan Africa typically include large state-owned enterprises (SOEs): 90% of the power projects analysed in IEA (2016) are being contracted and constructed by Chinese SOEs (see Table 10.2). The remaining 10% of these projects are being constructed by private Chinese developers specialising in large-scale infrastructure, construction, and civil-engineering projects in the energy sector.[4] As already mentioned, under EPC contracts, Chinese developers are responsible for all aspects of the project, from the initial feasibility stage via plant engineering and the subcontracting of components and related services to the plant's final commissioning. EPC is thus instrumental in selecting technology providers.

iii) *Technology providers* – Given an increasingly saturated domestic market and fierce competition in the European and US markets, Chinese technology-producing companies, such as those mentioned in Table 10.2, have increasingly moved into sub-Saharan Africa (Shen, 2020).

Table 10.2 draws on the available data to show the changes in exports of renewable-energy technology from China to sub-Saharan Africa over two five-year

TABLE 10.1 Key Chinese financial institutions, EPC contractors, and technology suppliers involved in the green energy sector in sub-Saharan Africa

	Finance	*EPC contractors*	*Technology suppliers*
Hydro	• China Export-Import Bank (China Exim Bank) • Chinese Development Bank (CDB) • Sinosure	• Sino Hydro • PowerChina Resources • Three Gorges Corporation	• Dongfang Electric Corporation • Harbin Electric Corporation • Shanghai Electric Power
Wind	• Industrial and Commercial Bank of China (ICBC) • Bank of China (BoC)	• CGC Overseas Construction Group • Hydro China • Longyuan Power Group	• Goldwind • Sany • Sinovel
Solar		• China Jiangxi Corporation • Powerway • Beijing Xiaocheng	• JinkoSolar • Yingli • JA Solar

Source: authors, adapted from Chirambo (2018), Shen and Power (2016), and Tan-Mullins, Urban, and Mang (2017).

TABLE 10.2 Exports of hydro, wind, and solar equipment from China to Africa 2006–2016 (USD million)

	2006–2010	*2011–2016*	*Total*
Hydro*	2,647	9,824	12,471
Wind	1,807	532,189	533,996
Solar	41,706	393,058	434,764
Total	46,16	935,071	981,231

Source: authors, based on COMTRADE (HS codes: 841011, 841012, 841013, 850231, and 854140). *Export of hydraulic turbines and water wheels from China to Africa.

periods. There have been massive increases in exports since 2010 in all three sectors. Hydro-technology exports and imports are relatively low compared to wind and solar because core technology only constitutes a relatively small share of the overall capital expenditure in hydro projects. However, China-Africa trade in hydro-technologies like turbines more than tripled in the second five-year period when compared to the first. Nonetheless this increase is nothing like as dramatic as the increase in wind and solar, both of which are growing exponentially. These data show how recent a phenomenon the trade in renewable energy from China to Africa is and how quickly it is growing.

To summarise, the increasing influence of China in the renewable-energy sector in sub-Saharan Africa can be observed across the three renewable-energy sub-sectors analysed in this article. Interestingly, we see a tendency for Chinese investors and contractors to supply projects on a turnkey basis delivered as a bundled package comprising a considerable representation of Chinese investors, engineering companies, and technology suppliers. A possible reason for the development of this Chinese model may be the nature of China's funding-support requirements, which stipulate that investors are eligible for export credits only if the equipment used is manufactured in China.

The following section draws on primary data to examine the key factors and indicators developed for this analysis (in the next section). Three sub-sections describe each of the case-study projects in turn. Table 10.3 provides an overview of the key actors in these three projects across both the *stage of infrastructure delivery* (engineering, procurement, construction, and various sub-tasks) and the *stage of service delivery* (operation, maintenance, and distribution). These are preceded by an *initiation* stage focusing on entrepreneurial development, and the negotiation stage, which is important because it defines the nature and scope of the subsequent steps.

TABLE 10.3 Overview of the projects: actors and roles

Capacity and ownership		*Adama*	*Bui*	*Garissa*
Ownership and capacity	*Owner/sponsor*	Ethiopian Electricity Power	Bui Power Authority	Kenya Rural Electrification Authority
	Energy source	Wind	Hydro	Solar PV
	Size	204 MW	400 MW	55 MW
	Cost	USD 462 million	USD 560 million	USD 135 million
Infrastructure delivery	*EPC and project management*	HydroChina and CGC Overseas Construction Group	Sino Hydro	China Jiangxi Corporation for International Economic and Technical Cooperation
	Finance	Export-Import Bank of China (85%) and Government of Ethiopia (15 %)	Export-Import Bank of China (89%) and Government of Ghana (11%)	Export-Import Bank of China (100%)
	Front-end and detailed engineering	HydroChina and CGC Overseas Construction Group	Coyne et Bellier (France) and Sino Hydro	CJIC and Maknes Consulting
	Core technology supply	Goldwind (China) and Sany (China)	Produced in China by Alstom (France)	JinkoSolar (China) and BYD (China)
Service delivery	*Electricity distribution*	Ethiopian Electric Services (EES)	ECG/Gridco	Kenya Power and Lighting Company
	Plant operation	Ethiopian Electricity Power	Bui Power Authority	Kenya Electricity Generating Company and CJIC
	Plant Maintenance	Ethiopian Electricity Power	Sino Hydro	KECG

Source: authors

Findings from the case studies

Adama wind power project

The Adama wind power project consisted of two phases of planning and construction by a joint venture between Chinese turnkey EPC contractor HydroChina and the CGCOC group, a Chinese construction company, for Ethiopian Electricity Power (EEP), the project owner. The first phase included the installation of 51 MW of wind power and was finalised in 2012. The second phase, Adama II, included the installation of 153 MW and was commissioned in 2015.

The types of *jobs* created in the Adama project are directly linked to the financing agreements, which specified that Chinese technology was to be used in the project. The turnkey contract held by HydroChina-CGCOC covered the majority of the value chain for the project, from its design and construction to handover training. Local jobs in project construction totalled 1000 across the two phases compared to approximately 400 jobs held by Chinese employees. The contract between EEP and HydroChina stated that unskilled labour should be recruited locally and that using staff and skilled labour from sources within Ethiopia was to be encouraged. However, the large number of Chinese employees involved during this phase suggests that the job types varied, and that project management was to a large degree carried out by Chinese nationals. The key project-management personnel included approximately 13 Chinese staff for phase II, ten of whom had already worked on phase I.

Local content in the project was limited to the minimum involvement of local firms in the supply of construction materials such as concrete, while the state-owned shipping company was involved in the transportation of wind-turbine components. All imported equipment, materials, and construction equipment were exempt from customs duties, value added tax, additional taxes, and the withholding tax. Furthermore, there was only minimal involvement by local communities in respect of deciding compensation for the temporary and permanent loss of farmland in order to build the wind farms and the necessary access roads. Beyond the access roads and water pumps, other social development projects were not deemed to be required. HydroChina held multiple information sessions and seminars to educate local residents on the impacts of wind farms.

In respect of *technological learning*, the investment model, designs, and blueprints for the project were developed independently by HydroChina and CGCOCC. All permanent equipment for the project was sourced and imported from Chinese companies – the unit transformer, 33KV cabinet, main transformer, circuit breaker, grounding transformer, SCADA system, and communication equipment – which constrained local learning. However, a team of 17 employees from Ethiopian universities was engaged by EEP to monitor implementation of the project during the construction stage and administer the contract. These employees were engaged to carry out a number of supervisory tasks, including reviewing micro-siting and layout designs, supervising the civil infrastructure, construction and erection of the wind turbines, and preparing acceptance certificates and project manuals. The university consultancy arrangement was the result of a national strategy to involve universities in projects in order to facilitate technology transfers and capacity-building. The project owner's knowledge accumulation was focused on Operations & Maintenance (O&M) related technological learning, while the university consultancy was specifically tasked with acquiring knowledge in project management, the implementation of construction contracts and ultimately building capacity for the manufacture of the components of wind-power technologies.

In *summary*, the Adama project is a case of medium co-benefit creation, with moderate local job creation in low-skilled construction and O&M, some local sourcing of peripheral services, and the critical involvement of actors in the local knowledge system. There was some technological learning, but it was still rather restricted. Most learning was confined to service delivery domains, with little to no learning in the infrastructure delivery domain. The main explanation for the economic co-benefits observed here is to be found in the semi-strategic stance adopted by the Ethiopian government, with a deliberate and explicit effort to obtain useful knowledge from the project implementation process. The nature of the technology adopted and the absence of a corresponding local supply base meant that there were few possibilities for local inclusion in the manufacturing chain, but there was a possibility for further inclusion in services, such as plant construction, turbine assembly, and installation. However, the project was undertaken mainly as a 'bundled' model with end-to-end services delivered by the Chinese consortium. This model was chosen through non-competitive and direct negotiations between the local government and the Chinese developers. Policy was the most decisive factor in securing some benefits, but it was not extended beyond involving key knowledge actors, so that further potential economic activities were not localised.

The Bui Dam hydropower project

Construction of the Bui Dam – with Sinohydro, a Chinese state-owned enterprise that is the world's largest dam-builder with a global market share of more than 50% in charge of its execution – started in 2006. The contract with Sinohydro was a turnkey or EPC contract, which meant that Sinohydro was only in charge of its construction and not its operation. The Bui Dam, a roller-compacted concrete (RCC) gravity dam in Ghana with a capacity of 400 MW, was completed in 2013, the entire dam (including turbines, powerhouse etc.) and its operation being turned over to the Bui Power Authority (BPA)[5] upon completion of the project.

Formally, strategic oversight of the project lay with the Ghanaian Ministry of Energy (MoE) and the operational oversight with the Bui Power Authority (BPA). A nuanced understanding of mega-dam construction is needed to fulfil such oversight duties sufficiently (Flyvbjerg, Holm, and Buhl, 2002). However, various interviewees suggested that Sinohydro's reporting to the MoE and the BPA was relatively sporadic and at times incomplete, due to limited capacity within MoE.

In respect of *jobs*, of the 1,836 workers employed at the Bui Dam construction site, as many as 91% were Ghanaian, the project thus providing 'temporary employment for roughly one out of 20 workers in the Tain District' where the project is located. On-the-ground management of the project, however, was exclusively Chinese. Informants suggested that importing relatively low-skilled construction workers from faraway China instead of hiring them locally, with only the little training required by them, would increase the project's costs. Around 50 Ghanaian staff, employed by BPA, are now involved in the operation and maintenance of the project.

With regard to *local content*, most material-processing content and associated sourcing needed for the dam, mostly concrete, were sourced locally. The exact percentage of local content going into this project is difficult to establish, but one informant estimated that at least 60% of this project consisted of local content. This high share of local content was to some extent policy-driven, as a clear local-content policy guides investment in the country. While overall local content provision was significant, it is also clear that the more sophisticated provision of products and services was retained by Sinohydro, which, for example, procured three 133 MW hydro turbines from the French company Alstom's factory in China.

With respect to *technological learning*, we distinguish between learning related to construction and to operation. While the construction of a large dam is a complex endeavour, with hydropower dams completed post-2000 facing an average cost overrun of 33% and an average schedule overrun of 18%, its operation is relatively uncomplicated. BPA was expected to be able to operate the dam upon its completion. However, this turned out not to be the case. Sinohydro was re-engaged to ensure that major maintenance was carried out (also reported by GhanaWeb, 2017). This suggests that little technological learning took place on the Ghanaian side in connection with the project's maintenance when it was constructed. Also, Sinohydro did not transfer any significant knowledge and expertise regarding the technology to the Ghanaians. Therefore, despite the local employment during the O&M stage, the locals struggled to carry out the various maintenance tasks due to which Sinohydro were rehired again for capacity-building.

To *summarise*, this is a case of low co-benefits, with employment of workers in Bui district during construction, but with little national impact. Limited technological learning took place, mostly confined to the operations part of service delivery and not including maintenance or construction, but there was a significant degree of local sourcing of construction materials. The main explanation for the identifiable economic co-benefits is the nature of the technology, where project management is highly complex, where only a few steps in infrastructure delivery in the value chain can be carried out remotely, and where construction needs to be localised. However, due to the absence of independent local firms, in Ghana these steps were carried out by Chinese firms. The project contract was directly negotiated between the Ghanaian government and the Chinese developers. In the absence of a strategic vision on the part of the government, the EPC's full-package provision left very little room for localisation and learning in this deal. The core insight from the Bui Dam case with regard to co-benefits from the perspective of the Ghanaian stakeholders is thus that the most crucial long-term co-benefit, technological learning, was not facilitated by Sinohydro. However, those co-benefits that are frequently discussed in the popular press, namely local content, local participation, and job creation, were more substantial.

The Garissa Solar PV project

The Garissa PV project is the first grid-connected solar project in Kenya, with a capacity of 54 MW. It was conceived by China (the Jiangxi Province representatives – JPR), along with Kenya (Ministry of Energy – MoE). The lead project developers (in particular, JPR) also facilitated securing the full project finance via China's Exim Bank, provided as a concessional loan. The project, commissioned in 2016, is administered and owned by the Rural Electrification and Renewable Energy Corporation (REREC).[6] While there is a FiT in place in Kenya to attract private investment and standardise tariffs, this was circumvented, and direct negotiations were used instead.

The choice of technology suppliers for the Garissa project was determined by the tied financing agreement, which mandated the use of Chinese technology. The JPR recruited their own state-owned enterprise, the China Jiangxi Corporation for International Economic and Technical Cooperation (CJIC), as the lead EPC and signed a contract with Jinko Solar to supply panels, and with Byd for inverters. CJIC also subcontracted two Chinese companies for project design and civil works. After the project's completion, there was a brief handover period from CJIC for the O&M, with a service agreement of two years, to the Kenya Electricity Generation Company (KenGen), responsible for undertaking O&M at the plant and contracted by REREC.

While there was no explicit strategy, the priority for *local jobs* was subject to a verbal agreement between REREC and CJIC. The overall project management was carried out by Chinese nationals, while nearly 85% of the workers employed during the project's construction were Kenyan nationals. However, most of them were hired on a casual basis, without formal contracts. During the construction period, nearly 300 to 350 Kenyan workers were employed. Of this, a majority took on low-skill tasks as carpenters, masons, drivers, manual lifters, and security guards, and they were involved in constructing the office buildings, lifting solar panels, and performing such manual tasks. The rest were engaged in semi-skilled tasks, including the installation of solar panels, electrical work, and steel work. In this period, nearly 75 Chinese employees were engaged in preparing steel structures, supervising tasks, operating JCB machines, and performing electrical tasks. During the operational phase, nine O&M engineers are employed on a contract basis, of whom five are Kenyan nationals and four are Chinese, forming a team working in a similar capacity.[7]

The bundling of finance with an EPC contract left relatively limited scope for *local content*. The sub-contractors included mainly Chinese companies. For civil works, a local Kenyan company was contracted to provide manpower during the construction phase. While Kenya has a sizeable number of solar PV companies, they are focused mainly on off-grid systems and small-scale PV installations. A few companies are gradually scaling up in the hope of obtaining sub-EPC contracts for large projects, but there are limitations still pertaining to project design, sizing systems optimally, and handling O&M tasks.

In terms of *local technological learning*, there was only a limited transfer of core technological knowledge, since all the permanent equipment for the project was imported as embodied knowledge from China, including solar panels, accessories, electrical equipment, the control system, and construction tools. Some construction equipment was sourced locally in Kenya, including electrical cabinet boxes, switch boxes, and circuit breakers. While core technological learning was limited, there was learning in other areas, including 'systems' design and operations. REREC engaged a Kenyan firm, Maknes Consulting Engineers, to oversee technical activities in the project. Maknes played a supportive role in reviewing the project drawings and O&M manuals, supervising the installation work, and overseeing technical progress. Reportedly, the tasks carried out by Maknes in the Garissa project were similar to those undertaken in other projects, albeit not on this scale. In other words, local knowledge acquisition regarding large-scale PV was deliberately designed into the project, which may be relevant to future projects.

To *summarise*, this is a case of low co-benefits. Although local job creation was significant (of the three projects, the highest per megawatt installed), local equipment provision and skills and knowledge transfer were limited and peripheral. Although one local engineering firm became involved in the infrastructure delivery process, gaining experience relevant to project execution, local learning was mainly confined to O&M. The main explanation for the limited economic co-benefits that were observed in this case are to be found in the institutional arrangements surrounding the project, with limited strategic intent evoked by local policymakers in relation to its organisation. The project was directly negotiated and involved a consortium model involving Chinese firms, contractors, and financiers with limited involvement by local actors. Although local solar firms could arguably have taken responsibility for parts of the project's construction, this was precluded by the 'tied finance' underpinning the project.

Economic co-benefits and their determinants

The three projects differ significantly in their technical nature, but it is relevant to bring them together for analysis and comparison. In this final analytical section, we start by providing an overview of the identification of co-benefits before proceeding to an explorative discussion of the determinants of these benefits.

The nature of inbound flows of capital and technology from China

The *nature and flows of capital and technology* were important influencing factors when it comes to the realisation of co-benefits. They differ greatly with respect to labour intensity, capital requirements and complexity involved in each of the projects, a finding aligned with the previous literature, which highlighted that industry localisation effects are highly technology-specific (Schmidt and Huenteler, 2016). For example, the relatively high degree of local content in the Bui case can be explained by the high transportation costs of cement for construction and the need to produce the cement on site.

However, in none of the three cases was the choice of technology for the project rooted in such deliberations or overall national energy plans (with the partial exception of the Bui Dam). On the contrary, the interviews suggest that technology selection was heavily influenced by the Chinese *lead agents involved*, who had their own technological preferences, in conformity with the previous literature (Ajakaiye and Kaplinsky, 2009; Kaplinsky and Morris, 2009). The analysis suggests that benefits are constrained by a dominant pattern of 'tied financing' associated with such chains, and it confirms the role of *the nature of finance*.

The case of Garissa showed how Jiangxi Province initiated the discussions and favoured its own state-enterprise, CJIC, while sourcing finances from China's Exim Bank. Similarly, the Adama case showed how the major actors in the project, EEP and HydroChina/CGOCCC as the EPC contractors, negotiated the contract and all contingent decisions. In the Bui case, the Chinese technology suppliers and EPC contractors also followed the Chinese investors in a tied-finance agreement. It was a requirement that investors had to produce the equipment in China in order to be eligible for export support. A non-Chinese contractor (Alstom) also received economic benefits because the equipment used in the project had been produced in China. Moreover, the contractual arrangements for this project, using an EPC contract, could have been more advantageous to the Ghanaian stakeholders, with the MoE and BPA likely to benefit much more from a build-operate-transfer (BOT) contract, which would have legally obliged Sinohydro to build the capacities needed for BPA to maintain the Bui Dam.

Local institutional and economic conditions

The analysis also suggests that local conditions – local deployment models, industrial policies, the domestic supply base, and local capabilities – significantly influence the nature of both the project and associated co-benefits. It is relevant to note that the projects analysed were negotiated in the context of weak institutional regimes, or even 'institutional voids' (Silvestre, 2015), when it comes to the *host economy deployment policy model* for renewable energy. This meant that projects were negotiated 'ad hoc' even when there was a FiT policy in place, which was eventually circumvented (the Garissa case), or there were initially intentions regarding local content, which ultimately could not be met (the Bui case).

The policy stance is a key variable and can make the difference between 'naturally occurring co-benefits' and 'induced co-benefits'. The majority of identified co-benefits are of the former type (e.g., sourcing local cement in the case of hydro), but some case material also points to the latter occurring.

The *industrial policy approach* also influences the associated co-benefits, confirming insights in the existing literature (Baker and Sovacool, 2017; McCrudden, 2004; Power et al., 2016). A more deliberate and strategic form of engagement means a greater likelihood of local capacity-building. The best example of this is the wind project in Adama, where explicit attention was paid to technological, learning, and supply-chain development during the contracting stage. As a counterpoint, the Garissa project was implemented in the context of a laissez-faire regime that entailed limited local jobs in the supply chain, limited suppliers, and hardly any engagement with a local university or research institute. In this case, the project could be viewed as a missed opportunity that REREC could have utilised specifically to focus on enhancing local skills and technical capacities, and/or supported synergies with local universities and similar repositories of knowledge to develop local capacities and strengthen the linkages of local industries. A locally active policy stance and the application of existing bargaining power, even if low, is key. It is interesting to note that Kenya has subsequently adopted a more active policy approach and has embedded local content ambitions into the newly passed energy bill (Kingiri and Okemwa, this volume).

Furthermore, the three cases emphasised the importance of the relative strength of the domestic supply base and how this needs to be considered in relation to the choice of technology (as discussed above). Our findings are aligned with the argument that co-benefits depend significantly on the capabilities of local firms engaged in green-technology manufacturing (Lema, Iizuka, and Walz, 2015). The manufacturing of most core technologies and components might be unlikely to take place in sub-Saharan Africa. However, there are a range of assembly tasks, as well as many services, that are being undertaken locally in the case of all three technologies examined here.

Investment decisions may benefit from a bottom-up approach to the selection of projects and technologies, considering first the range of activities that can easily be supplied locally (e.g., peripheral components such as solar-panel racks or wind-turbine foundations) and secondly those activities that are in the zone of proximate development, that is, where realistic capability-stretching may enable localisation (e.g., assembling turbine panels). However, the three cases all suggest that local involvement in strategic services, not least project management, is strategically important because it creates greater scope for influencing decisions concerning supply chains. Hence, the politically negotiated initiation stage of projects, where negotiations around financing may specify roles and responsibilities during the project-execution stage, is key (Hanlin, Okemwa, and Gregersen, 2019; Kirchherr and Urban, 2018). This may involve choice of technology and technology provider, as well as specifying the role of local actors and other conditions, which have a direct bearing on the creation of co-benefits.

The nature and organisation of the investment project

Our research showed that project organisation has important implications for economic co-benefit creation. In terms of the *contractual arrangements*, as mentioned already, the nature of tied finance had the knock-on effect of creating 'bundled projects' organised by Chinese EPCs. In Adama, the technical specifications of this project were quite clearly designed and influenced by the project developers, the financing, and the EPC contractors' terms. The origin of the technology was defined by the majority financial investment from China Exim Bank, while the lists of suppliers and technical equipment illustrate the preference for Chinese suppliers. Further favourable conditions were granted to the importation of equipment, with exemptions from both customs duties and taxes related to their import. However, negotiations on the part of the government of Ethiopia were designed to ensure local participation through the involvement of the universities and state-owned shipping companies.

Similarly, in the Bui project, the turnkey EPC contract that put Sinohydro in charge of its construction and operations had implications for the *project's organisation*. Some 60 relevant players were involved in the Bui Dam project overall, with Sinohydro responsible for its implementation and for organising its own supply chains.

In Garissa too there was a full-package provision of EPC contracts. Chinese developers made turnkey investments with significant imported content and frequent use of imported labour. The technical specifications of the project were designed and influenced by the project's EPC contractors and the financiers' terms and conditions. Further favourable terms were provided for any imported equipment with exemptions of both custom duties and taxes related to their import. To a large extent, the project was executed as a package 'parachuted' in from China, which limited the agency and influence that could be exerted by the national actors (Bhamidipati and Hansen, 2021).

The element of finance is significant because it shifts the relative bargaining power strongly in favour of the investor-contractor consortium. As a result, the co-benefits are largely dependent on the project developers that are engaged in making the key decisions concerning the project. However, there may be some scope for *planned capacity-building* in project negotiations. In the Garissa case, the project provided naturally occurring, learning-by-doing opportunities for skills development and for familiarising a host of Kenyan stakeholders with the solar PV technology, as well as with the processes entailed in designing and operating a utility-scale PV project. The beneficiaries included select REREC staff, Kenyan electricity firms (KPLC, KenGen, KETRACO), the Kenyan workers engaged with semi-skilled tasks, and the five Kenyan engineers hired for the O&M phase on a contractual basis. The engineers benefitted directly from the training and acquisition of relevant skills (including technical, electrical, IT, and safety-related skills). The unskilled Kenyan workers secured temporary jobs and incomes, but they also performed the sorts of tasks that are generic to most construction projects. Importantly, however, the engagement of Maknes Consulting was an important step because it created a 'vessel' for the transfer of local capabilities and lessons from one project to the next. Nonetheless the overall turnkey model of the project involving mainly Chinese contractors, the centralised nature of project delivery, and the limited planned efforts to increase local capacity-building limited the scope for co-benefits.

The government of Ethiopia utilised a similar strategy, but went further in its decision to give universities the mandate to act as the owner's consultants with the aim of increasing technology transfers, as knowledge transfer defined the unique organisational arrangements of the Adama case. Bringing in universities as important actors in this situation is interesting and suggests the intention to develop industry-university linkages. It emphasises how universities can act as recipients of knowledge transfers in the innovation system. It also accentuates universities' roles in innovation systems, where a heterogeneous group of actors that are not firms are important in contributing to capability accumulation in terms of innovation, sustainability, and long-term dynamism. However, in practice, further studies need to be conducted to assess the quality of knowledge and technology transfer, as all parties in the Adama project mentioned challenges in the collaborative arrangements.

Conclusions

This chapter has set out to examine the type and nature of the local economic co-benefits that may arise from Chinese renewable-energy investments in sub-Saharan Africa. It contributes to a small but growing body of empirical research on the economic opportunities of implementing green transformations in latecomer countries. The existing literature on such economic opportunities (i.e., the potential co-benefits) has mainly focused on large 'emerging economies' with established programmes for renewable energy, comparably strong production, and innovation systems, and the pre-existing potential for a high degree of localisation of green economic activities, and even for exports of green technologies (Binz et al., 2017; Lema, Fu, and Rabellotti, 2020). Much less attention has been paid to low and lower-middle income countries where strategies and policies for greening with renewables are much more recent and where practical implementation is dependent on significant inflows of capital and technology.

The chapter has sought to attend to this gap by focusing on specific renewable-energy investment projects in sub-Saharan Africa. Given the increasing Chinese involvement in renewable energy in this region, it was important to understand the extent, nature, and determinants of the resulting co-benefits when projects are organised by Chinese renewable-energy developers. Since this push for co-benefits, although increasing, is still in its infancy, its insights are to be derived mainly from case studies of pioneer projects.

Main findings and policy implications

The project-level analysis in this chapter suggests that the projects examined made some contributions to the local economies, but it is necessary to emphasise the highly restricted nature of the benefits we identified. Hence, we stress the need for caution when it comes to overly optimistic expectations of co-benefits arising from investments in renewable-energy infrastructure projects in sub-Saharan Africa.

In a broader perspective, the findings of this chapter highlight the significant challenges associated with the notion of green latecomer development and sustainable industrialisation in sub-Saharan Africa. In the context of latecomer development, such a strategy may be easier to achieve in upper-middle income 'emerging economies' compared to low income or lower-middle income countries with more limited institutional capabilities. This chapter has shed light on substantially different settings, where growth and development-enhancing objectives are rather difficult to achieve through large green infrastructure projects. This is not least because of the geographical separation, unequal distribution of capabilities, and skewed power relations between the users and producers of green infrastructure in Africa.

This does not mean that green latecomer development should be abandoned as a strategy in countries like Kenya, Ethiopia, and Uganda. On the contrary, it means that, at least in the context of the provision of green energy infrastructure, green latecomer development needs to be stepped up to become effective: an active and directed policy approach needs to be devised for maximising the co-benefits of further renewable energy investments in the future. To unfold this insight further, we connect insights from our findings with three pertinent policy issues.

First, while we find evidence of benefits, these benefits, however limited, did not emerge as automatic by-products of the investments. Every green investment decision needs to be preceded by exerting the full extent of the available bargaining power. Local bargaining power is often constrained, but it is not non-existent. This can ensure the maximum possible local content, jobs in knowledge-intensive tasks, and deliberately designed transfers of knowledge and capabilities from existing foreign suppliers of green infrastructure (Chinese or otherwise) to African users and associated local enterprises and organisations in local systems of production. While this point may seem obvious, there are indications that major investment decisions have been made mainly with the primary benefits in mind (i.e., reducing carbon emissions) and without paying sufficient attention to the strategic opportunities to achieve the associated economic co-benefits.

Second, these policies and strategies should focus deliberately on opportunities in the process of delivering these green infrastructure projects. There is a tendency to neglect this stage while focusing too much on the processes of delivering sustainable energy. For example, the cases analysed show that, while there were quite significant transfers of knowledge through training and overseas secondment related to operations and routine maintenance (i.e., the service delivery process), there was no correspondingly significant and deliberate transfer of capabilities related to the preceding infrastructure delivery process. Accordingly, the ambition needs to take the form of the gradual building of local capabilities related to the latter. If the greening of local energy systems is to be beneficial to local economic development, it is not sufficient to say, as is sometimes done in investor and climate change circles, that it does not matter who creates the infrastructure as long as it is green and cost-efficient. Our findings indicate that significant co-benefits will only arise with substantial local involvement in the high value-adding and more knowledge-intensive stages of the infrastructure delivery process.

Third, green energy infrastructure should not be treated in isolation in this respect. While these types of projects could become important learning and development platforms, the attainment of infrastructure project execution capabilities is relevant outside this specific domain, that is, in building roads, ports, electricity distribution systems etc. as well. Interestingly, in all three cases, independent local entities were assigned to the role of the owners' consultants. These entities could become important vessels for local transfers of lateral capabilities from one project to the next. However, due to the strategic importance of these capabilities and their national public-good nature, they may also need to be located in government offices.

Acknowledgements

The authors gratefully acknowledge the Danish Ministry of Foreign Affairs (Grant: DFC 14-09 AAU) for providing funding support for the IREK research on which this chapter is based. The authors would also like to thank all the participants of the IREK Workshop, held in Nairobi in 2019, in particular Rob Byrne and Helene Ahlborg for their invaluable comments and inputs.

Notes

1 A longer article version of this book chapter is currently under review in the Journal World Development.
2 As Shen (2020) emphasises, it is difficult to obtain a precise estimate of the size of and trends in Chinese activities in the power sector in sub-Saharan Africa. This reflects a larger problem regarding data shortcomings on funding from China because China has not released a breakdown of its lending activities.

3 The RE projects were chosen in three SSA countries: the Adama project in Ethiopia (wind energy), the Bui Dam project in Ghana (hydro energy), and the Garissa project in Kenya (Solar PV). The core of our analysis thus builds on primary data obtained at the project level. This information was used for micro-level analyses exploring inbound flows, local conditions, the characteristics of organisational arrangements, and the three main types of co-benefit. The main sources of information for these case studies are site visits at each project and a total of 38 in-depth interviews with project organisers and key informants. Given the lack of existing studies, the chapter provides a first exploratory attempt to analyse the co-benefits and their determinants in Chinese projects.
4 Five of these companies combined are responsible for three-quarters of the total generating capacity between 2010 and 2015 in SSA (IEA, 2016).
5 Previously, the French consulting firm Coyne et Bellier had produced the dam design, and the British consultancy Environmental Resources Management (ERM) had conducted the Environmental and Social Impact Assessment (ESIA).
6 REREC is a government organisation mandated to spearhead and drive renewable energy development along with rural electrification in Kenya.
7 Furthermore, additional local employment during O&M is to be generated in the form of security guards, solar-panel cleaners, and general cleaners for the project site spread over 85 hectares.

References

Ajakaiye, O. and Kaplinsky, R. (2009) 'China in Africa: A relationship in transition', *European Journal of Development Research*, 21(4), pp. 479–484. https://doi.org/10.1057/ejdr.2009.30

Baker, L. and Sovacool, B.K. (2017) 'The political economy of technological capabilities and global production networks in South Africa's wind and solar photovoltaic (PV) industries', *Political Geography*, 60, pp. 1–12. https://doi.org/10.1016/j.polgeo.2017.03.003

Bhamidipati, P.L. and Hansen, U. (2021) 'Unpacking China-Africa relations and local agency: Frictional encounters in the transition to solar power in Kenya', *Geoforum*, 119, 206–217.

Binz, C., Gosens, J., Hansen, T. and Hansen, U. (2017) 'Toward technology-sensitive catching-up policies: Insights from renewable energy in China', *World Development*, 96, pp. 418–437. https://doi.org/10.1016/j.worlddev.2017.03.027

Chirambo, D. (2018). Towards the achievement of SDG 7 in sub-Saharan Africa: Creating synergies between Power Africa, Sustainable Energy for All and climate finance in-order to achieve universal energy access before 2030. *Renewable and Sustainable Energy Reviews,* 94, 600–608.

Flyvbjerg, B., Holm, M.S. and Buhl, S. (2002) 'Underestimating costs in public works projects: Error or lie?', *Journal of the American Planning Association*, 68(3), pp. 279–295. https://doi.org/10.1080/01944360208976273

Ghana Web. (2017) 'Bui dam generates $335.7m revenue', *Ghana Web*, 12 December. Available at: https://www.ghanaweb.com/GhanaHomePage/business/Bui-dam-generates-335-7m-revenue-608901 (Accessed: 09/10/2020).

Hanlin, R., Okemwa, J. and Gregersen, C. (2019) 'Building competences and capabilities through projects: Examples from Kenya's renewable energy sector', IREK Working Paper No. 8. Available at: https://www.irekproject.net/wp-content/uploads/IREKPaper8.pdf

International Energy Agency (IEA). (2016) *Boosting the Power Sector in Sub-Saharan Africa: China's Involvement*. Paris: International Energy Agency.

IEA (2020) *Africa Energy Outlook*. Paris: International Energy Agency.

International Renewable Energy Agency (IRENA) (2013) *Africa's Renewable Future: The Path to Sustainable Growth*. Abu Dhabi: International Renewable Energy Agency.

IRENA (2019) *Scaling Up Renewable Energy Deployment in Africa*. Abu Dhabi: International Renewable Energy Agency.

Kaplinsky, R. and Morris, M. (2009) 'Chinese FDI in Sub-Saharan Africa: Engaging with large dragons', *The European Journal of Development Research*, 21(4), pp. 551–569. https://doi.org/10.1057/ejdr.2009.24

Kirchherr, J. and Urban, F. (2018) 'Technology transfer and cooperation for low carbon energy technology: Analysing 30 years of scholarship and proposing a research agenda', *Energy Policy*, 119, pp. 600–609. https://doi.org/10.1016/j.enpol.2018.05.001

Lema, R., Iizuka, M. and Walz, R. (2015) 'Introduction to low-carbon innovation and development: Insights and future challenges for research', *Innovation and Development*, 5(2), pp. 173–187. https://doi.org/10.1080/2157930X.2015.1065096

Lema, R., Fu, X. and Rabellotti, R. (2020) 'Green windows of opportunity? Latecomer development in the age of transformation towards sustainability', *Industrial and Corporate Change*, 29(5), pp. 1193–1209. https://doi.org/10.1093/icc/dtaa044

McCrudden, C. (2004) 'Using public procurement to achieve social outcomes', *Natural Resources Forum*, 28(4), pp. 257–267. https://doi.org/10.1111/j.1477-8947.2004.00099.x

Power, M., Newell, P., Baker, L., Bulkeley, H., Kirshner, J. and Smith, A. (2016) 'The political economy of energy transitions in Mozambique and South Africa: The role of the Rising Powers', *Energy Research and Social Science*, 17, pp. 10–19. http://dx.doi.org/10.1016/j.erss.2016.03.007

Sanfilippo, M. (2010) 'Chinese FDI to Africa: What is the nexus with foreign economic cooperation?' *African Development Review*, 22, pp. 599–614. https://doi.org/10.1111/j.1467-8268.2010.00261.x

Schmidt, T. S. and Huenteler, J. (2016) 'Anticipating industry localization effects of clean technology deployment policies in developing countries', *Global Environmental Change*, 38, pp. 8–20. https://doi.org/10.1016/j.gloenvcha.2016.02.005

Shen, W. (2020) 'China's role in Africa's energy transition: A critical review of its intensity, institutions, and impacts', *Energy Research and Social Science*, 68, p. 101578. https://doi.org/10.1016/j.erss.2020.101578

Shen, W. and Power, M. (2016) 'Africa and the export of China's clean energy revolution', *Third World Quarterly*, 6597(July), pp. 1–20. https://doi.org/10.1080/01436597.2016.1199262

Silvestre, B.S. (2015) 'Sustainable supply chain management in emerging economies: Environmental turbulence, institutional voids and sustainability trajectories', *International Journal of Production Economics*, 167, pp. 156–169. https://doi.org/10.1016/j.ijpe.2015.05.025

Sperling, F., Granoff, I. and Vyas, Y. (2012) *Facilitating Green Growth in Africa*. Tunis: African Development Bank.

Tan-Mullins, M., Urban, F. and Mang, G. (2017) 'Evaluating the behaviour of chinese stakeholders engaged in large hydropower projects in Asia and Africa', *The China Quarterly*, 230, pp. 464–488. https://doi.org/10.1017/S0305741016001041

Wegenast, T., Krauser, M., Strüver, G. and Giesen, J. (2019) 'At Africa's expense? Disaggregating the employment effects of Chinese mining operations in sub-Saharan Africa', *World Development*, 118, pp. 39–51. https://doi.org/10.1016/j.worlddev.2019.02.007

11

LOCAL CONTENT AND CAPABILITIES

Policy processes and stakeholders in Kenya

Ann Kingiri and Josephat Mongare Okemwa

Abstract

This chapter presents a historical analysis of Kenya's policy process in renewable electrification (RE) from 1999 to 2019. The policy process reflects some efforts to promote industrialisation in the energy sector, which may be a positive step towards attainment of inclusive and sustainable development in this subsector. It is not, however, clear how these efforts translate to meaningful learning and local capabilities that are critical to building a sustainable industrialisation in emerging economies. Using novel data from a survey and secondary data, the chapter investigates changes in RE policies as far as capability development and Local Content Requirements (LCRs) are concerned and stakeholders' views about this. With the term local content, we mean the use of Kenyan local expertise, goods, and services, people and business for systematic development of national capacity, and capabilities geared towards the enhancement of the economy. The analysis of the energy policy process through the lens of capabilities identifies key policy and practice gaps that require policy support in developing requisite capabilities for support to sustainable industrialisation. From the stakeholders' survey, the majority of the stakeholders were of the opinion that deployment of technologies for RE provides opportunities for building local technical capabilities and other skills like project management.

Introduction

Interrogation of diffusion dynamics of technologies for RE has attracted attention because they are perceived to have political, social, environmental, and economic attributes, which are critical in achieving a holistic and transformational sustainable industrialisation (Schmitz and Scoones, 2015). To this end, the

DOI: 10.4324/9781003054665-11

consideration of LCRs is becoming an important industrial policy tool which is largely expected to translate to local employment and private sector development. LCRs are provisions that bind foreign investors and companies to a minimum threshold that must be procured or purchased locally in order to support nascent local industries (Hansen et al., 2015; Johnson, 2015; Advisors, 2013). In Kenya there has been a change in policy dynamics to embrace industrial development in the energy sector as evidenced by inclusion of local content elements in the GoK (2019). It is however not clear how these dynamics relate to learning and capabilities that are critical to a sustainable industrialisation. This chapter interrogates the policy process resulting in the enactment of the GoK (2019). It is guided by the following overarching research questions:

i) How are industrial capacities and capabilities reflected in the Kenyan RE policy process and what are stakeholders' perspectives about this?
ii) What are the opportunities for making policies that promote capabilities related to RE in the future?

The study contributes to the on-going scholarly discussion about the prospects of a green energies revolution in contributing to sustainable industrialisation in emerging economies. The chapter is structured as follows: first the analytical framework is presented, followed by the methodology that has informed the study. Next is a critical review of capabilities and LCRs dynamics around deployment of technologies in RE based on secondary materials and perspectives of stakeholder groups (policy makers, private sector, researchers, and nongovernmental organisation – NGOs). The study concludes by drawing policy and practice-oriented lessons for an expanded discussion about inclusive renewable electrification in Kenya.

Analytical framework

The conceptual framing of this chapter benefits from the general literature on capabilities and learning as it relates to technology transfer and the structural aspects of a national innovation system (NIS). This is complemented by the industrial policy literature, primarily on the LCRs. The analytical framework looks at some basic understanding around policy making involving a dynamic technological innovation process. These theoretical concepts are used to investigate stakeholders' understanding about the key policy processes that enhance or characterise local technical and non-technical capabilities in the uptake of renewable technologies in Kenya with particular focus on wind and solar PV.

Renewable technologies and capabilities

There has been a significant growth in the number of off-grid energy firms arising from RE efforts in many countries. This has opened up research in the innovation and development studies scholarly fields, particularly research that focuses on low carbon technologies in emerging economies (Ockwell and Byrne, 2015). The capacity to enhance adoption of new technologies has long been known to be a key determinant of technical change in developing countries (Nelson and Pack, 1999). Further, technological capabilities are critical components of NIS (Fagerberg and Srholec, 2008), which in turn are determined by systematic policy interventions. However, the context within which sustainability technologies are being advanced and the potential for their uptake depends largely on the social and institutional ecosystem that supports the deployment process including accumulation of capabilities (Foxon and Pearson, 2008; Geels, Hekkert, and Jacobsson, 2008; Lema et al., 2018). As new technologies are introduced to a new context, the process of adaptation is determined by many institutional and structural factors including local and organisational arrangements (Sovacool, 2014; Bell, 2012). This contributes significantly to the development of a local innovation ecosystem that is critical for technological deployment more generally. Experiences of emerging countries with renewable energy technologies show that focused institutional support strengthens technology capability development (Ru et al., 2012). This then implies that paying attention to local capabilities-building associated with deployment of technologies is important for strengthening respective NIS. Lema et al. (2018) argue that attention should be given to how technological capabilities are acquired. This is because there are complex and multiple user-producer interactions that demand critical thought about how technological and experiential learning occurs and how this contributes to local capabilities.

A key question for the RE efforts in Kenya has been how local capability can be strengthened in a way that promotes inclusive and sustainable industrialisation. Capability issues are important because promoting growth of specific sectors would enable a country like Kenya to create and strengthen its value addition activities including ability to design, build, and operate in these sectors. Empirical research has shown that introduction of solar PV in Kenya initially failed because of lack of local capability for appropriate installations and maintenance (Ondraczek, 2013). Other scholars have also attempted to understand the global nature of technologies for RE and what this means for local capabilities (Lema et al., 2018). These scholars note that local capabilities in wind and solar PV technologies are perceived to be important in addition to the degree to which capabilities are sought from outside Kenya and utilised by local actors. This chapter attempts to expand this scholarly field. It does so through a qualitative study carried out to understand capabilities and local content issues in the renewable electrification policy process in Kenya and stakeholders' views on this. As mentioned elsewhere, the stakeholder groups as system builders can strategically shape the technological field in which technologies in RE can develop and diffuse. This implies that they are critical in advocating for policies that could support a functional system more generally (Ockwell and Byrne, 2015).

LCRs as an industrial policy tool

The motivation for policy makers in developing countries to promote technologies for RE is spurred by a number of factors. There is an increasing urge for RE technologies to meet not only the political, socio, and economic goals but also the transformational and sustainable goals including inclusive development (Schmitz and Scoones, 2015). The benefits emanating from renewable energy are multifaceted and include mitigating carbon emissions, which is directly connected to economic development, job creation, and other technological benefits (Lewis, 2014). Besides technical and economic benefits, another motivation relates to the ability of these technologies to enhance localised technological and innovation capabilities (Baker and Sovacool, 2017; Ockwell et al., 2018).

According to Johnson (2015), the main motivation for LCRs in developed countries is job creation. In developing countries, including sub-Saharan Africa (SSA), the motivation is mainly related to protection of nascent industries and creation of opportunities for local companies to benefit from local and foreign firms' large-scale investments (ibid). Other motivations include technology transfer, creation of new industries, and providing jobs for local people. LCRs aim to provide protection to local young firms while they accumulate capabilities to a level where they can compete with local and international firms (Hansen et al., 2015). In the RE sector, countries have used LCRs to support domestic industries. However, politics, market failures, and weak institutions are perceived to be barriers to promotion of local content (Johnson, 2015; Hansen et al., 2015). LCRs may also not create a level playing ground for global trade due to manipulation and may also lead to global trade conflicts (Ettmayr and Lloyd, 2017). When applying the LCRs' elements to specific technological innovations, Rodrik (2004) note the importance of paying attention to incentivising the very activities that allow for an expansion of the economy by generating new areas of comparative advantage. He further adds that focusing on activities across sectors has the potential to enhance interactive learning or what he refers to as cross-cutting opportunities.

Policies to promote local capabilities may vary across contexts. In South Africa, the energy regulatory framework has provided an enabling environment for creation of technological capability to independent power producers. Consequently, this has enabled many local communities to have an equity share in the renewable energy projects as well as making sure that the LCRs are followed (Baker and Sovacool, 2017). In this regard, LCRs are – in many cases – becoming an important industrial policy tool that have the potential to enhance domestic capabilities of local manufacturing (Johnson, 2015).

Qualitative data is used in this chapter to understand how LCRs are incorporated into RE policy instruments. Further, stakeholder views have been sought, specifically to receive more detail on the policy process and examples of policy windows to affect policy change.

Policy process and policy windows

A policy process entails policy making and policy implementation and both have multiple elements that influence the ultimate efficiency of a policy instrument (Richardson, 1982). Policy making refers to the procedures and institutional arrangements that shape the policy process (Nilsson et al., 2012). It covers all stages of the policy cycle including problem identification, agenda setting, policy formulation, evaluation or assessment, and policy adaptation (Rogge and Reichardt, 2013). Policy implementation is where responsible authorities devise strategies to ensure efficient enforcement of specific policy instruments (Nilsson et al., 2012). It also entails actual implementation and how this is influenced by actors, for instance public service workers or street level bureaucrats who have discretion and power in policy implementation (Gilson, 2015). Efficiency may entail collaboration amongst different value chain stakeholders, provision of resources, and capacity building. Further, because of the functional dynamism of a technological system, capabilities-building at different levels including that of policy makers may be necessary (Jacobsson and Bergek, 2011). While the two aspects of the process are important, the style of policy processes is critical as a 'standard operating procedures for making and implementing policies' (Richardson, 1982, p. 2). It may provide among other things flexibility in terms of implementation of different aspects of the policy process (Rogge and Reichardt, 2013).

It is acknowledged that policy processes are complicated by the fact that many actors are involved, all with different values and interests. Indeed, policy processes in dynamic technological innovation systems like RE are associated with uncertainties linked to different expectations and interests. The solution to different vested interests may be a consideration of a policy mix whereby new policy instruments supporting particular niches may be added to already existing regimes instead of replacing them altogether (Kern and Howlett, 2009). The policy mix concept points towards mobilisation of incentives and support in a policy process while also countering potential challenges that may hinder policy support in a social technical system (Edmondson, Kern, and Rogge, 2018). This chapter takes as a point of departure the argument that the bridge between policy making and policy implementation is building requisite capabilities (institutions and human; hard and soft). These include policy innovation, managerial skills, project implementation skills, and negotiation and advocacy skills, amongst others. Debatably, policy aspects that address capabilities must be considered in the policy mix.

A better understanding of policy windows presents opportunities for learning and potential policy change. Kingdon (2003) describes how windows of opportunity for policy change can unexpectedly create a short-lived opportunity for uptake of relevant knowledge evidence which has been previously ignored. Rose et al. (2017), inspired by Kingdon's work, further note that the policy windows concept is linked to agenda setting in policy processes and may explain why certain policy issues gain attention in the list of policy maker's agenda. Timing is also important if research is to influence policy. In the environmental science domain, Rose et al. (2017) outline recommendations that can help both environmentalists and researchers to engage with the concept of policy windows and increase the likelihood of knowledge uptake for policy change. These include: i) capacity to foresee emergent windows like policy organisations and knowledge brokers that provide platforms for engagement, ii) capacity to respond to opening windows, iii) capacity to frame research in line with appropriate windows, and iv) capacity to persevere in closed windows which may call for arguing for incremental policy change.

This study interrogates Kenya's RE policy process to understand occurrence of policy windows in a bid to expose the extent to which capabilities as a concept has become more or less central in this process.

Methodology

The chapter builds on primary and secondary material, with empirical data having been collected as part of the IREK project. The data has provided an opportunity to generate evidence around capabilities and collaboration on renewable electrification in Kenya. Apart from the IREK stakeholder survey in 2016, the chapter draws on other sources of data, notably an analysis of policy documents thereby tracking the policy process in RE from 1999 to 2019 and documentation of IREK project interaction with key stakeholders.

In mid-2016, the IREK project undertook a survey involving stakeholders across the renewable energy sector comprising policymakers, energy professionals, academia, and NGOs. Respondents were asked to answer questions relating to their current perceptions, attitudes, and knowledge of:

- The use and practices of wind and solar technologies in Kenya.
- Current policies for solar PV and wind energy in Kenya.
- Current barriers to diffusion of technologies in these fields.

The survey placed a specific emphasis on:

- The type and extent of collaborations (local and international) within the industry that foster and enhance diffusion of solar PV and wind technologies.
- The types of capabilities/capacity building that are needed to ensure that these technologies can be effectively introduced and utilised in Kenya.

The survey was complemented by other secondary activities that were sources of data, namely:

- Analysis of stakeholders' recommendations emanating from IREK stakeholder workshops (IREK, 2015, 2018b) and insights from selected critical projects whereby interviews were conducted between 2017 and 2019.

- Review of relevant energy policies and strategies with respect to capabilities and local content provisions.
- Interaction with Kenya's parliamentary committee on Energy, 2018 during IREK's project submission on 15 March (IREK, 2018a).

Additional primary data was generated via follow-up interviews with selected policy makers and stakeholders including those targeted for critical projects or case studies. These were conducted in 2017 in Nairobi, Kenya (see also IREK, 2017).

The survey respondents and interviewees were drawn from government institutions, research and educational institutions, think tanks, industrial associations, donors, NGOs, and the private sector in the Kenyan energy sector. Figure 11.1 presents the important actors that responded to the IREK survey as well as those interviewed at different times as part of the case studies. Together they constitute the participating stakeholders in this study.

The mapping out of all stakeholders involved in this study as outlined in Figure 11.1 serves two purposes. First, it helps us to categorise them into the different stakeholder groups they belong to. The majority of the study respondents were mainly from the research and educational institutions, private sector, and NGOs. Second, the mapping informs the discussion in the subsequent sections.

The analyses of the different sets of data followed different stages. The first stage of the data analysis focused on the perspectives of Kenya's stakeholder groups about capabilities in solar and wind subsectors based on the survey data, interviews, and case studies. The second stage entailed the analysis of data from secondary sources used to triangulate findings from the primary data. The final step entailed interrogation of the emerging perspectives using the theoretical framework adopted for this chapter.

Results: prospects for building capabilities through renewable electrification

This section documents the results of the data analysis and relevant discussion. The first part summarises the outcome of the review of key energy policies and strategies. The last parts detail the perspectives of stakeholders and how these inform policy and practice for promotion of RE local capabilities in Kenya. The latter sections discuss the results drawing on both survey data and interview data.

The renewable energy policy processes in relation to local capabilities

This section documents the energy policy processes with a view to exposing some underlying barriers and opportunities associated with capabilities and related policy change. In addition, what has changed in the RE policy landscape is also explored. The analysis pays attention to specific elements of a policy

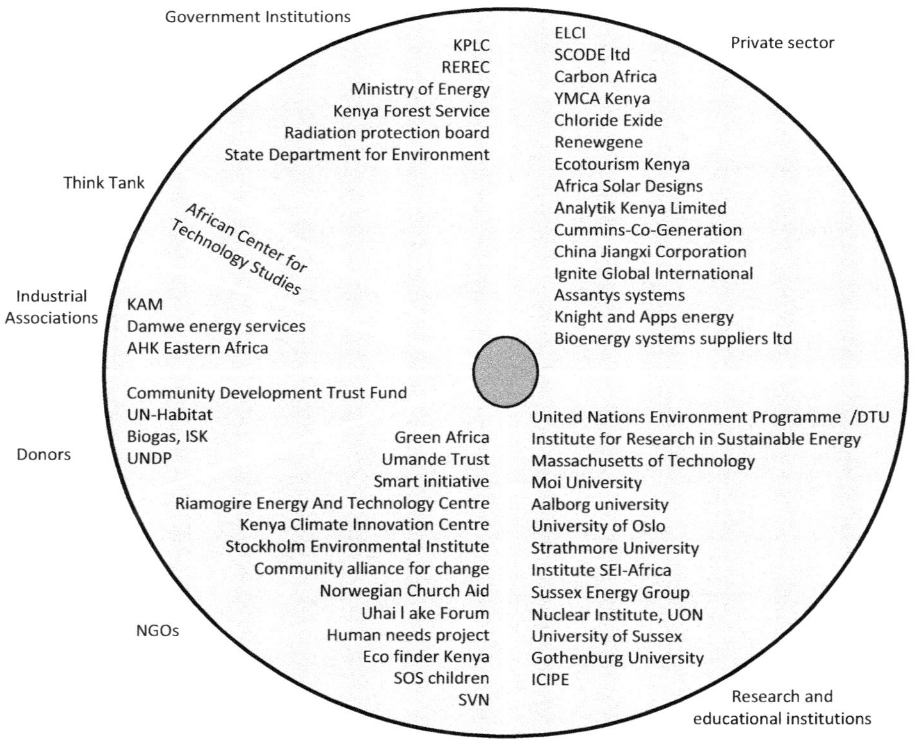

ACRONYMS

REREC	Rural Electrification and Renewable Energy Corporation
RERAC	Renewable Energy Resources AuthorityCommittee
SOODE	Sustainable Community Development services
EPPRA	Energy and Petroleum Regulation Authority
ELCI	Environmental Liaison Centre International
LCPDP	Low Cost Power Development Plan
KPLC	Kenya Power and Lightning Company
UNDP	United Nations Development Programme
KAM	Kenya Association of Manufacturers
UN	United Nations
ICIPE	International Centre of Insect Physiology and Ecology
MPID	Mainyoito Pastoralists
O	Integrated Development Organization
DTU	Technical University of Denmark
YMCA	Young Men Christian Association
SNV	Stichting Nederlandse Vrijwilligers (Netherlands Development Organisation)

FIGURE 11.1 Important actors that responded to the IREK survey as well as those interviewed at different times as part of the case studies. *Source: authors.*

process in line with the analytical framework adopted for the chapter. The analysis also looks at the Energy Bill formulation process and the roadmap to inclusion of LCRs in the enacted policy instrument, the GoK (2019). Other relevant policy instruments that demonstrate policy related efforts to build local capabilities in Kenya's energy sector are also explored.

Local content provisions in the energy policy instruments

The energy policy formulation process has been advanced over the years mainly through the Ministry of Energy (MoE). Ogeya et al. (2022; this volume) outline the history of the different energy policy instruments that have been introduced from early 2000 to 2019. Figure 11.2 summarises these key policy instruments and the respective formulation process in relation to key actors, agenda setting, and which aspects of capabilities were given attention in each of the policies and strategies analysed.

In terms of agenda setting (circle 3 from X, Figure 11.2), the motivation for building capabilities for RE has progressively advanced, driven by different factors. In the early 2000s, the driver of requisite policy processes was the need to diversify sources of energy and renewables targeting the off-grid population. Since then, efforts towards promotion of RE in the country have gained momentum particularly in the policy arena (Byrne, 2009). According to Makara (2019), the agenda has slowly shifted to general economic, social, and environmental goals, and more recently the promotion of local manufacturing firms as a contribution to Kenya's Big Four Agenda. This government Agenda prioritises food security, manufacturing, universal healthcare, and affordable housing, all of which depend significantly on availability of energy (ibid). Both the public and private actors have dominated the capabilities and requisite skills building efforts in different ways and times as shown in Figure 11.2, circle 2 from X. Although the RE policy roadmap may have commenced earlier, this study pays attention to the policy-making window between 1999 and 2019. For a detailed account of the policy evolution process with respect to actors (circle 2, Figure 11.2) and respective policy instruments (inner circle, Figure 11.2), see Ogeya et al. (2022; this volume).

With respect to capabilities (outer circle, Figure 11.2), there are some historical elements of capabilities-building being included in the early energy policy documents. However, the focus on capability-building and capability requirements has been limited until 2014 when the draft Energy Bill was presented to parliament and included a section on local content. The GoK (2014) and later the GoK (2018b) were consequently drafted. Both the GoK (2015a) and the GoK (2019) have local content provisions which are explored next. The GoK (2014) regulations were intended to apply to operations of all energy sources that included non-renewables and renewables. The regulations provide for a local content plan execution that includes technical and non-technical skills and capacity building targeting domestic or indigenous

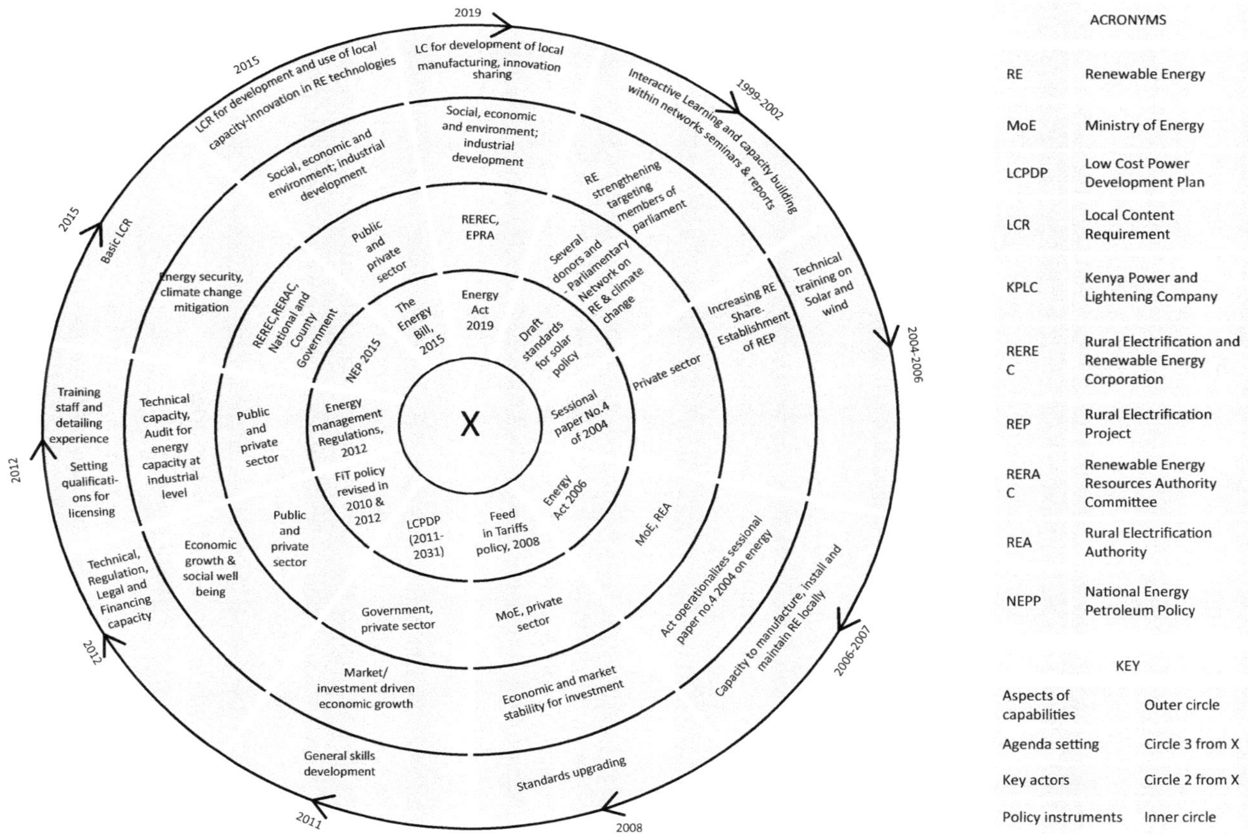

FIGURE 11.2 Selected policy instruments in the Kenyan energy policy process for capabilities in renewables (until 2019). *Source: authors.*

firms and prioritising local Kenyan citizens for training and employment. Local content was further popularised in 2015 when discoveries of oil were made in Kenya. The post-oil discoveries era triggered development of laws and policies that would guide oil production. The local content aspect is therefore articulated in the GoK (2015b) primarily to develop local talent and retention of value created from energy resources through appropriate policies. The policy notes that the main challenge of developing local content is lack of relevant skills and capacity. The policy further recommends capacity building in all sectors of the economy through training and international partnerships.

The drafting of the GoK (2014) may have informed the local content framing in the subsequent version of GoK (2015a) and later the GoK (2019). The GoK (2015a) strives to promote local content (LC) provisions entailing general use of local capabilities for economic growth, promotion of appropriate local capacity for the manufacture, installation, maintenance, and operation of basic renewable technologies and proposes a Renewable Energy Feed-in Tariff System to stimulate innovation in technologies for RE. The local content in the GoK (2015a) is defined as 'use of local expertise, goods and services, people, businesses and financing for development of national capabilities and for enhancement of the economy'. The LC provisions under the GoK (2019) include developing and procuring capabilities of local workforce, services, and supplies, for the sharing of accruing benefits. Further, the Act strives to promote development of appropriate local capacity for the manufacture, installation, maintenance, and operation of basic renewable technologies and as a one stop shop for information and guidance to investors on RE projects. In the GoK (2019), local content is referred to as 'added value brought to the economy from energy related activities'. There is a clear elaboration in the Act of what forms of capabilities may be anticipated in the execution of this clause and the inclusion of benefits sharing. Arguably, this may denote some flexibility in the implementation of the capability and capacity development and considers the importance of inclusivity.

The local content regulations outlined in this section point towards use of local personnel and materials. They also promote the creation of backward linkages in the value chain through requiring companies involved to utilise local firms for security, transport, clearing, and forwarding when conducting projects (see Hanlin and Okemwa, this volume).

Other capabilities provisions in the energy policies and other relevant policies

There are rules and regulations on qualifications and experiences provided for in the policy instruments that are relevant for working in the sector. There are also government minimum standards for training and skills motivated by the government intention to create an enabling environment for building individual skills. Kenya's Energy and Petroleum Regulatory Authority (EPRA) has been mandated to license all engineers working on solar projects. EPRA is the sole entity that regulates all electrical contractors and all electrician contractors who work on solar projects in ensuring that they are all licensed and/or that a company working on these projects has at least one licensed electrical contractor. The Kenya Bureau of Standards regulates the standards of all solar equipment that is manufactured and imported into Kenya.

In the GoK (2012), enacted under the GoK (2006), qualifications and experience for licensing range between High Diploma to Masters in Technology and Engineering related fields and a minimum of two years' work experience in energy operations or maintenance or planning. In an effort to actualise these provisions, EPRA has outlined various requirements for solar PV technicians categorised into three classes, T1, T2, and T3, before they are licensed. To be licensed for T1, one is required to be a holder of Kenya Certificate of Secondary Education (KCSE), electrical government trade test 2, and basic two years solar training experience. For T2, one is required to be a holder of KCSE, diploma, or electronic and intermediate solar training with 2 years solar installation experience; or a degree in electrical or relevant degree or higher national diploma with one-year solar installation experience. For T3, one is required to be a holder of KCSE, diploma, or electronic and advanced solar training with four years solar installation experience; or a degree in electrical or relevant degree or higher national diploma with 2 years solar installation experience.

The GoK (2019) states that a person who wishes to carry out the generation, exportation, importation, transmission, distribution, and retail supply of electricity must apply for a licence. Further, the Act requires that a person who wishes to carry out electrical installation work must be licensed as an electrical contractor by the Authority. In addition, a licence for electrical installation work shall be issued for a term of three years and may be renewed for a similar term upon expiry, subject to the licensee having undertaken any required additional training. It is important to note that, to be an electrical contractor, a person must be a certified electrical worker; or have in his employment a certified electrical worker. In addition, the Act stipulates the need for specific certification of solar system installation contractors. The two types of licensing that the Act refers to are the general electrical licensing types, namely Class A, B, and C, and solar PV technician certificates, Class T1, T2, and T3 (see description outlined above).

Other regulations that have been gazetted to assist the local renewable electrification sector include the GoK (2018a) which provides for tax exemptions for the supply or importation of equipment and materials for development and generation of energy for solar and wind energy technologies. Similarly, the First Schedule to the GoK (2013), (revised 2018) focuses on specialised equipment for the development and generation of solar energy, including deep cycle batteries which use or store solar power.

Insights from stakeholders: prospects for building capabilities through renewable electrification in Kenya

This section documents capabilities in the RE policy process in Kenya and stakeholders' views on the same.

Perspectives of stakeholder groups about capabilities in solar and wind subsectors in Kenya: insights from survey and interviews

This study and the broader IREK project was motivated by an underlying hypothesis that promoting (technological) capabilities in certain industries enables a country to create and strengthen its value addition activities in the relevant subject area; in this case in the ability to design, build, and operate solar PV and wind projects. The 2016 survey and follow-up interview questions were designed to elicit perspectives about whether and how RE efforts in Kenya contribute to raising different forms of technical and managerial capabilities. The majority of the stakeholders noted that promotion of renewable technologies (usually from abroad) significantly increases the ability for local technical and managerial capabilities to be built. This implies that capabilities are built from the interactions or collaborations within and outside the projects. The majority of stakeholders argued that mechanisms for knowledge generation and diffusion associated with different forms of capabilities in both wind and solar subsectors were weak when the survey was conducted in 2016. This calls for strategies to strengthen the linkages between knowledge supplies and users in a way that promotes learning and consequently local capabilities in these sectors.

Respondents to the 2016 survey were further asked to identify areas where capabilities are strong or lacking in the wind and solar PV technology sectors in Kenya. They were required to select from i) equipment manufacturing, ii) project development, iii) construction and installation, iv) operations and maintenance, and v) training. The study shows that equipment manufacturing capability is significantly low in both subsectors (Figure 11.3). Respondents argued that this may be the main reason why local projects rely heavily on foreign firms for acquisition and supply of energy generating equipment. The international transfer of technology is arguably an indication of top down transfer flow of knowledge. This is attributed to the undeveloped stage of Kenya's local manufacturing sector (Lema et al., 2018). Capability in operations and maintenance was a bit more developed, particularly in the solar subsector, which may denote opportunity for local capability deployment at the lower level of technology manufacturing and the deployment chain. This was corroborated by follow-up interviews. One entrepreneur dealing in solar PV noted that 'solar does not require frequent maintenance and it is easy to identify trained solar PV technicians all around Kenya … and in the case of wind, technicians with knowhow and experience are lacking' (interview on 13 February 2017).

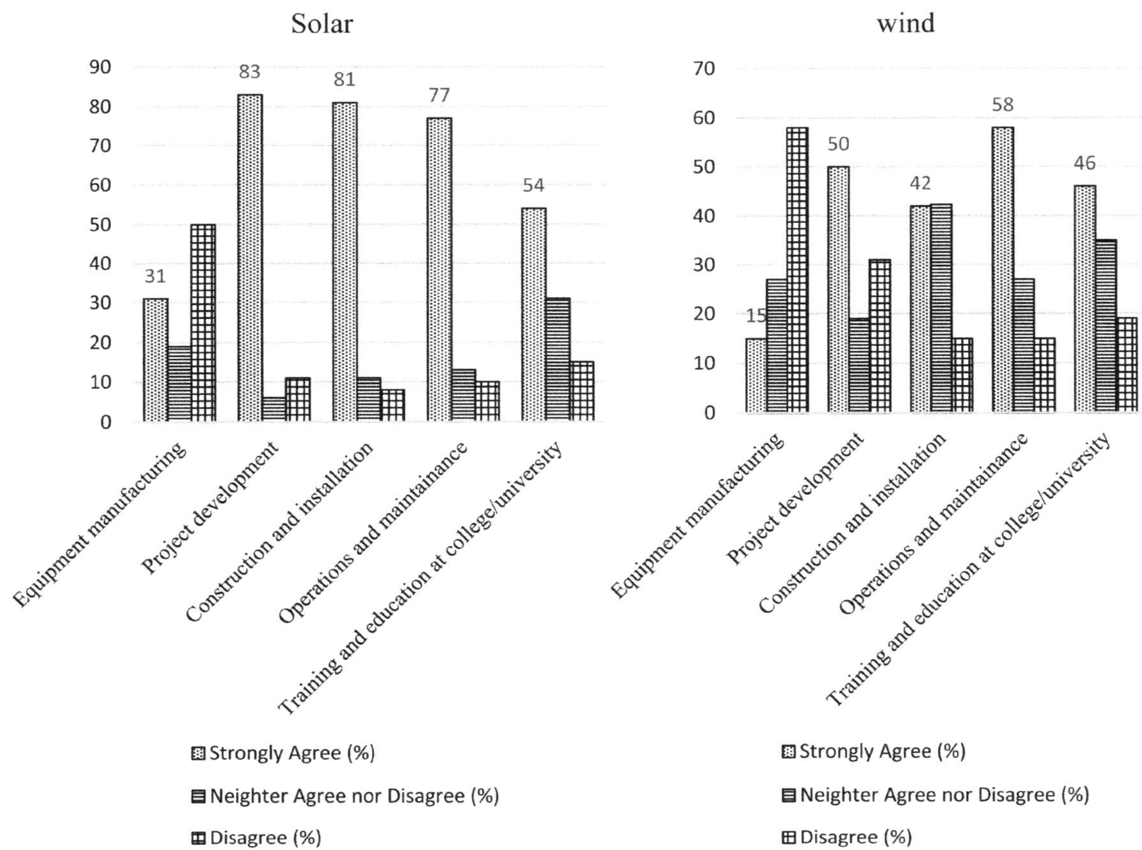

FIGURE 11.3 Respondents understanding of capability strengths in wind and solar (%). Percentages are used here to enable comparison between respondents' answers to the solar and wind questions when differing numbers of respondents answered these questions. *Source: authors.*

The respondents noted that lack of local training and education in these subsectors has confounded the weak local capabilities situation. Arguably, training and education at college level may contribute to basic knowledge as is the case with solar technicians but this may not guarantee attainment of practical knowledge that is relevant for sustained local capability. The majority of the respondents identified the importance of capacity for managing networking including international relationships to enhance successful technology transfer. This is a non-technical skill that needs to be built at the local level. Lema et al. (2018) contend that both manufacturing and deployment chains of solar PV and wind turbines may promote interactive learning, but that such options are frequently ignored in informing requisite policy and practice.

Perspectives of stakeholders about capabilities: insights from selected case studies

The IREK project has been studying some critical projects whereby qualitative data involving interviews with key persons on these projects further exposes mixed reactions about policy support for building local capabilities (see also Hanlin, Okemwa, and Gregersen, 2019). A number of these reactions are reported next.

A supportive policy environment is critical to capabilities-building

Government actions and policy initiatives have the power to create an enabling environment through which capabilities can be built. For instance, some respondents from selected case studies perceived that the Kenyan government and other institutions have supported renewable energy sector development through formulation of conducive policies. One programme director from a solar project noted that 'the environment is now favourable as policies governing the sector are in place as compared to when the project had no policy guidelines during the earlier stages of project installation' (interview on 7 February 2019). The respondent argued that this provides an opportunity for building of capabilities, collaborations, and job creation more generally. An engineer in one of the case study solar projects under the IREK research project argued that the policy environment and regulations provided guidance to the recruitment of engineers and technicians with the right qualifications including recruitment of project staff and contractors in the supply chain (interview on 4 March 2019). This was corroborated by a key staff member of a firm that was contracted to design, supply, and undertake installations works including electrical for power distributions (interview on 18 January 2019). He noted that existing regulations required the engineers and technicians from the contracting firm to be licensed by the government through REREC (formerly known as Electricity Regulatory Commission – ERC). The regulations further required the firm to develop an organisational policy that would ensure staff undertake training to acquire additional new skills. These sentiments were echoed by a senior manager working with the government, who asserted that the government is partnering with stakeholders to establish specialised academies or centres of excellence for specialised training in specific areas of renewable energy (interview on 28 February 2017). An interview with a co-founder of one of the wind firms affirmed government policy support by noting that 'the government has been very supportive and proactive in spite of the bureaucracy and lack of coordination'. He added that 'the wind firm has survived under different governments and as such, local knowledge on how to manage and develop the project has been acquired to date' (interview on 6 February 2017). These findings seem to suggest that government support is critical in ensuring a favourable policy environment to build local capabilities.

Lack of commitment to support policies that are important for capabilities-building by stakeholders

Others had mixed feelings about the policy support emanating from the government, especially for relatively small projects. One pessimistic respondent from a solar project cited lack of commitment to support policies for collaborative small projects, which he considered to be major contributors to economic and sustainable development. He remarked 'a big ten megawatt project will help few people get jobs, however small scale projects will create tens of thousands of jobs but the government is not having a vision to do that' (interview on 22 January 2019). He further noted that the net metering policy has not been implemented and small projects have not benefitted. He called for the national and county governments to collaborate in the implementation of relevant policies that support RE technologies for the poor. This purported lack of policy support to the grassroots projects was confirmed by a policy regulator who lamented 'there is lack of proper incentives from the government to support renewable energy industry … raw materials and equipment are not locally available and the cost of power is high'. The regulator further added 'generally, there is little support from the government to support the local fabrication' (interview on 23 January 2019).

Sentiments relating to another fourth solar project identified collaboration issues. A county official informed that the local beneficiaries had not been involved in major decisions pertaining to this project's launch as well as benefits sharing. The other issue was lack of clear policy guidelines on skills development targeting local staff. A technician who was interviewed on this project mentioned that a training contract between the government and the contracted company was drawn. This provided for Kenyan local engineers to receive hands-on training to acquire skills for day-to-day monitoring of the unit which at the time of the interview had not been carried out. The respondent perceived that this had not been forthcoming because foreign staff preferred working at night and as such, local engineers were unable to learn these skills (interviews with three people between 13 and 15 February 2019). This may be interpreted as lack of commitment to support policies that are important for capabilities-building.

Prospects for building local capabilities through RE: summing up the stakeholders' views

Overall, the majority of the stakeholders argued that mechanisms for knowledge generation and diffusion associated with different forms of capabilities in both subsectors are currently weak. This calls for strategies to strengthen the linkages between knowledge supplies and users in a way that promotes learning and consequently local capabilities. We note and argue that local universities and colleges have a role to promote capabilities at both the manufacturing and deployment chains through appropriate education and training. However, combining or integrating practical knowledge in the formal technical training is an issue for policy and requires critical thought. The study further identified the complexity around technology transfer and local capability-building. Debatably, the expected knowledge transfer from international or foreign firms should be complemented with requisite local training to enhance skills that can contribute to sustainable uptake of renewable technologies.

Stakeholders acknowledge that Kenya's policy process in renewable energy has been shaped by multiple actors who are critical in providing direction for the manufacturing and deployment of solar and wind technologies in Kenya. Academic literature has also credited stakeholders for their role in creating a conducive regulatory and policy environment and for undertaking dynamic system-building processes (Jacobsson and Bergek, 2011; Bergek et al., 2008). Further, depending on the nature of interactions, different actors in the value chain stimulate learning through the functions they undertake. This provides an opportunity for local capability-building (Lema et al., 2018).

Discussion

Advocating for capabilities-building in RE policy and practice

The literature on policy windows discussed earlier in this chapter demonstrates how to promote evidence-informed policy. As part of policy research under IREK, the research team got an opportunity to make an oral and written submission on the Energy Bill to the Energy parliamentary committee in early March 2018 (IREK, 2018a). The submission to the policy makers was supported by case studies and examples of success in respective counties. One of the parliamentarians reacting to the presentation expressed lack of awareness about one case study project under IREK that purportedly has great potential for building domestic skills. The IREK recommendations to the committee emphasised the need to incorporate commitments for capabilities-building for enhanced renewable energy uptake in the Energy Bill and the need to launch renewable technology studies. This may be perceived to be an unexpected 'window of opportunity' for IREK research to influence policy change. Arguably, this intervention may be linked to inclusion of flexible, capabilities-oriented, local-content elements in the GoK (2019). Seizing upon a policy window by publishing a compilation of relevant knowledge on capabilities is in agreement with recommendations to develop specific tips for approaching policy windows (Rose et al., 2017).

The IREK policy research has also demonstrated the importance of policy mixes and how this might influence renewable technological uptake and the embedded local capabilities-building in the Kenyan context. Depending on context, policy mixes may be a prerequisite for a positive policy outcome in the field of sustainability transitions (Edmondson, Kern, and Rogge, 2018). Interviews conducted by IREK team members confirm this. For instance, a respondent from a critical solar project under IREK, who previously served as a technician and a manager, remarked 'the government from one of the counties in Kenya issued us with an allotment letter for the land where the project is installed' (interview on 26 February 2019). The national policy was not enough for the implementation of the project and the county policy was complementary. The policy mix in relation to capabilities is inconclusive from this study and hence warrants further research.

Prospects for building local capability through LCRs

It is important to note that the local content idea is not new in Kenya's energy policy process. Byrne (2009, p. 148) noted that in the early 2000s when discussions about renewable energy were gaining currency, the formulation of standards for solar PV then considered local content, citing the importance of protecting local manufacturers. The formulation team was made up of mainly policy makers and technical experts under the Ministry of Energy then. Strangely, this motivation was not long lived because the GoK (2006) does not have local content provisions. These, however, are included in the GoK (2015a) and finally in the enacted GoK (2019). Despite politics, the latter years of the energy policy process may be perceived to have been more participatory and inclusive in terms of agenda setting. Following a public invitation to make submissions related to the Energy Bill, subsequent activities of strategic actors like the IREK team and other stakeholders may have influenced the policy-making process. Capacity to seize a window of opportunity (Rose et al., 2017) to underline the importance of local capabilities may arguably have been critical.

There have been subtle motivations to invest in local capabilities as revealed by this study. According to a respondent from a private sector it should be a normal practice to give priority to local companies or to engage labour from locals in projects (interview on 8 February 2017). However, this has not been always the case. Previously, before the enactment of the GoK (2019), we may conclude that local capacity building efforts by different stakeholders were unstructured and ad hoc. A senior respondent from a government institution asserted that 'the Chinese do not use local subcontractors but because of local (government) push, this is changing'. He further added that in an effort to encourage knowledge transfer through LCRs, 'we try to enforce training and local employment' (interview on 8 February 2017). Another interviewee from a private company with a 15-year maintenance contract in a wind company corroborates this purportedly changing trend. He informed us that

> *during the initial stages of installation, we had 5% local and 95% foreign engineers. However, this has changed over time with more locals now taking the role. In addition, frequent maintenance due to wear and tear of turbines allows for local training.*
> *(Interview on 8 February 2017)*

One foreign entrepreneur interested in developing local capacity in the wind sector lamented that in a previous tender with the government, the signed contract did not provide for equipment maintenance, which resulted in system failure due to negligence (interview on 8 February 2017).

The local content policies in Kenya have been criticised for having a limited impact in the energy sector including mining or mineral extraction (Mwendwa, 2019). This notwithstanding, the Energy Act (2019) creates 'a window of opportunity' for stakeholders in renewable energy to ensure that the implementation of LCRs takes into cognisance the need to develop capabilities of local manufacturers. Realising the opportunities will however require collaboration between the local and international actors as well as the policy actors at the local level.

Conclusion and recommendations

This chapter focuses on capability-building and local contents issues in the RE policy process in Kenya between 1999 and 2019 and set out to answer the following questions: i) how are industrial capacities and capabilities reflected in the Kenyan RE policy process and what are stakeholders' perspectives about this?; and ii) what are the opportunities for making policies that promote capabilities related to RE in the future? Through the lens of stakeholders in the energy sector and analysis of key secondary documents, the study identifies interesting dynamics that help us answer these questions.

To start with, the findings from the stakeholders' survey indicate that deployment of technologies for RE has provided opportunities for building local technical capacities and managerial capacities. In addition, the analysis of the policy process has shown that there has been a steady process relating to inclusion of capabilities provisions in the energy policy instruments. At the early phases (in reference to GoK, 2006), capabilities-building was synonymous with human resource development, i.e., developing skills and competences of technicians and engineers to deal with the RE technologies. In later phases, there was more emphasis on local content issues and capability development in policies and strategies.

The analysis has identified key policy and practice gaps that require policy support in developing requisite capabilities for support to sustainable industrialisation of the sector.

The results show that there is partial application of the capability concept in the process and in particular the application of LCRs for solar PV and wind technologies for Kenya's RE, especially in the early phases of the process that culminated into the enactment of the GoK (2019). The results further demonstrate that there were different factors and actors that influenced the agenda setting and ultimate policy outcome. The policy process at the early phases was motivated by the urge to increase the share of RE in the country's energy mix. The agenda progressively shifted to embrace local economic oriented interests like industrial development and global environmental goals. The inclusion of local content provisions in the GoK (2015a) and ultimately in the GoK (2019) may be perceived to be a protracted path towards industrialisation through promotion of local capabilities for economic growth.

The study has established that although capabilities in the solar sector are more advanced than wind, overall capabilities are inadequate in both sectors to steer a sustainable industrialisation. The stakeholders put emphasis on building capabilities in the operations and maintenance field, especially in the wind subsector, and a corresponding perception that Kenya does not have to rely on foreign firms for operations and maintenance expertise. From the broader IREK work and as recorded in Hanlin, Okemwa, and Gregersen (2019) and in other chapters in this book (for instance Hanlin and Okemwa, 2022; this volume; Karjalainen and Byrne, 2022; this volume), there is evidence that Kenya is progressively building capacities in operation, maintenance, and service skills for both wind and solar PV. However, more support is needed in this area and as a first step there is a need for more research in this regard.

The survey in particular shows that Kenya does not have huge manufacturing capacity and given the likelihood of an upward trend in increased influx of technologies for RE in the future, there is a need for policy makers to rethink where to direct requisite policy support. For instance, should they focus on building an industrial manufacturing base and try to become the source of solar PV or wind equipment manufacturers whose products are then exported? Or should the attention be on increasing access to energy, thereby reducing energy poverty and working towards national climate change targets? The former fits with the focus on longer-term job creation and economic growth that was emphasised by so many of the survey respondents. However, so too does a focus on ensuring 100% access to electricity (whether grid connected or otherwise), at least in the short and medium term until the country has been fully electrified through jobs building renewable power plants and maintaining them. A key question that requires more research by academics and more consideration by policy makers is whether focusing on increasing access to energy utilising foreign technology creates transferable skills and opportunities for employment and viable operations and maintenance businesses beyond the wind and solar PV subsectors.

Training is one way to enhance future opportunities for requisite capacity building and policy change in RE. With regards to training needs in the solar and wind subsectors, the study raises a key need for more assessment of the current situation to be conducted. This requires action not just by the Government of Kenya but also firms and other actors involved in solar PV and wind projects in Kenya to ensure that they have adequately trained staff. More research on the level of in-house training that takes place within firms and organisations involved in this field would provide interesting additional context. The education system should also investigate more on what specific training may be useful to develop at Kenyan universities and Technical Vocational Education and Training Institutions (TVET). A review of county level vocational training efforts, particularly in counties that have large or large numbers of renewable energy projects, would be helpful in this regard. All these issues relate to lack of clarity on factors that motivate the policy makers to prioritise capability-building.

Both private and policy actors have influenced the energy policy process but largely without taking cognisance of holistic capabilities, including technical and managerial capabilities, as well as capabilities along the local manufacturing value chain. This study has shown that stakeholder groups (policy makers, private sector, researchers, and NGOs) as system builders can strategically shape the technological field in which technologies in RE can develop and diffuse. Consequently, it is noted that windows of opportunities for learning and building capabilities at different scales in RE are maximised when stakeholders collaborate and where there is policy support. Further, stakeholders have a role to play in influencing the direction of policy change but only if they develop the capacity to seize windows of opportunity and use practical evidence for training and advocacy.

Based on the results of this study, policy and practice recommendations should build on the following lessons learned:

- Stakeholders have different understanding of capabilities. This can impact negatively on capabilities-building efforts. There is therefore a need for continued awareness creation amongst different stakeholders, especially policy makers and academics.
- Evidence is critical in influencing the perspectives of policy makers about the need for a policy change, in this case the importance of capabilities in RE.
- In addition to evidence, timing of engagement during the policy formulation process is paramount. Researchers engaging in policy processes have a high probability of influencing policy outcome where a window of opportunity arises.
- A comprehensive framework for review of energy policies should guide in the subsequent implementation. There should also be consideration for policy mix because the energy sector is dynamic and attracts interest in other non-energy policy domains.

Overall, the findings suggest the importance of building a strong capacity in the area of solar PV and wind energy in Kenya. The stakeholder survey identifies areas where the capacity is low or may be needed. Arguably, RE for sustainable industrialisation can only be realised if there is a conducive policy environment that supports training opportunities, uptake of RE technologies and investment, as well as collaboration. This notwithstanding, Kenya needs to find its own way forward based on in-depth analysis of its natural resources and capacity for change (see for instance Schmitz, 2007; Schmitz and Scoones, 2015). This should then inform strategic direction in the implementation of the enacted Energy Act (2019) and in particular the local content aspects in order to enhance sustainable industrialisation. To this end, the government could proactively work to boost production capability of local actors for both wind and solar technologies.

Acknowledgement

This research is supported by the Danish Ministry of Foreign Affairs, Grant: DFC 14-09AAU. The authors acknowledge this support.

References

Advisors, W.T.I. (2013) 'Local content requirements and the green economy' in *Ad Hoc Expert Group Meeting on Domestic Requirements and Support Measures in Green Sectors: Economic and Environmental Effectiveness and Implications for Trade*, June, pp. 13–14. Available: at: https://acf2017.africacarbonforum.com/sites/default/files/documents/DITC_TED_13062013_Study_WTI.pdf (Accessed: 2/10/2021)

Baker, L. and Sovacool, B.K. (2017) 'The political economy of technological capabilities and global production networks in South Africa's wind and solar photovoltaic (PV) industries', *Political Geography*, 60, pp. 1–12. https://doi.org/10.1016/j.polgeo.2017.03.003

Bell, M. (2012) *Low-Carbon Technology Transfer. Low-Carbon Technology Transfer: From Rhetoric to Reality*. London: Routledge. https://doi.org/10.4324/9780203121481

Bergek, A., Jacobsson, S., Carlsson, B., Lindmark, S., and Rickne, A. (2008) "Analyzing the functional dynamics of technological innovation systems: A scheme of analysis", *Research Policy*, 37(3), pp. 407–429. https://doi.org/10.1016/j.respol.2007.12.003

Byrne R.P. (2009) *Learning Drivers: Rural Electrification Regime Building in Kenya and Tanzania*. DPhil Thesis. Brighton, UK: University of Sussex.

Edmondson, D.L., Kern, F. and Rogge, K.S. (2018) The co-evolution of policy mixes and socio-technical systems: Towards a conceptual framework of policy mix feedback in sustainability transitions', *Research Policy*, 48 (10). https://doi.org/10.1016/j.respol.2018.03.010

Ettmayr, C. and Lloyd, H. (2017) 'Local content requirements and the impact on the South African renewable energy sector: A survey-based analysis', *South African Journal of Economic and Management Sciences*, 20(1). https://doi.org/10.4102/sajems.v20i1.1538

Fagerberg, J., and Srholec, M. (2008) 'National innovation systems, capabilities and economic development', *Research Policy*, 37 (9), pp. 1417–1435. https://doi.org/10.1016/j.respol.2008.06.003

Foxon T. and Pearson, P. (2008) 'Overcoming barriers to innovation and diffusion of cleaner technologies: Some features of a sustainable innovation policy regime', *Journal of Cleaner Production*, 16 (1), pp. 148–161. https://doi.org/10.1016/j.jclepro.2007.10.011

Geels, F.W., Hekkert, M.P. and Jacobsson, S. (2008) 'The dynamics of sustainable innovation journeys', *Technology Analysis & Strategic Management*, 20 (5), pp. 521–536. https://doi.org/10.1080/09537320802292982

Gilson, L. (2015) 'Michael Lipsky's street-level bureaucracy: Dilemmas of the individual in public service' in: Balla, Steven J., Lodge, Martin, and Page, Edward (eds.), *The Oxford Handbook of Classics in Public Policy and Administration*. Oxford: Oxford University Press.

Government of Kenya (GoK) (2006) 'The Energy Act 2006, Legal Notice No. 102 of 2006'. http://www.eisourcebook.org/cms/Kenya%20Energy%20Act,2006.pdf (Accessed: 10/01/2020)

GoK (2012) 'The Energy (Energy Management) Regulations 2012'. https://www.epra.go.ke/download/the-energy-energy-management-regulations-2012/ (Accessed: 10/10/2019).

GoK (2013) 'Value Added Tax Act, (2013)'. Chapter 476. Revised Edition 2012 [1993]. www.kenyalaw.org (Accessed: 10/01/2020).

GoK (2014) 'Energy (Local Content) Regulations, 2014'. Available at: https://energy.go.ke/?p=237 (Accessed: 10/10/2019).

GoK (2015b) *Draft Energy Bill*. 154p. Republic of Kenya: GoK.

GoK (2015a) 'Draft national energy and petroleum policy (NEPP)'. http://www.erc.go.ke/images/docs/National_Energy_Petroleum_Policy_August_2015.pdf

GoK. (2018a) 'Finance Act (2018)'. Special Issue, Kenya Gazette Supplement No. 121 (Acts No. 10).

GoK (2018b) 'Local Content Bill (2018)'. http://www.parliament.go.ke/node/10677 (Accessed: 10/01/2020).

GoK (2019) 'The Energy Act, 2019', no. Acts No. 1 (2019).

Hanlin, R., Okemwa, J. and Gregersen, C. (2019) 'Building competences and capabilities through projects: Examples from Kenya's renewable energy sector', *IREK Working Paper No. 8*. http://irekproject.net (Accessed: 10/10/2019).

Hanlin, R. and Okemwa, J. (2022) 'Interactive learning and capability-building in critical projects', in *Building Innovation Capabilities for Sustainable Industrialisation: Renewable Electrification in Developing Economies*. New York: Routledge. https://doi.org/10.4324/9781003054665-7

Hansen, M.W., Buur, L., Kjær, A.M. and Therkildsen, O. (2015) 'The economics and politics of local content in African extractives: Lessons from Tanzania, Uganda and Mozambique', *Forum for Development Studies*, 43 (2), pp. 201–228. https://doi.org/10.1080/08039410.2015.1089319

Innovations and Renewable Electrification in Kenya (2015) 'Drivers for uptake of renewable electrification in Kenya', *IREK Briefing Note No. 1*, Feb. Available at: http://irekproject.net (Accessed: 10/10/2019).

IREK (2017) 'Insights from joint research interviews', *Briefing Note No. 3*, Feb. Available at: http://irekproject.net (Accessed: 10/10/2019).

IREK (2018a) 'Embedding capabilities in energy policies for effective deployment of renewable technologies: IREK project policy intervention', *Briefing Note No. 5*, March. Available at: http://irekproject.net (Accessed: 10/10/2019).

IREK (2018b) 'Towards a functional renewable energy sector in Kenya: Recommendations from stakeholders', *Briefing Note No. 6*, March. http://irekproject.net (Accessed: 10/10/2019).

Jacobsson, S. and Bergek, A. (2011) 'Innovation system analyses and sustainability transitions: Contributions and suggestions for research', *Environmental Innovation and Societal Transitions*, 1(1), pp. 41–57. https://doi.org/10.1016/j.eist.2011.04.006

Johnson, O. (2015) 'Promoting green industrial development through local content requirements: India's National Solar Mission', *Climate Policy*, 16 (2), pp. 178–195. https://doi.org/10.1080/14693062.2014.992296

Karjalainen, J. and Byrne, R. (2022) 'Moving forward? Building foundational capabilities in Kenyan and Tanzanian off-grid solar PV firms', in *Building Innovation Capabilities for Sustainable Industrialisation: Renewable Electrification in Developing Economies*. New York: Routledge. https://doi.org/10.4324/9781003054665-9

Kern, F. and Howlett, M. (2009) 'Implementing transition management as policy reforms: A case study of the Dutch energy sector', *Policy Sciences*, 42(4), pp. 391–408. https://doi.org/10.1007/s11077-009-9099-x.

Kingdon, J. (2003) *Agenda, Alternatives, and Public Policies*. 2nd ed. New York: Longman Press.

Lema, R., Hanlin, R., Hansen, U.E. and Nzila, C. (2018) 'Renewable electrification and local capability formation: Linkages and interactive learning', Energy Policy, 117, pp. 326–339. https://doi.org/10.1016/j.enpol.2018.02.011

Lewis, J.I. (2014) 'The rise of renewable energy protectionism: Emerging trade conflicts and implications for low carbon development', *Global Environmental Politics*, 14 (4), pp. 10–35. https://doi.org/10.1162/glep_a_00255

Makara, M. (2019) 'Affordable and reliable energy pivotal to Kenya's delivery of its 'Big Four Agenda': Kenya's energy sector can enhance the country's economic growth', *Africa Business Insight, Maritz Africa*. 12 June. Available at: https://www.howwemadeitinafrica.com/affordable-and-reliable-energy-pivotal-to-kenyas-delivery-of-its-big-four-agenda/63351/ (Accessed: 20/02/2020).

Mwendwa, D. (2019) 'The development of local content in the oil and gas industry in Kenya: The need for negotiations'. https://www.researchgate.net/publication/335243350_THE_DEVELOPMENT_OF_LOCAL_CONTENT_IN_THE_OIL_AND_GAS_INDUSTRY_IN_KENYA_THE_NEED_FOR_NEGOTIATIONS (Accessed: 23/06/2020).

Nelson, R. and Pack, H. (1999) 'The Asian miracle and modern growth theory', *The Economic Journal,* 109 (457), pp. 416–436. http://www.jstor.org/stable/2565712 (Accessed on 2/10/2020).

Nilsson, M., Zamparutti, T., Petersen, J.E., Nykvist, B., Rudberg, P. and McGuinn, J. (2012) 'Understanding policy coherence: Analytical framework and examples of sector-environment policy interactions in the EU', *Environmental Policy and Governance*, 22 (6), pp. 395–423. https://doi.org/10.1002/eet.1589

Ockwell, D. and Byrne, R. (2015) 'Improving technology transfer through national systems of innovation: Climate relevant innovation-system builders (CRIBs)', *Climate Policy* pp. 1–19. https://doi.org/10.1080/14693062.2015.1052958

Ockwell, D., Byrne, R., Hansen, U.E., Haselip, J. and Nygaard, I. (2018) 'The uptake and diffusion of solar power in Africa: Socio-cultural and political insights on a rapidly emerging socio-technical transition', *Energy Research and Social Science*, 44, pp. 122–129. https://doi.org/10.1016/j.erss.2018.04.033

Ondraczek, J. (2013) 'The sun rises in the east (of Africa): A comparison of the development and status of solar energy markets in Kenya and Tanzania', *Energy Policy*, 56, pp. 407–417. https://doi.org/10.1016/j.enpol.2013.01.007

Richardson, J. (1982). Policy Styles in Western Europe (Routledge Revivals) (1st ed.). London, UK: Routledge. https://doi.org/10.4324/9780203082010

Rodrik, D. (2004) 'Industrial policy for the twenty-first century', *SSRN Electronic Journal*. Available at SSRN: https://ssrn.com/abstract=617544 or http://dx.doi.org/10.2139/ssrn.617544

Rogge, K.S. and Reichardt, K. (2013) 'Towards a more comprehensive policy mix conceptualization for environmental technological change: A literature synthesis', *Working Paper Sustainability and Innovation*, No. S3/2013, Fraunhofer ISI, Karlsruhe. http://hdl.handle.net/10419/77924 (Accessed on 2/ 10/2020).

Rose, D.C., Mukherjee, N., Simmons, B.I., Tew, E.R., Robertson, R.J. Vadrot, A.B.M., Doubleday, R. and Sutherland, W.J. (2017) 'Policy windows for the environment: Tips for improving the uptake of scientific knowledge', *Environmental Science and Policy*, 113(November 2020), pp. 47–54. https://doi.org/10.1016/j.envsci.2017.07.013

Ru, P., Zhi, Q., Zhang, F., Zhong, X., Li, J. and Su, J. (2012) 'Behind the development of technology: The transition of innovation modes in China's wind turbine manufacturing industry', *Energy Policy* 43, pp. 58–69. https://doi.org/10.1016/j.enpol.2011.12.025

Schmitz, H. (2007) 'The rise of the East: What does it mean for development studies?', *IDS Bulletin*, 38(2), pp. 51–58. https://doi.org/10.1111/j.1759-5436.2007.tb00351.x

Schmitz, H. and Scoones, I. (2015) 'Accelerating sustainability: Why political economy matters', *IDS Evidence Report* 152. Brighton, UK: Institute of Development Studies. https://opendocs.ids.ac.uk/opendocs/handle/20.500.12413/7077 (Accessed on 1/10/2020).

Sovacool, B.K. (2014) 'Energy studies need social science', *Nature* 511, pp. 529–530. https://doi.org/10.1016/j.jeem.2008.02.004

12

RENEWABLE ELECTRIFICATION PATHWAYS AND SUSTAINABLE INDUSTRIALISATION

Lessons learned and their implications

Rasmus Lema, Margrethe Holm Andersen, Rebecca Hanlin, and Charles Nzila

Abstract

What economic opportunities are involved in the process of renewable electrification in developing economies? This is a central question running through this book. Achieving these opportunities requires appropriate policy. Therefore, in this last chapter, we examine the evidence collected throughout the book and discuss the policy implications. We are particularly focused on the co-benefits that are attained from renewable electrification efforts in developing economies. Specifically, how these can be used to build long term learning and capabilities that have broader relevance for the economy than simply through the provision of green electricity. We start by setting out the motivation, framing and key research themes addressed in this book. We then present the findings regarding when and how renewable electrification can enhance local sustainable development outcomes. Our key argument is that it matters how green economic activities are organised. We, therefore, bring out the evidence from the various chapter of the book concerning the three key themes of the book, (a) projects design, organisation, and linkages, (b) deployment model and choice of technology, and (c) policies and political actors at the national and international level. We conclude the chapter by bringing out the implications, providing key pointers for policy action.

Introduction

The green transformation comes with important opportunities for economic development, both in advanced and developing economies. This is a widely held argument in policy circles as well as in the popular press. In developing economies, it can set in motion processes of economic development and transformation

DOI: 10.4324/9781003054665-12

referred to as 'sustainable' or 'green' industrialisation (UNIDO, 2016). Yet, it is widely acknowledged that the link between the green transformation and sustainable industrialisation is not automatic and that it comes with considerable constraints in developing economies. The underlying motivation for the research brought out in this book was to gain a better understanding of the circumstances that can help to turn the opportunities into reality in low and lower-middle-income countries.

The central theme running through the chapters in this book is that *it matters how green economic activities are organised*: they can be designed in ways that are more or less conducive to the long-term development of local industry. Our key starting assumption was that sustainable industrialisation achievements often depend critically on the local availability of capabilities to change (i.e., 'innovate') existing relationships along the chains that connect users and producers in the creation and deployment of green technologies.

To explore this, we have focused on a particular type of green technology, i.e., renewable energy, and a particular setting, i.e., the use of such technology in low and lower-middle-income countries to expand and transform existing electricity systems. We refer to this as renewable electrification. It has the dual purpose of creating increased access to electricity while greening the overall electric power regime.

We adopt a dynamic perspective, conceptualising the relationship between renewable electrification and related industry development over time. On the one hand, as mentioned above, the capabilities going into the renewable electrification process are critical to the realisation of green industrialisation outcomes. On the other hand, the successive steps in the renewable electrification process each provide a platform for learning, i.e., accumulation of capabilities. These can either enhance the sustainable industrialisation outcomes of subsequent steps in the renewable electrification process or undermine them, if learning is limited. It is this contingent process of *building innovation capabilities for sustainable industrialisation* which we have sought to unpack in this book.

The first chapter (Hanlin et al., 2022, this volume) sets the scene and defines our questions while the second chapter (Andersen and Lema, 2022; this volume) provides our conceptual framework. The remaining nine chapters examine different aspects of the process of building innovation capabilities for sustainable industrialisation. The purpose of this final chapter is to summarise the main insights of the book and bring out their implications.

The chapter is structured as follows: the next section reiterates the key research themes addressed in this book in order to situate the findings in wider debates about pathways to sustainable industrialisation. The following section forms the bulk of the chapter. It conveys the findings regarding the relative importance and dynamics of different aspects of renewable electrification for enhancing capability outcomes. It is structured around aspects related to the three key themes of the book, (a) projects design, organisation, and linkages, (b) deployment model and choice of technology, and (c) policies and political actors at the national and international level. The final section brings out the implications for policy. We summarise key findings arising from the research presented in this book, with the hope that the pointers for policy action can prove useful for more sustainable pathways that combine primary energy benefits and secondary localised economic co-benefits in the context of renewable electrification and the green transformation more broadly.

Pathways to sustainable industrialisation

The central theme in this book concerns pathways leading towards systems change in the way green infrastructure development is conceived in policy and designed, developed, and implemented in practice so local economic benefits are achieved. We pay particular attention to benefits with long-run potential for sustainable industrialisation gains. In this respect, we draw on the notion of transformative pathways characterised by directionality, distribution, and diversity (Leach, Scoones, and Stirling, 2007) in order to envisage pathways in the renewable electrification processes towards sustainable industrialisation.

Sustainable industrialisation

The overall objective is to assess whether and how renewable electrification can lead to inclusive structural change. Before addressing how the chapters in this book answer this question and generate new insights, it is useful to recapitulate what we mean by sustainable industrialisation and how different pathways of renewable electrification might contribute to it.

Our notion of sustainable industrialisation is three-dimensional and implies that structural change meets economic, environmental, and social goals simultaneously. It means that the process of industrialisation increases the greening, durability, and inclusiveness of economic activities (Figure 2.3, Andersen and Lema, 2022; this volume).

The contribution of renewable electrification to *greening* of economic activities is obvious: it changes the structure of energy systems by adding new green energy sources. It has the potential to become a force of 'creative destruction' by dismantling existing high-carbon models as new energy innovations are created and diffused (Lema, Iizuka, and Walz, 2015). For many years, scholars, politicians, and practitioners thought that such creative destruction should and would be spearheaded in high-income economies because greening was costly. Low- and middle-income countries would therefore adopt a 'cleaning-up later' industrialisation model as envisaged by the notion of an environmental Kuznet curve (Pegels and Altenburg, 2020). However, there is increasing recognition that a 'greening now' model can be viable because early greening can bring economic co-benefits that can make positive contributions to the industrialisation process itself. As discussed in Hanlin et al. (2022; this volume), however, such positive contributions from renewable electrification are far from automatic in many low- and middle-income countries. This is because the localisation of benefits depends on pre-existing capacities and capabilities and because the industrial development potential of the greening process through renewable electrification is largely absent in policy debates.

The *durability* of the economic activities involved depends on whether or not they lead to relevant forms of 'upgrading', defined here as an increase in the overall skill content of products and services and the creation of dynamic capabilities in organisations and production systems. Upgrading helps to ensure that skills, assets and systems do not become redundant beyond their immediate purpose, in the context of fast technological change and globalisation. In the absence of upgrading (or in the case of outright downgrading), local economies may experience premature deindustrialisation and immiserising or welfare worsening growth (Kaplinsky, Morris, and Readman, 2002). It is important to recognise that durability of activities does not necessarily mean that these activities are (only) concentrated around manufacturing activities. Rather, this aspect of the sustainability of industrialisation means that it involves positive structural change overall, involving a shift in production towards higher-knowledge intensity of products and services reflecting underlying assets based on higher knowledge and skilled labour (Ciarli et al., 2018).

Sustainable industrialisation is a form of structural change that also involves *inclusiveness* when it comes to increasing the participation of firms and workers in the economic activities and their involvement as active contributors to change. Given the other aspects inherent to sustainable industrialisation – greening and durability – such participation may often be challenging because it involves entirely new skills and processes. The international dimension is therefore of central importance: one crucial aspect is that more local firms and workers are involved as opposed to only or mainly foreign firms that have superior capabilities and capacities. Within countries, the transition to renewables is often intertwined with overall efforts of electrification, involving service delivery to rural communities that have tended to be marginalised from processes created by typical industrial development strategies. A key aspect of ensuring increased inclusiveness in sustainable industrialisation is thus inclusion of local firms and workers into the *learning processes* involved in low carbon development (Johnson and Dahl Andersen, 2012).

Sustainable industrialisation is structural change in which economic activities are green, durable, and inclusive at the same time. There can be significant tensions between these three aspects of sustainable industrialisation (i.e., 'trade-offs'). However, it is important to recognise that such mutually opposed pressures can be mitigated and brought about by strategies and policies of the key stakeholders involved. It is precisely this transformation of tensions into synergies, that should take centre stage in policy discussions: making the most of current renewable electrification efforts by re-shaping their pathways towards sustainable industrialisation.

Transformative electrification pathways

Pathways describe the particular directions in which new technologies – in this case green electrification technologies – are shaped during the process of their creation, adoption and use. The nature of such shaping has direct links to the realisation (or not) of the multidimensional sustainability outcomes discussed above. As discussed earlier, research focused on the direction, distribution, and diversity dimension of pathways provides a useful entry point to analysing renewable electrification (Stirling, 2009).

Direction implies that different innovation pathways towards electrification are considered and that trajectories, which support multidimensional sustainability are supported. In this book, the issue of directionality is addressed in particular through the focus on deployment models and choices of technology. There is a great need to redirect deployment pathways in directions that favour types of technology that are more appropriate and models that meet a larger array of sustainability outcomes. While directionality is often considered at the systems level (e.g., energy systems), it is also important to consider how individual parts of the system (e.g., projects) take shape. In the context of this book, this includes the direction of technological capabilities being used and built in projects, specifically, the degree to which there is learning which implies a movement from capabilities to use green energy technologies towards capabilities that help develop wider skills and competences related to design and manufacturing (Bell, 2009). This requires dispersed innovative capabilities that are deeply and pervasively embedded in local renewable energy systems.

The aspects of *distribution* imply a focus on how renewable electrification can eventually provide positive gains to local workers and firms. This requires an assessment of winners and losers of the renewable electrification process when it comes to the appropriation of the co-benefits involved. It is about 'just sustainabilities' (Agyeman, 2008; Scoones, 2016) in the domain of economic co-benefits and their distribution. In this respect, a key aspect addressed in the book is about the distribution of economic gains from green electrification in sub-Saharan Africa (SSA), between local and foreign actors. Hitherto these benefits have been heavily skewed in favour of the latter. This includes both the economic dividends from investments and the enterprise profits arising from infrastructure delivery. As will be discussed further below, several chapters in this book show how foreign firms typically occupy almost every step in the electrification value chains and in particular the most profitable and learning-intensive ones. Immediate jobs and local demand stimuli along backward linkages are important (to the extent they arise), but access to learning and capability-building is critical in the longer run. Many capabilities involved in renewable electrification are lateral in nature and can be used in other infrastructure settings.

The need for *diversity* of pathways is essentially about diversity in the accumulation of knowledge and experiences (Scot and Steinmuller, 2018) and an acknowledgement of the unequal distribution of capabilities and relations of power between actors in this process (Ockwell and Byrne, 2016). In the context of this book it relates strongly to the need to rebalance capabilities to generate 'greater intensity and diversity of localised innovation that complements the role of technological imports' (Bell, 2009, p. 12). Our book highlights the difficulties of having structured and systematised diversity with different technologies and modes of organisation, because powerful (typically foreign) actors often create lock-in to certain pathways. Diversity needs to be deliberately designed into projects to circumvent this. The notion of 'sustainability experiments' discussed in Andersen and Lema (this volume) could be a vehicle to change this. Deliberate efforts to meet the multidimensional sustainability criteria and placing more emphasis on locally designed projects is a way forward, for example, as in China where experiments ('demonstration programmes') were designed to crowd-in local players in successive rounds of direction-changing projects. This does not entail that imports are entirely substituted or that foreign advice from consultants are abandoned. There will certainly still be a need for openness in such locally grounded projects, but foreign inputs would have a supporting role rather than being in the driving seat. In the area of renewable electrification, this emphasises the need to create learning spaces to enhance the local participation in the innovation and economic processes and not always to focus only on modes of organisation that are designed largely by foreign companies.

Lessons learned

By putting capabilities centre stage in the exploration of the connection between renewable electrification and sustainable industrialisation we are standing on the shoulders of a long research stream. This established body of research has emphasised the centrality of organisational capabilities in the analysis of local innovation (Ely et al., 2013; Lema, Izuka, and Walz, 2015; Vidican, 2014), technology transfer and collaboration (Bell, 2012; Ockwell and Mallett, 2013), and innovation systems (Altenburg and Pegels, 2012; Johnson and Dahl Andersen, 2012; Ockwell and Byrne, 2015) in the context of low carbon development. This body of research has helped to extend conventional focus on the socio-economic benefits arising directly from access to green energy to those that arise indirectly as potential co-benefits of the renewable electrification process itself, in particular the potential 'learning benefits' (see Hanlin et al., 2022; this volume; Andersen and Lema, 2022; this volume). In the following we start each of the three subsections with the broad research-guiding 'propositions' which are informed by this literature and which have helped steer our empirical work. We then discuss the key findings we can draw out from the research as lessons learnt.

Project design, organisation, and linkages

We started our research with the idea that the way renewable energy projects are designed and organised, and the type and quality of linkages they include, matters significantly for capability development in low and lower-middle-income countries. Specifically, we assumed that learning-by-doing would be limited because the 'doing' would tend to be done by foreign firms and workers and because knowledge transfer would not occur as an automatic byproduct of project activities. In extension, we assumed that whether or not learning occurs depends on the degree to which it is specifically planned for in the design phase. These features of project design are, in turn, also the choice of the 'project model' such as whether it is organised as a relatively closed turnkey project coordinated by a single lead agent or a more open and loosely coordinated project involving several collaborators. Here the assumption was that a more open project would provide opportunities for building local capabilities. We explore these assumptions, summarising our evidence, in the following.

The various studies included in this book suggest that the renewable electrification projects observed typically revolve around foreign-dominated modes of project organisation. Bhamidipati et al. (this volume) examining three major projects found that in terms of the flows and nature of the projects: benefits are constrained by the dominant pattern of full-package provision of EPC contract. The authors show that local staff were involved in the construction phase of some projects (e.g., 70–90% of total project employees in one case), but most highly skilled activities were mainly carried out by foreign nationals, in these cases Chinese nationals. In the operation and maintenance phase of the three projects, fewer local staff were employed but for longer time. In some cases, handing over to local staff was not fully achieved because locals were not always sufficiently trained to take over. There were some backward linkages, e.g., from provision of local services and manufacturing inputs from local firms, but these tended to be limited.

Hanlin and Okemwa (2022; this volume) provide case studies of four different renewable electrification projects with the aim to examine interactive learning and capability-building in critical projects. In all four cases studied, foreign Engineering, Procurement, and Construction (EPC) firms have played a central role either across the project or for key parts of it. The EPC firms have tended to structure tasks and linkages in these projects. However, these case studies provide grounds for cautious optimism. They showed that valuable capabilities, particularly project management and linkage (or networking) capabilities, were being built. Specifically, the authors showed that these case studies reveal increasing possibilities for local companies to gain an EPC role, especially in small solar PV projects, and that the way contracts are formulated relating to project management has a bearing on the likelihood of backward linkages and capabilities-building in projects of all sizes.

The study by Karjalainen and Byrne (this volume) goes deeper by providing a categorisation of solar photovoltaic (PV) companies in Kenya and Tanzania according to their levels of innovativeness, focusing specifically on firms operating in the domain of off-grid electrification. They show that foreign PV companies operating in Kenya and Tanzania have more innovative capabilities than the local ones. At the same time, however, local firms involved in innovative off-grid electrification activities have learned novel techniques through joint ventures and partnerships with foreign firms. Many local innovative firms have had active international linkages.

The insights unearthed by Nzila and Korir (2022; this volume) provide a systematic comparison across five different renewable energy technologies and five different steps in the value chain drawing on a comprehensive survey undertaken in Kenya. The study shows that capabilities for renewable electrification deployment in the country are relatively high on average but with noticeable bottlenecks. The findings also indicate that while management capabilities are generally rated high, there are a number of areas that require improvements, in particular with respect to the capabilities to identify, assess, negotiate, and finalise project financing terms as well as capabilities within maintenance. It was further revealed that the overall capability levels are highest in the solar PV domain. These findings serve to reinforce the premise that (for a developing economy) learning, capability development and benefits of renewable electrification are not obvious across the renewable energy (RE) value chain since to a large extent they might be constrained by the (mostly external) dominant actors. Hence there is need for interventions geared towards harnessing the full benefits of renewable electrification, including directing a new paradigm: from continued dependence on external actors in most steps of the RE value chain to the targeted development of local capabilities.

The main insight, when cutting across these studies, is that local actors, while involved in all projects observed, tend to be marginally involved in renewable electrification projects. Overall, there is only little local content provided and it tends to be concentrated in (a) the provision of auxiliary inputs into the construction phase such as provision of e.g., rackings for solar PV projects or cements for the foundations of wind turbine projects or (b) the provision of routine operation and maintenance tasks such as cleaning of facilities. The modes of project organisation observed – typically the EPC model involving a foreign main contractor – have left negligible scope for more technology-intensive functions such as overall project design or provision of critical inputs such as core technologies. This has typically meant that linkages are concentrated in areas with limited learning opportunities, insofar as provision of marginal inputs provide marginal access to the critical capabilities to design and orchestrate similar project in the future. This is important because the capacity to take on the role of project organisers locally may enhance project design that includes overall local 'secondary benefits' in terms of jobs, more critical local content, and further learning. In short, the prevalence of the EPC model in current green electrification means that local firms experience learning mainly in routine tasks but make little progress in tasks that may provide further opportunities for backward and lateral linkage creation.

This finding is important to the sustainable industrialisation debate and takes centre stage in the insights provided by this book. Meanwhile, it is important both to bring out further the material that provides clear-cut support for it and also the exceptions and boundaries and to explain some of the key factors that help to explain this state of affairs.

Deployment model and choice of technology

Project designs, including their internal organisation and external linkages, do not exist in a vacuum. They are dependent on overall renewable 'deployment models' that may vary significantly between countries, but they are dynamic, changing over time, and several paths may develop simultaneously (Lema et al., 2018). Key dimensions of such models are overall choice of renewable energy sources (the balance of hydro, solar, wind, etc.) versus sources, project size (e.g., many solar panels versus few or small versus large hydro) and technology sizes (e.g., small versus large wind turbines). They also include typical ownership structures and degree of international openness. In this book we have paid particular attention to centralised versus decentralised models and associated choices of technology where technology is understood in a broad sense. It is important to recognise that there are significant trade-offs in the realm of primary benefits – between e.g., on the one hand, economies of scale, speed of deployment, and so-called proven business models, and, on the other hand, geographical reach (in particular to rural communities and integration with new business models involving mobile money). But our focus has been on the secondary benefits in terms of direct economic activities (jobs and local content) and associated capability-building. Below we bring together key insights arising from the chapters in this book.

Several chapters in the book emphasise that the overall choice of core technology is crucial for learning opportunities. The majority of chapters are focused on wind and solar technologies, but some chapters also include analyses of geothermal energy and biomass powers (Nzila and Korir, 2022; this volume; Ogeya et al., 2022; this volume). The chapter by Nzila and Korir (2022; this volume) is key in this respect because it includes a systemic comparison across technology types. They found that local capabilities were acquired in PV and small hydro to a greater extent than in biogas, geothermal, and wind energy. In the latter group, capability development constraints were particularly pronounced in the areas of servicing and maintenance capability. Other chapters reinforce this finding, in particular with respect to solar PV. This technology, as compared to other renewables, has a lower share of upfront capital cost in overall project costs, meaning that the share of cost activities related to installation (and associated peripheral equipment), operation, and maintenance is higher. These activities provide opportunities for local involvement and learning.

Hansen et al. (2022; this volume) make the argument that although the nature of core technology has implications for secondary benefits, it is rather the 'size', which is critical. It is more in this sense that the choice of technology is crucial and has huge implications, for instance for how projects are organised and hence also for options for learning and capability-building. The same theme is picked up by Wandera (this volume) who argues that small wind has particular benefits compared to large wind as this technology exhibits features of appropriate technology such as simplicity, low capital cost, and ease of maintenance. Wandera does lend cognisance to the fact that small wind is only viable in areas with high surface wind speeds, and further argues that solar-wind-battery-hybrids have significant but untapped potential in Kenya and in other SSA countries (Johannsen, Østergaard, and Hanlin, 2020; Wandera, 2020). However, Hansen et al. (2022; this volume) emphasise, in particular, that large differences in learning and local economic development opportunities are related to whether projects are deployed on or off the national grid. They support the small is beautiful argument in the sense that there are more local learning possibilities in off-grid renewable projects which are small in size compared to projects feeding the grid.

These findings are echoed by Hanlin and Okemwa (2022; this volume) who provide in-depth insights into the process of learning across projects of different sizes. As mentioned above, they emphasise the observation that capability-building is constrained in EPC projects but there is evidence of local firms doing more engineering, procurement, and construction in small-scale projects. There is limited evidence of 'strategic' project management roles, even in small-scale projects as well as in the provision of technology. Local tasks are mainly 'routine' roles in construction, operations, and maintenance whereas foreign firms tend to cater for project design and coordination of construction.

Karjalainen and Byrne (2022; this volume) suggest that 'foundational capabilities' are built that could support sustainable industrialisation in a small-scale decentralised setting. Focusing on the off-grid solar PV sector in Kenya and Tanzania, they trace learning over 30 years and show that firms emerging in the 2010s and operating in Kenya and Tanzania have become globally leading. While most highly innovative companies are of foreign origin, this creates opportunities for local firms to strengthen their learning and potentially enter an 'early latecomer' phase where they could build increasingly complex capabilities, including for manufacturing. Models for forming use of technology appear more important than choice of technology/core technology. Innovations are focused largely on business models, but they also show that cash-sales of solar PV remain important and may be more valuable for local firms compared to the (centralised) PAYG model.

Another aspect of 'pathway models' regards their international openness. It is well-known that electrification efforts in SSA are quite heavily influenced by international actors across the value chain, from finance over equipment provision to project execution (Lema et al., 2018). The precise degree of external dependence is difficult to specify, but as earlier mentioned, foreign lead firms tend to take on coordinating roles. Interestingly, Gregersen and Gregersen (2022; this volume) explore 'learning spaces' in foreign-dominated projects in large scale wind, one European and one Chinese project. Focusing on how interactions between different stakeholders in wind power 'megaprojects' can lead to the accumulation of technological and managerial capabilities, they show that both formalised and tacit knowledge interaction can occur, even in the megaproject setting, but it has limits.

The chapter by Bhamidipati et al. (2022; this volume) also examines large projects, focusing on three Chinese projects, with the aim to examine the realisation of co-benefits (i.e., secondary benefits). They show that the choice of (core) technology is decided by key decision-makers in each of the three projects as a part of financing deals. The element of finance is significant because it shifts the relative bargaining power strongly in favour of the investor-contractor consortium, making co-benefits largely dependent on the actors engaged with making key decisions with regard to the project. Core components and equipment are almost exclusively imported from China or alternatively sourced from specialised suppliers in advanced economies. Apart from the primary benefits of electricity provision, local socio-economic benefits arising from the electrification process itself (secondary benefits) ranges between 'very few' and 'extremely few'.

Collectively, the chapters in this book show that patterns of learning capabilities and outcomes differ markedly between types of technology and between deployment models. Most of the evidence unearthed supports the argument that wider socio-economic benefits can be achieved more easily in small-scale decentralised models, but even here these benefits depend on key contingencies. We address these in the next section.

Policies and political actors at the national and global level

The following section brings out the key implications for policy arising from the research presented in this book. In this present section our focus is different: we aim to bring out insights about how policies and political actors influence capability-building directly and through influencing deployment models and modes of project organisation. Some of the chapters in the book address policy issues as their key focus while others deal with the policies as a part of their overall analysis.

The chapter by Kingiri (this volume) is explicitly focused on the role of policy and goes some way in explaining the prevalence of observed deployment models and modes of project organisation. It presents a historical analysis of Kenya's policy process in renewable electrification from 1999 to 2019 with a particular emphasis on the efforts to promote industrialisation in the energy sector. The chapter draws on a stakeholder survey of key policy gaps that require policy support in developing requisite capabilities for support to sustainable industrialisation. Although local content issues, as a result of pressure from various stakeholders, are becoming more pronounced in the discourse and agenda, it is still not fully implemented and reinforced. A key issue highlighted in the chapter is that this is difficult due to both domestic factors, such as limited manufacturing capabilities, and due to external factors, such as high levels of competition from other countries (China included) as far as production of solar PV and other equipment is concerned. In an illustration of this, the chapter by Karjalainen and Byrne (2022; this volume) shows how attempts to develop the local manufacture of battery charge regulators for solar home systems failed because of better quality and cheaper products imported from China.

In other words, while local content policies exist on paper, they are difficult to implement in reality, because of pressure/conditions from providers of finance combined with a 'mismatched' or uncompetitive local supply base. The latter is also related to the education system where specific training efforts in some Kenyan universities and Technical Vocational Education and Training Institutions (TVETs) do not fully meet the production and innovation needs associated with the renewable electrification process. As a result, one chapter in the book (Ogeya et al., 2022; this volume) highlights the existence in Kenya of a 'locked-in' system where change is difficult to achieve due to existing institutions and political structures.

Bhamidipati et al. (this volume) address similar issues and seek to examine 'local institutional and economic conditions' as key determinants of learning, linkages, and local content (economic co-benefits) arising from three green energy projects in Kenya, Ethiopia, and Ghana respectively. They find that in all three cases, there were local content policies in place but these policies were largely circumvented in project contracts. They highlight both policy effort and inequalities in bargaining power between external consortia (bringing together finance, equipment, and services) and local stakeholders, in particular governments but also other firms and organisations involved in the project.

This aspect is also clearly illuminated by Gregersen and Gregersen (2022; this volume). In the Ethiopian case, the government has gone beyond production-system thinking and involved knowledge and innovation-system building elements to ensure more local learning in and around projects. While still with several shortcomings, the Ethiopian Government has taken an active role in design of projects to ensure maximum local learning, by ensuring that professional users are more involved in the project execution. This, according to the authors, explains that while there was some local learning in both the Kenyan and the Ethiopian cases in the field of O&M and how to add more renewable energy to the national grid, the Ethiopian Adama case involved slightly more learning about how to design a (large-scale) renewable energy project.

Another important aspect concerns the policy support for traction in the different dimensions of the deployment model. Hansen et al. (2022; this volume) show that a disaggregated level of analysis is important for understanding (and ultimately designing) policies in the renewable electrification field. Hence the overall focus on policies for 'renewables' needs to be decomposed in ways that are sensitive to different types of renewable technologies and, in particular, the different *versions* of these renewables that are supported, e.g., standalone versus grid-connected solar. As argued above, these have very different potential outcomes in terms of benefits and their distribution. Governments support different renewables and deployment models that can imply benefits at different scales and in different contexts (e.g., urban/rural). Overall, the political economy dynamics tend to favour 'proven' models with powerful lead firms and strong economies of scale in large projects.

In continuation, Wandera (2022; this volume) argues that despite – or perhaps because of – the huge attention to large wind in Kenya, there has been limited support and a formation of a very weak (if not absent) system of small wind deployment in Kenya. When it comes to rural and decentralised deployment, rooftop solar PV has received more attention than hybrid mini-grids that are more technically demanding and have higher upfront investment costs. However, this is slowly changing and more conducive policies, e.g., for more wind used in hybrid off-grid systems, may alter the role of small wind turbines (SWT). Yet, a recent donor-funded experimental project to increase small wind (Kenya's Miniwind project) has been discontinued due to the complexities involved in cross-function collaboration and development of viable business models.

Karjalainen and Byrne (2022; this volume) tell the other side of this story, showing how multiple donor projects in the late 1990s and early 2000s supported the deployment of solar PV, through technical skills training and awareness activities. These were highly successful in rolling out solar home systems (SHS), but economic arrangements have been left to the market forces and have not helped individual companies to catch up and arrange for the localisation of products and services.

In sum, this book shows that socio-economic outcomes of renewable electrification – their contribution to sustainable industrialisation – are influenced by disparate policy fields, chiefly energy policy and industrial policy, that are rarely brought together to make the most of ongoing economic activities. In particular, deliberate policies towards capability development are strikingly absent. Although important opportunities have been identified in and around the economic activities involved in the electrification process, they are rarely formulated let alone implemented. In the following section we set out an interpretation of our findings against key normative objectives as specified in Hanlin et al. (2022; this volume), before proposing key suggestions for policy.

Conclusions and pointers for policy action

We started this chapter by observing that the green transformation comes with opportunities and constraints for economic development, but that these differ markedly between countries. In many high-income economies, the objective to maintain and expand high levels of economic development and welfare with this transformation is embedded in the notion of 'green growth' (Bowen and Fankhauser, 2011). Emerging economies such as China and India seek to exploit changes in global markets and technologies to increase the pace of latecomer development (Pegels and Altenburg, 2020). But while such strategies have been effectively crafted in advanced and in emerging economies, what might the windows of opportunity look like in low and lower-middle-income countries? And what will it take to realise such options?

In this book we have sought to address these questions by focusing on renewable electrification as a case in point to examine its contribution to sustainable industrialisation. If renewable electrification in lower-middle-income countries does not contribute effectively to sustainable industrialisation, there is no reason to assume that other elements of the green transformation (energy efficiency, mobility, etc.) would.

We have drawn on a conceptual framework focusing on learning, capabilities, and outcomes. As set out in Chapter 2 (Andersen and Lema, 2022; this volume), we have put forward the argument that sustainable industrialisation outcomes depend on the local learning opportunities involved in the underlying economic activities. Hence, a more specific question has driven our work: what economic opportunities are involved in the process of renewable electrification in developing economies and how can policy help to ensure that such opportunities are realised in a way that makes maximum contributions to structural change? So far there are surprisingly few insights regarding this question, for two main reasons.

The first reason is that some analysts and global policy makers still consider these questions superfluous because the primary (green energy) benefits should, in their view, take precedence. Hence, the argument goes, renewable electrification – even without noticeable local economic benefits – should be pursued due to its climate change and energy security advantages. They reason that the green transformation is cheaper overall when 'delivered' by actors with pre-existing technologies and capabilities, as opposed to when it is delivered by those firms who need to pay the additional costs of moving along a technological learning curve.

In this book, we argue that this viewpoint is critically flawed. First, delivery of green energy by external firms and agents may create path-dependency and long-run lock-in to external solutions. Second, overly depending on external actors in the provision of green energy entails big risks in terms of 'technological fit' of new facilities and in terms of operation, maintenance, and sustainable use of the renewable energy systems already installed. Third, the long-term viability of the green transformation depends on popular legitimacy and support, which requires localisation of economic benefits. It needs to be supported by realistic prospects of green economic development in low- and middle-income countries and not just by 'green growth' in advanced economies. There are strong links between the first and the third of these arguments to the recognition of the overall importance of equity in climate change policy (Klinsky et al., 2017), in particular when it comes to climate change mitigation as a cornerstone of the green transformation.

The second reason is that although some of these counterarguments have been taken on board and the discourse has (to some extent) expanded from primary benefits to economic co-benefits, this discourse has been based on assumptions that do not hold. A key conclusion arising from our work is thus that communities and policymakers need to beware of overly optimistic expectations about the secondary (economic) benefits that renewable electrification projects may generate in low- and middle-income countries. These benefits tend to be limited and they are not automatic by-products of the expansion and modification of energy systems with renewables. More generally, a major argument in our book is that the specificities of green development in low- and middle-income countries are insufficiently understood and are also insufficiently incorporated into the global and local policy discussions.

So, whereas the strategies of green growth may work in advanced economies, the evidence unearthed in this book suggests that it has much less potential in the low- and middle-income countries, in particular if conducted as business-as-usual. In order to make it 'work', a number of major changes are required in policy. In the remainder of this chapter, we bring together and discuss new kinds of policy action which could be explored in different contexts. We outline key policy deliberations across national and global levels of policymaking and a summary of these is provided in Box 12.1. We call them key 'pointers' for policy action because they need to be carefully interpreted, assessed, and shaped depending on circumstances, local industrial context, specific features of the technologies in question etc.

However, these suggestions are all underpinned by one central idea running through this book, namely the need to make economic co-benefits a requirement of green transformations in developing economies. This necessitates persistent expansion of the policy focus from primary benefits to economic co-benefits. As discussed in this book, it also entails that production and innovation capability development is put centre stage in all aspects of green transformation policy, such as those dealing with local firms, projects, and organisations relevant for renewable electrification.

BOX 12.1 SUMMARY OF KEY POINTERS FOR POLICY ACTION

1. Combine plans of energy system greening with industrial development and technological development strategies. This requires that policy domains that typically develop separately – i.e., the energy-environmental and industrial development domains – are aligned, co-designed, and developed in conjunction.
2. Ensure frameworks for project selection, such as auction systems, and increase accountability and selection criteria across a broader set of industrial development goals as opposed to just energy production.
3. Make local co-benefits a key criterion for selection of projects. Devise and use impact assessments for skilled jobs, local content, and capability development prior to any project decision.
4. Re-balance the emphasis on capability development in energy projects away from the conventional focus on renewable energy project service delivery (operation and maintenance) to pay more attention to renewable energy project infrastructure delivery (particularly project design and execution).

5. Create in-depth maps of renewable energy supply chains and focus on capacity and capability-building in 'zones of proximate development' (capabilities that are within reach, but not yet acquired locally) in both the manufacturing and deployment chains of sustainable energy projects.
6. Create learning spaces such as experimental projects (sustainability experiments), that try out not only different types of technologies, but also different new types of project management, localised supply, and community involvement. Document and use the experience in revising project selection and design criteria.
7. Create national agencies that can function as vessels of domain expertise, enable systematic learning, and facilitate knowledge transfer between different successive and otherwise unconnected projects.
8. Create a network of national 'centres of excellence' in universities and vocational training institutions and make sure to insert national education institutions into renewable energy projects as partners/learning consultants.
9. Help national consortia to bring together finance from impact investors with local and global companies for projects that meet the multidimensional sustainability criteria and related learning objectives.
10. Build multi-stakeholder global coalitions to define and implement mission-oriented innovation programmes with the aim to use greening transformation initiatives to foster structural change. Make finance from progressive institutional investors a cornerstone.

At the national level, plans of energy system greening must be persistently combined with industrial development and technological upgrading strategies. This requires that policies that typically develop in separate policy domains need to be co-created across the energy-environmental and industrial spheres. For example demand-driven initiatives to facilitate energy system greening, such as feed-in tariffs or auctions, need to develop in alignment with active industrial policy and appropriate measures to ensure the appropriate localisation of economic activities (Landini, Lema, and Malerba, 2021). As shown in several chapters in this book, the thorough analysis of past provision of green energy facilities can enable the identification of value chain activities that can feasibly be undertaken by local firms, thereby substituting currently imported capacities and capabilities. Such tailored strategies should be developed and adjusted dynamically to identify zones of proximate development in the upgrading process (developing capabilities that are within reach) and involve the targeted development of support systems, in particular the provision of relevant knowledge infrastructures for the cultivation of relevant design, engineering, and management capabilities.

In other words, the co-designing of policies needs to be closely coordinated with long-term national energy plans and nationally appropriate mitigation actions (NAMAs). This should also ensure that national development plans determine green transformation activities, not the other way around. Systemic frameworks, such as energy auctions, that assess and approve project selection should be pro-actively designed and implemented to raise investment and ensure favourable tariffs. Importantly, they should also ensure that projects are not developed on an ad-hoc basis promoted by specific consortia of finance and technology supply (see Bhamidipati et al., this volume). Ad-hoc project approval weakens the bargaining power of governments and typically comes along with informal 'foreign content requirements' tied to external sources of finance. Frameworks for project selection and approval must be focused on a broader range of goals to address the issues of directionality, diversity, and distribution discussed above. In this respect, the capacity of local authorities to design and manage according to 'guiding visions' is key (Lema et al., 2018).

When it comes to crafting such guiding visions, it is important to keep in mind that renewable electrification essentially is a process of successive provisions of discrete green infrastructure projects. As discussed above, localisation and learning does not arise automatically in these renewable energy projects. Therefore, deliberate and upfront planning of local content provision and capability development is needed for every single project and throughout the various stages of decision-making. The required measures do not work when they are supplementary add-ons provided after other technical and financial specifications are defined. Hence, these concerns need to move centre stage at the point of initial impact assessment and feasibility reporting to be able to select the projects with the largest scope for creating economic benefits, and they need to be included in the process of project design in order to make sure that this potential is realised. The stage of project design is where the key supply chain decisions are made, and hence this is where appropriate local rooting can be ensured.

Related to this, local content can only be built into projects if there are capabilities on the ground. A key problem of policy-design for renewable electrification is the temporal and specialised nature of projects, in particular for large-scale projects and pathways. Major projects may only be constructed a few times every decade. Incidentally, this is an additional argument for intensifying the exploration of small-scale, decentralised pathways. These are more frequent and have lower barriers to entry for local firms. Mid-size mini-grids may well be constructed by developers from low- and middle-income countries. Innovative business models developed in East Africa, as described in this book (see e.g., Karjalainen and Byrne (2022; this volume) may provide a platform for the attainment of market shares in this segment.

Whether they are small or large-scale projects, there is no guarantee of repeated involvement of particular national firms in a context of competitive bidding. Therefore, it is particularly important to create government agencies that can function as vessels of domain expertise, enable systematic learning, and facilitate knowledge transfer between different successive and otherwise discrete and unconnected projects. It is important to recognise, moreover, that several lateral capabilities are involved in the provision of green energy, i.e., capabilities that are involved in other types of infrastructure provision activities, be it ports, bridges, or railway lines. Government agencies can thus be supported by centres-of-excellence in infrastructure project management, located in key national universities, research institutes, and technical training and vocational institutes, which need to be invited on board as project partners involved in assessment and design as well as capability development and knowledge acquisition and collaboration.

Such requisite organisational entities, in particular the key government agencies, need to facilitate a diversity of capabilities, ranging across different domains from project management to more specific technological capabilities. They also need to accumulate experiences relevant to different sizes of projects, e.g., both large and small scale. Crucially, they also need to function as climate-relevant innovation system builders (Ockwell and Byrne, 2015) taking on the central role of forging relationship between firms and between firms and supporting institutions. In this way, virtuous circles of co-development of specific renewable energy projects with enhanced long-term capabilities in sustainable industrialisation may be generated.

Focused efforts to create learning spaces such as experimental projects or 'sustainability experiments' are important in this respect (Berkhout et al., 2010). One area where our work has highlighted the need for more experimentation is in project design, project management, and infrastructure delivery (execution) more generally. There is a need to re-balance the emphasis on capability development in energy projects away from the conventional focus on renewable energy project service delivery (operation and maintenance) to pay more attention to renewable energy project infrastructure delivery (particularly project design and execution). Therefore, these sustainability experiments should try out not only different types of technologies, but also different new types of project management, localised supply, and community involvement. These efforts must be documented and reviewed so the experiences can be taken on board and utilised in future project design. Crucially, the learning should be considered by those involved in project selection too – i.e., that more learning is built into the policy process; to close the loop in the policy process from implementation back to planning and designs.

These points have profound implications for the design of policies and support schemes of global institutions and global bilateral programmes. There is a need to coordinate and align objectives across global policy domains as well. Hence, international decision-makers need to work together to create policy packages, cutting across climate policy, and trade and development policy, for locally beneficial renewable electrification pathways. The raison d'être of such policy packages should be to create the economic co-benefits necessary to incentivise the exploration of new pathways of sustainable industrial development locally. It is necessary to increase the scope of national decision-making and introduce deliberate local learning in all elements of renewable energy interventions and related climate actions. It is not sufficient to devise such schemes based on the notion that cost-effective climate change mitigation trumps the localisation of economic benefits, such as technological and organisational learning as well as backward linkages to locally rooted supply chains. Increased awareness of what it takes to design projects in a manner that take into account needs for local learning and involvement is an obvious area for deepened government-to-government collaboration and mutual capacity building.

The principles above require a global system, which works actively to substitute imported capabilities with local ones, where in practical terms such substitution is achieved by assisting to build capabilities in low and lower-middle-income countries. In turn, it depends on a contentious but necessary acknowledgement that the economic interests of powerful global stakeholders need to be curbed and brought into alignment with national priorities for example in sub-Saharan Africa, thereby increasing fairness and popular legitimacy to support global support for the green transformation. This requires a disassociation between soft loans or aid for green energy and infrastructure and contracting decisions favouring global vested interests. Global agreements and standards are needed to ensure such separation. Moreover, every financing decision must include economic co-benefits from the initial point of project negotiations, not in exceptional cases but as a requirement instituted in global policy frameworks. They should be included in public Official Development Assistance (ODA) guidelines and in private environmental, social, and corporate governance (ESG) indicators.

In sum, considerable efforts are required in projects and at national and global levels to ensure that the possible economic co-benefits of renewable energy projects for sustainable industrialisation are realised. These need to be at the heart of current debates about 'leapfrogging' to avoid the fossil-fuel based energy scenarios of high-income countries. In other words, there is a need for entirely new visions for development pathways around renewable electrification which brings together the objectives of energy system greening, access to energy, and economic development. It requires a recasting of sustainability transition narratives, broadening the scope to incorporate a wider range of multidimensional sustainability criteria – to include economic and social dimensions – explicitly and centrally into multilateral and national policy frameworks.

References

Agyeman, J. (2008) 'Toward a 'Just' Sustainability?', *Continuum*, 22(6), pp.751–56.

Altenburg, T. and Pegels, A. (2012) 'Sustainability-Oriented Innovation Systems – Managing the Green Transformation', *Innovation and Development*, 2(1), pp. 5–22.

Andersen, M.H. and Lema, R. (2022) 'Towards a Conceptual Framework: Renewable Electrification and Sustainable Industrialisation', in *Building Innovation Capabilities for Sustainable Industrialisation: Renewable Electrification in Developing Economies*. New York: Routledge. https://doi.org/10.4324/9781003054665-2

Bell, M. (2009) *Innovation Capabilities and Directions of Development*. Vol. 33. Brighton: STEPS Centre.

Bell, M. (2012) *Low-Carbon Technology Transfer*. London: Routledge.

Bhamidipati, P.L et al. (2022) 'Chinese Green Energy Projects in Sub-Saharan Africa: Are There Co-Benefits?', in *Building Innovation Capabilities for Sustainable Industrialisation: Renewable Electrification in Developing Economies*. New York: Routledge. https://doi.org/10.4324/9781003054665-10

Bowen, A. and Fankhauser, S. (2011) 'The Green Growth Narrative: Paradigm Shift or Just Spin?', *Global Environmental Change*, 21(4), pp. 1157–59.

Ciarli, T., Savona, M., Thorpe, J. and Ayele, S. (2018) *Innovation for Inclusive Structural Change. A Framework and Research Agenda*. Brighton: Science Policy Research Unit.

Ely, A,, Smith, A., Stirling, A., Leach, A. and Scoones, I. (2013) 'Innovation Politics Post-Rio+20: Hybrid Pathways to Sustainability', *Environment and Planning C: Government and Policy*, 31(6), pp. 1063–81.

Gregersen, C. and Gregersen, B. (2022) 'Interactive Learning Spaces: Insights from Two Wind Power Megaprojects', in *Building Innovation Capabilities for Sustainable Industrialisation: Renewable Electrification in Developing Economies*. New York: Routledge. https://doi.org/10.4324/9781003054665-8

Hanlin, R. and Okemwa, J. (2022) 'Interactive Learning and Capability-Building in Critical Projects', in *Building Innovation Capabilities for Sustainable Industrialisation: Renewable Electrification in Developing Economies*. New York: Routledge. https://doi.org/10.4324/9781003054665

Hanlin, R., Andersen, M.H., Lema, R. and Nzila, C. (2022) 'Renewable Electrification and Sustainable Industrialisation', in *Building Innovation Capabilities for Sustainable Industrialisation: Renewable Electrification in Developing Economies*. New York: Routledge. https://doi.org/10.4324/9781003054665-1

Hansen, U.E. et al. (2022) 'Centralised and Decentralised Deployment Models: Is Small Beautiful?', in *Building Innovation Capabilities for Sustainable Industrialisation: Renewable Electrification in Developing Economies*. New York: Routledge. https://doi.org/10.4324/9781003054665-4

Johannsen, R.M., Østergaard, P.A. and Hanlin, R. (2020) 'Hybrid Photovoltaic and Wind Mini-Grids in Kenya: Techno-Economic Assessment and Barriers to Diffusion', *Energy for Sustainable Development*, 54, pp. 111–126.

Johnson, B. and Dahl Andersen, A. (2012) *Learning, Innovation and Inclusive Development : New Perspectives on Economic Development Strategy and Development Aid*. Aalborg, Denmark: Aalborg University Press.

Kaplinsky, R., Morris, M. and Readman, J. (2002) 'The Globalization of Product Markets and Immiserizing Growth: Lessons from the South African Furniture Industry', *World Development*, 30(7), pp. 1159–77.

Karjalainen, J. and Byrne, R. (2022) 'Moving Forward? Building Foundational Capabilities in Kenyan and Tanzanian Off-Grid Solar PV Firms', in *Building Innovation Capabilities for Sustainable Industrialisation: Renewable Electrification in Developing Economies*. New York: Routledge. https://doi.org/10.4324/9781003054665-9

Kingiri, A. and Okemwa, J. (2022) 'Local Content and Capabilities: Policy Processes and Stakeholders in Kenya', in *Building Innovation Capabilities for Sustainable Industrialisation: Renewable Electrification in Developing Economies*. New York: Routledge. https://doi.org/10.4324/9781003054665-11

Klinsky, S., Roberts, T., Huq, S., Okereke, C., Newell, P., Dauvergne, P., O'Brien, K., Schroeder, H., Tschakert, P., Clapp, J., Keck, M., Biermann, F., Liverman, D.,

Gupta, J., Rahman, A., Messner, D., Pellow, D. and Bauer, S. (2017) 'Why Equity Is Fundamental in Climate Change Policy Research', *Global Environmental Change*, 44, pp. 170–73.
Landini, F., Lema, R. and Malerba, F. (2021) 'Demand-Led Catch-up: A History-Friendly Model of Latecomer Development in the Global Green Economy', *Industrial and Corporate Change*, 29(5), pp. 1297–1318.
Leach, M., Scoones, I. and Stirling, A. (2007) *Pathways to Sustainability: An Overview of the STEPS Centre Approach*. Brighton: STEPS Centre.
Lema, R., Hanlin, R., Hansen, U.E. and Nzila, C. (2018) 'Renewable Electrification and Local Capability Formation: Linkages and Interactive Learning', *Energy Policy*, 117(August 2017), pp. 326–39.
Lema, R., Iizuka, M. and Walz, R. (2015) 'Introduction to Low-Carbon Innovation and Development: Insights and Future Challenges for Research', *Innovation and Development*, 5(2), pp. 173–87.
Lema, R. et al. (2022) 'Renewable Electrification Pathways and Sustainable Industrialization: Lessons Learned and Their Implications', in *Building Innovation Capabilities for Sustainable Industrialisation: Renewable Electrification in Developing Economies*. New York: Routledge. https://doi.org/10.4324/9781003054665-12
Nzila, C. and Korir, M. (2022) 'Are the Capabilities for Renewable Electrification in Place? A Kenyan Firm-Level Survey', in *Building Innovation Capabilities for Sustainable Industrialisation: Renewable Electrification in Developing Economies*. New York: Routledge. https://doi.org/10.4324/9781003054665-6
Ockwell, D. and Byrne, R. (2015) 'Improving Technology Transfer through National Systems of Innovation: Climate Relevant Innovation-System Builders (CRIBs)', *Climate Policy*, 16(7), pp. 836–854.
Ockwell, D. and Mallett, A. (2013) 'Low Carbon Innovation and Technology Transfer' in Urban, F and Nordensvärd, J. (eds.) *Low Carbon Development: Key Issues*, pp. 109–28. Abingdon: Earthscan, Routledge.
Ogeya, M.C., Osano, P., Kingiri, A. and Okemwa, J. (2022) 'Challenges and Opportunities for the Expansion of Renewable Electrification in Kenya', in *Building Innovation Capabilities for Sustainable Industrialisation: Renewable Electrification in Developing Economies*. New York: Routledge. https://doi.org/10.4324/9781003054665-5
Pegels, A. and Altenburg, T. (2020) 'Latecomer Development in a 'Greening' World: Introduction to the Special Issue', *World Development*, 135, pp. 1–11.
Scoones, I. (2016) 'The Politics of Sustainability and Development', *Annual Review of Environment and Resources*, 41, pp. 293–319.
Stirling, A. (2009) *Direction, Distribution and Diversity! Pluralising Progress in Innovation, Sustainability and Development*. Brighton: STEPS Centre.
UNIDO (2016) '*Industrial Development Report 2016: The Role of Technology and Innovation in Inclusive and Sustainable Industrial Development*', Vienna.
Vidican, G. (2014) 'Challenges and Opportunities for Capturing Local Benefits', *Ökologisches Wirtschaften–Fachzeitschrift*, 29(4), p. 19.
Wandera, F.H. (2020) 'The Innovation System for Diffusion of Small Wind in Kenya: Strong, Weak or Absent? A Technological Innovation System Analysis', *African Journal of Science, Technology, Innovation and Development*, (in press) pp. 1–13.
Wandera, F. (2022) 'Understanding the Diffusion of Small Wind Turbines in Kenya. A Technological Innovation Systems Approach', in *Building Innovation Capabilities for Sustainable Industrialisation: Renewable Electrification in Developing Economies*. New York: Routledge. https://doi.org/10.4324/9781003054665-5

INDEX

Note: Page numbers in *italics* indicate figures, **bold** indicate tables and page numbers with "n" indicates the end notes in the text

For Product Safety Concerns and Information please contact our EU
representative GPSR@taylorandfrancis.com
Taylor & Francis Verlag GmbH, Kaufingerstraße 24, 80331 München, Germany

www.ingramcontent.com/pod-product-compliance
Ingram Content Group UK Ltd.
Pitfield, Milton Keynes, MK11 3LW, UK
UKHW020316270926
475702UK00050B/909
* 9 7 8 0 3 6 7 5 1 6 2 4 6 *